Contemporary Polish Ontology

Philosophical Analysis

Edited by
Katherine Dormandy, Rafael Hüntelmann,
Christian Kanzian, Uwe Meixner, Richard Schantz and
Erwin Tegtmeier

Volume 82

Contemporary Polish Ontology

Edited by
Bartłomiej Skowron

DE GRUYTER

ISBN 978-3-11-076371-3
e-ISBN (PDF) 978-3-11-066941-1
e-ISBN (EPUB) 978-3-11-066951-0
ISSN 2627-227X

Library of Congress Control Number: 2019946358

Bibliographic information published by the Deutsche Nationalbibliothek
The Deutsche Nationalbibliothek lists this publication in the Deutsche Nationalbibliografie;
detailed bibliographic data are available on the Internet at http://dnb.dnb.de.

© 2021 Walter de Gruyter GmbH, Berlin/Boston
This volume is text- and page-identical with the hardback published in 2020.
Printing and binding: CPI books GmbH, Leck

www.degruyter.com

Contents

Bartłomiej Skowron
Some Introductory Thoughts on Contemporary Polish Ontology —— VII

Tomasz Bigaj
On Essential Structures and Symmetries —— 1

Mariusz Grygianiec
Prospects for an Animalistically Oriented Simple View —— 25

Filip Kobiela
How Long Does the Present Last? The Problem of Fissuration in Roman Ingarden's Ontology —— 51

Zbigniew Król and Józef Lubacz
The Subject's Forms of Knowledge and the Question of Being —— 71

Andrzej Biłat
The World as an Object of Formal Philosophy —— 87

Urszula Wybraniec-Skardowska
Logic and the Ontology of Language —— 109

Krzysztof Śleziński
Benedict Bornstein's Ontological Elements of Reality —— 133

Janusz Kaczmarek
On the Topological Modelling of Ontological Objects: Substance in the *Monadology* —— 149

Krzysztof Wójtowicz
Does Mathematical Possibility Imply Existence? —— 161

Rafal Urbaniak
Neologicism for Real(s) – Are We There Yet? —— 181

Jacek Paśniczek
Possible Worlds and Situations: How Can They Meet Up? —— 205

Marek Magdziak
The Ontologic of Actions —— 219

Michał Głowala
"Physical Intentionality" and the Thomistic Theory of Formal Objects —— 245

Bartłomiej Skowron, Tomasz Bigaj, Arkadiusz Chrudzimski, Michał Głowala, Zbigniew Król, Marek Kuś, Józef Lubacz, and Rafał Urbaniak
An Assessment of Contemporary Polish Ontology —— 271

Author Index —— 295

Bartłomiej Skowron
Some Introductory Thoughts on Contemporary Polish Ontology

This book is a collection of articles authored by Polish ontologists living and working in the early part of the 21st century. Harking back to the well-known Polish Lvov-Warsaw School, founded by Kazimierz Twardowski,[1] we try to make our ontological considerations as systematically rigorous and clear as possible – i.e. to the greatest extent feasible, but also no more than the subject under consideration itself allows for. Hence, the papers presented here do not seek to steer clear of methods of inquiry typical of either the formal or the natural sciences: on the contrary, they use such methods wherever possible. At the same time, I would like to draw attention to the fact that despite their adherence to rigorous methods, the Polish ontologists included here do not avoid traditional ontological issues, being inspired as they most certainly are by the great masters of Western philosophy – from Plato and Aristotle, through St. Thomas and Leibniz, to Husserl, to name arguably just the most important.

The subject of the present volume is no single ontological issue, in that its purpose is to demonstrate the richness of ontology as currently practised in Poland. The articles contained here touch upon and range across the most important ontological issues: substance and dispositions, persons and knowledge, as well as language, time and mathematical objects – not to mention the ontology of action and the metaphysics of possible worlds. During the very first meeting of the Polish Philosophical Society in Lvov in 1904, Kazimierz Twardowski spoke the following words: "The one and only dogma of the Society will be the conviction that dogmatism is the greatest enemy of scientific work. Just as all radii of a circle, though originating from different points around its circumference, combine and meet in its center, so we wish all directions taken by the work and philosophical views of our Society to aim at just one goal: the illumination of the truth" (Twardowski, 1904, p. 241, trans. C. Humphries). The philosophical metaphor of a circle, in which various methods and issues, striving for true cognition, converge in the middle, fits well with the current book: the reader will encounter ontological analyses here

---
[1] For a general overview, see Woleński (2015).

Bartłomiej Skowron, International Center for Formal Ontology, Faculty of Administration and Social Sciences, Warsaw University of Technology, Poland.

https://doi.org/10.1515/9783110669411-001

that employ a wide range of methods, but which nevertheless all focus on seeking out ontological insights of the most perspicuously truthful kind.

Ontology in Poland, as in other countries, is diverse and heterogeneous – and it is surely good that it is so! This collection of articles brings together the work of ontologists who, in their deliberations, mostly make use of methods specific to either the formal sciences (mathematics and logic) or the natural ones (physics). Ontology of this kind, while approximating to a form of scientific inquiry, remains most definitely a philosophical discipline: it does not become just another part of science (even in the broadest sense of that term), but rather just uses scientific tools. This type of ontological approach is not the only one present in Poland, but it is one that in one way or another continues the tradition initiated by Kazimierz Twardowski, Stanisław Leśniewski, Roman Ingarden, Kazimierz Ajdukiewicz and Józef Maria Bocheński – one still enthusiastically continued in the opening decades of the 21st century by Bogusław Wolniewicz and Jerzy Perzanowski. We ourselves, while seeking to take this tradition still further in Poland, also form part of the increasingly popular direction known as "mathematical" philosophy (sometimes also referred to as "logical" or "formal" philosophy), which may be regarded as the sister of "analytical" philosophy in the strict sense of that term.

I shall now turn to the content of the articles that compose this volume. In Tomasz Bigaj's opening contribution, entitled "On Essential Structures and Symmetries", the author addresses a classic issue: do physical objects exist in isolation? That is to say, are they independent, or do they need larger wholes in order to exist, within which they must coexist in relations obtaining between them and other objects? Structuralists claim that structures are ontologically primary as beings, and it is only within such structures that individual objects can be identified and studied. Nevertheless, the question naturally arises of whether the objects that find themselves transformed within some given structure lose their identity after being transformed by it, or remain the same. Bigaj defends a version of structuralism in which the identification of individual objects in different possible worlds (that is, after the changes in question have taken place) is carried out by means of qualitative relational structures, which he calls "essential structures", thus defining his position as "essential structuralism". This solution allows for the identification of objects regardless of the fact of their being transformed across possible worlds.

In physics, the concept of symmetry plays a special role, because it is believed that what distinguishes real physical reality from that which is theory-dependent is precisely that which remains invariant across all variation at the level of physical theory itself. Bigaj discusses the concept of identity or sameness in relational structures in detail, pointing out that a standard model-theoretic approach to symmetry fails to meet the expectations of structuralists. He proposes a new concept of symmetry, fully compatible with structuralism, in which qualitative properties of objects (e.g., electron mass) are used to identify objects. Then he juxtaposes his own approach to symmetry with the standard symmetries that occur in physical theories to show how his new conception of it works. He considers discrete symmetries, such as permutations of quantum-mechanical states of multiple particles of the same type (e.g., electrons or photons), as well as continuous symmetries that include Leibniz shifts and Galilean boosts.

The next article touches upon the ontology of persons and, in particular, the problem of identity – the topic previously analysed in a slightly different context and with different methods by Bigaj. Mariusz Grygianiec, in his contribution "Prospects for an Animalistically Oriented Simple View", notes that the following two positions are treated as mutually incompatible: (1) that human beings are biological organisms (i.e. that every person is identical with a certain organism), and (2) that there are no non-circular criteria for personal identity such as would constitute necessary and sufficient conditions for personal identity. Such criteria can only ever be trivial, circular or uninformative. These views are contrasted in the literature, with the former called "animalism", the latter the "Simple View". In his text, Grygianiec offers three potential strategies for reconciling these seemingly contradictory positions.

Grygianiec presents the two positions in detail. He discusses the criteria and conditions of identity they involve, giving examples of such criteria, including the axiom of extensionality for sets and Davidson's criterion for the identity of events. He then presents three strategies for combining animalism and the Simple View. The first refrains from treating "human animal" as a natural-kind term, and consequently also from construing "human being" as such a term, where this results in there being no metaphysical criteria for the identity of persons. The second treats the terms "human animal" and "person" as both natural-kind terms, but rejects the possibility of formulating criteria for personal identity. The third strategy, preferred by Grygianiec, adopts an animalist stance while at the same time excluding the possibility of formulating criteria of identity for persons, in that the term "person" is open-ended and not a categorically well-defined concept, while also not being a natural-kind term – so that metaphysical criteria of identity cannot be associated with it. In this way Grygianiec reconciles animalism and the Simple View.

One of the properties of real objects is that they exist in time. However, what exactly does this mean, and what are its consequences for real existence? The real object is, one might say, somehow doomed to reside in one present. Its present being is always to be found between its past and its future. It cannot remain in its present – it must, so to speak, trail away into the past. Roman Ingarden characterized this suspension between past and future as a fissure-like existence. To some extent, this mode of existence can be overcome by human beings through conscious living.

They can go beyond the present moment: for example, by recalling the past or looking to the future with hope. Thanks to this, they are able to reduce the fragility of their existence. But can the limits of the present be extended infinitely in both directions, so as to cover both past and future? An object existing in this way would resemble an absolute being. These fundamental problems are taken up by Filip Kobiela in his contribution, entitled "How Long Does the Present Last? The Problem of Fissuration in Roman Ingarden's Ontology". Kobiela harks back to the tradition established by Ingarden, which has enjoyed a rich existence in Poland itself but remains insufficiently well-known elsewhere. He recalls the analyses of time put forward by St. Augustine, Bergson and Popper. He also skilfully quotes the analyses of two other Polish thinkers, the philosopher Bogdan Ogrodnik and the well-known writer Stanisław Lem.

As a result of their deliberations, ontologists often affirm certain claims. Roman Ingarden, to advert to him once again, asserted that if an object changes its way of existence, it must lose its identity, because it is not possible to change its way of existence and preserve its identity as an object. An epistemologist listening to such considerations will immediately be prompted to ask of the ontologist various questions. How did you come to know that? On what basis do you claim it? How can you attain such knowledge at all? Could you have allowed yourself to be mistaken in your deliberations and arrive at a false conclusion? From the very dawn of philosophy there has been a dispute over the relative priority of ontology and epistemology. "The Subject's Forms of Knowledge and the Question of Being", a joint contribution by Zbigniew Król and Józef Lubacz, concerns the ways in which a subject can acquire knowledge – especially ontological knowledge.

Contrary to popular philosophical trends, Król and Lubacz seek to show that hidden assumptions, non-act-like components of cognition, and – something that rarely appears in epistemological reflections – trust play an important role in the acquisition of knowledge. Definitions of knowledge often rest on the conviction that to know something is to be in possession of a justified true belief. One of the conclusions presented by the authors is the opposite statement: it is possible that a subject can genuinely know something that is false.

Another contribution dealing with the struggle for the primacy of ontology over epistemology is Andrzej Biłat's text, entitled "The World as an Object of Formal Philosophy". Biłat adopts and defends an ontological paradigm of philosophy. The starting point for philosophy, according to the latter, is the question "What exists?", which differs from the Cartesian paradigm (represented here by the article of Król and Lubacz just discussed), for which the very first question is "What can I know?". Biłat, inspired by the intuitions of Plato and Aristotle, presents a formal analysis of concepts basic to the ontological paradigm of philosophy: namely, the concept of the world as an extensional whole (i.e. a non-empty class including all classes of beings), the concept of the real world (i.e. the non-empty class of all temporal beings), and that of the world of nature (a concretum occupying the entirety of space at any given time). After giving a precise definition of these concepts, he outlines the position of extensional metaphysical realism, seeking to show that this position is entailed as a metaphysical consequence of contemporary logic (to be precise, monadic second-order logic) and the methodology of the natural sciences (including cosmology).

Biłat's contribution is a good example of a formal approach inspired by the thoughts of the great classics of philosophy: Plato, Aristotle and others. The approach in question is one that seeks to address issues central to philosophy in ways that make use of modern developments in the formal and the empirical sciences, including logic, mathematics and physics. From the perspective of such an approach, formal philosophy – the term "logical philosophy", as propagated by Jerzy Perzanowski, is also widely used in Poland to refer to this – is not only a certain method to be encountered in contemporary philosophy, but also something that has been an important component of philosophical thinking from the very beginnings of philosophy. Formal ("logical") philosophy in Poland is not detached from the Western tradition of philosophy, and its inspirations reach right back to ancient philosophy, as Biłat demonstrates – as well as taking in modern philosophy, as the reader will shortly discover when we come to consider the contribution of Janusz Kaczmarek to this volume.

Urszula Wybraniec-Skardowska, in her paper "Logic and the Ontology of Language", which discusses ontological aspects of language, presents a theory of language in which she uses the tools of logic and set theory. She harks back to the logical conception of language proposed by Ajdukiewicz, according to which it consists of a vocabulary and a syntax (i.e. rules for how to build new expressions), a semantics (assigning meaning and denotation to linguistic expressions) and a pragmatics (governing relations between signs and users). The logical approach to language makes language an ideal object, which is no longer the natural language we use on a daily basis. While both logicians and linguists deal with language in their research, it should be noted that they are not really concerned with the same object: a language viewed in terms of logic is a much simpler, idealised construction, and should not be confused with a living natural language composed of concrete expressions. The author of this article exploits this ontological heterogeneity with respect to language, characterizing it in terms of the ontological duality of linguistic expressions: on the one hand such expressions can be construed as concrete objects, like inscriptions on a blackboard, on the other as theoretical objects, such as a class or type. Naturally, the question then arises of which linguistic expressions are to be deemed primitive: is it the case that expression-tokens are ontologically fundamental (as in concretism), or do expression-types have priority (as in platonism)?

Wybraniec-Skardowska constructs a formal syntactic theory for both platonism and concretism, from which she then derives some ontologically interesting conclusions: it turns out that a syntax built on expression-tokens is logically equivalent to one built on expression-types, where this carries the entailment that a syntax can be constructed without positing abstract objects – something that would certainly be music to the ears of every concretist! However, as Jerzy Perzanowski noted when commenting on this particular author's research, one should also take into account the fact that the proposed equivalence is articulated with the help of set theory, while set theory is itself by no means ontologically neutral. (Even assuming the existence of an infinite set amounts to a strong ontological commitment.) Further to this, the author anyway argues that when we consider matters locally, as internal to the structures she herself proposes, logic favours neither concretism nor platonism. In order to offer readers a fuller picture of her proposed formal ontology of language, Wybraniec-Skardowska also constructs a formal semantics and pragmatics, where these serve to complement other elements within the whole construction.

In the first half of the 20th century, apart from such great figures of Polish philosophy as Kazimierz Twardowski, Stanisław Leśniewski, Alfred Tarski and Roman Ingarden, Benedict Bornstein, who is less well-known, was actively involved in philosophy in Poland. He had the ambition of creating an all-encompassing metaphysical system, which he called the "architectonics of the world". As analytical philosophers often avoided metaphysics, his work was not met with enthusiasm from Kazimierz Twardowski's students. For this reason his achievements, overshadowed as they have been by those of philosophers of the Lvov-Warsaw School, have not had a great impact on the development of ontology in Poland. Nevertheless, Bornstein did forge boldly ahead, and his thinking now looks to have been several decades ahead of his time.

In the course of his ontological investigations Bornstein employed mathematics, including geometry and topology (of a sort fitting their state of development at that time) – something by no means in vogue then, when formal philosophy was principally focused on logical concerns. Elements of his ontological system are presented in the context of an introductory discussion of his work by Krzysztof Śleziński, in his paper "Benedict Bornstein's Ontological Elements of Reality". Bornstein, who, as was already noted above, is not yet well-known as a philosopher, has certainly suffered from the fact that his work is too advanced to be accessible to a wide range of readers of philosophy, in that it requires familiarity with contemporary mathematics. Nevertheless, his ideas were without a doubt moving in much the same direction as those that have since provided the basis for both the dynamic development of mereotopology and the emergence of spatial logic more generally that we are witnessing today. His philosophical (not mathematical) considerations also count as pioneering in relation to the philosophy of category theory (as introduced by Mac Lane and Eilenberg), where logic and geometry meet again. In his article here, Śleziński points to the directions of research into Bornstein's thought waiting to be undertaken by contemporary ontologists – especially those with a mathematical training and a strong metaphysical temperament. There is certainly much to be done in this area. It is also worth adding that the basic assumptions of Bornstein's conception of metaphysics as a mathematical science go hand in hand with contemporary mathematical philosophy as promoted, for example, by Hannes Leitgeb (cf. Leitgeb 2013).

In the next article, "On the Topological Modelling of Ontological Objects: Substance in the *Monadology*", Janusz Kaczmarek uses topology to model the notion of substance in the context of a Leibnizian approach. This is an example of work in the field of formal ontology, where certain mathematical structures are deployed in order to arrive at an enhanced understanding of issues central to metaphysics. These structures are by no means accidental, as Kaczmarek uses spatial structures (topology being a generalization of geometry) just as Bornstein tried to. Kaczmarek defines substance as an ordered collection of topologies (monads) together with a dominant topology (dominant monad). Such a substance resembles the notion of a system as investigated by Bocheński, Bunge and Ingarden. Kaczmarek highlights this similarity and draws significant conclusions from it. One of the consequences of his definition of substance is the claim that in a substance with more than two elements, other substances can be distinguished. He also provides a criterion of identity for substances. He points out that each substance contains only one dominant monad, and goes on to introduce formal equivalents of certain classical metaphysical concepts: e.g., perception (in many variants) and appetition.

Kaczmarek's paper ends by setting out some currently unresolved problems in the field of topological ontology. In this matter, as in the case of Bornstein's works, there is still much work to be done. Although he does not mention it himself, Kaczmarek's method of analysing philosophical problems remains close to Ajdukiewicz's method of paraphrase: Ajdukiewicz used formal logic, while Kaczmarek suggests using topological structures instead, but both set themselves the task of analysing traditional problems by these means, with Ajdukiewicz rejecting transcendental idealism and Kaczmarek seeking to elucidate the Leibnizian conception of substance.

Mathematical theorems are often taken to furnish a model instance of necessary truth. Yet what does it mean, exactly, for a theorem to be of necessity true? The standard answer from Leibniz states that a necessary theorem is one that is true in all possible worlds. And well, the universe of possible worlds is very rich indeed, containing as it does everything logically possible. Because of this, the notion of necessity as truth in all possible worlds is often linked to a thesis asserting that for mathematical objects to exist is equivalent in principle to their just being logically non-contradictory. In the universe of mathematics, existence just means logical possibility: the creative power of mathematics is limited only by logical constraints. In his article "Does Mathematical Possibility Imply Existence?", Krzysztof Wójtowicz analyses the thesis that the sheer logical possibility of a mathematical object (i.e. its non-contradictoriness) entails its existence. This realist stance can be expressed in brief as follows: all possible mathematical objects exist. If we reduce mathematics to some kind of basic object, for example sets, then this thesis can be expressed as the claim that all mathematical objects are simply sets. Wójtowicz does not agree with such an ontological reduction: he defends the thesis that mathematical objects exist *per se*, regardless of their multiple representations. In his argumentation he appeals to both advanced results from the foundations of mathematics and ontological intuition, while also referring to mathematical practice.

The next paper – also dealing with mathematical objects – is Rafał Urbaniak's contribution, entitled "Neologicism for Real(s) – Are We There Yet?". The author discusses the ontology of real numbers. In the context of investigations of the foundations of mathematics we encounter various kinds of ontological foundation being proposed for real numbers: in particular, their reconstruction in set theory, which Urbaniak invokes in the third section of his own article. In the main part of his paper, he reviews the project of reconstructing the real numbers in logic, trying to show that it is possible to defend an improved version of the traditional position in the foundations of mathematics, namely logicism, whose main thesis is that mathematics can be derived from logic. The modern version of logicism is – unsurprisingly – known as neologicism. Its main idea is to apply appropriate abstraction principles to the reconstruction of real numbers. A classic example of such an abstraction principle is the one that Frege used to define the notion of extension: F and G have the same extension if F and G apply to the same objects. To put it more formally: $\{x: Fx\} = \{x: Gx\} \equiv \forall x(Fx \equiv Gx)$. This type of principle engendered Russell's paradox, however, so the application of such principles is subject to a degree of cognitive risk. Urbaniak presents different principles of abstraction and shows how they can be used to reconstruct real numbers. In particular, he presents the reconstructions of real numbers proposed by Peter Simons, Stewart Shapiro and Bob Hale. He then criticises each of these proposals, pointing out their weaknesses: it turns out to be by no means easy to alight upon an abstraction principle that will be neither too strong nor too weak. In his contribution, he also takes up important threads of both an ontological and an epistemological kind that pertain to abstraction principles. For example, he considers the question of whether such abstraction principles lead to the creation of new abstract objects, and what the principles of abstraction might be for true statements.

One highly popular concept in contemporary logic and ontology is that of possible worlds, which is often explored and entertained alongside that of situations. Jacek Paśniczek explores relations between possible worlds and situations in his paper "Possible Worlds and Situations: How Can They Meet Up?". A possible world is commonly defined as a maximally consistent object, which means that for any proposition A, either A or $\neg A$ can be true in this world, but not both. However, such notions of possible worlds have proved inadequate in the context of certain areas of research – for example, those involving non-classical logic. Thus, the concept of impossible worlds – i.e. worlds in which, for some proposition A, both A and $\neg A$ can be true – has been coined. Such impossible worlds are referred to as "non-normal" worlds, or "non-standard" possible worlds. Paśniczek calls them "n-worlds". In his own contribution, he points out the weaknesses of the set-theoretic approach to n-worlds, and presents an algebraic proposal, built on a suitably enriched De Morgan lattice. Structurally, the set-theoretic and algebraic approaches are equivalent, but it is precisely this equivalence that shows that the concept of a set does not play the most important role in the ontology of n-worlds. Paśniczek's findings indicate that such an ontology can be described algebraically, independently of any language.

Jerzy Perzanowski was one of the most eminent Polish ontologists working at the turn of the 20th and 21st centuries. Perzanowski (1988, p. 87–88) distinguishes three aspects of ontology: *ontomethodology*, *ontologic* and *ontics*. *Ontomethodology* deals with ontological methodology, and its results may take the form of ontological principles, such as Ockham's razor. *Ontics*, meanwhile, is the conceptual analysis of ontological issues, resulting in a description of the ontological universe. (Perzanowski pointed to Roman Ingarden as a typical Polish "onticist".) *Ontologic*, on the other hand, is the study of the logical foundations of the ontological universe in question, with a typical ontologist being the well-known Polish philosopher Stanisław Leśniewski, the initiator of the foundational systems known according to his own terminology as Protothetic, Ontology and Mereology.[2] (We may mention in passing here that Leśniewski remarked of his doctoral students that they were all geniuses. This was in fact so, as his only doctoral student was Alfred Tarski!) Returning to our main story, what results from *ontologic* is the logic of the ontological universe.

---

[2] For a review, see Simons (2015).

The next text in the present volume is Marek Magdziak's paper entitled "The Ontologic of Actions", in which the author sets out to construct a specific logical and ontological basis for ethics. Following Georg Henrik von Wright, Magdziak views actions as intentionally brought about changes to the world: i.e. as the production or omission of states of affairs. Both actions and the states of affairs produced and destroyed by them are particularly important for ethics, as they are – at least arguably – the quintessential bearers of our ascriptions of goodness or badness. (For instance, we find it natural to say such things as "in failing to keep your promise, you did the wrong thing", or "it's good you were evaluated fairly by your teacher".) Magdziak, in his axiomatic system, which is a multi-modal propositional logic, analyses the omission, performance and possibility of actions, together with the production and destruction of states of affairs by a given action; he also introduces and analyses the ethical operators "it is bad that..." and "it is good that...". He presents a semantics for his proposed ontologic of action, this being a slight modification of the standard relational semantics for normal modal propositional logic.

The ontologic of action presented by Magdziak articulates a specific ontological basis for ethical reasoning: i.e. reasoning whose conclusion is some action itself. Once again, Magdziak's work is in line with the current of logical philosophy still being vibrantly pursued in Poland today. Such philosophy, though very much alive in Poland, is rather less popular amongst students than contemporary social or political philosophy, for the simple reason that it requires from readers a deep level of concentration combined with an unhurried approach to study. This is something readers can certainly experience for themselves – in, I would venture to add, a wholly positive way – should they be prepared to devote the time and energy required to do justice to Magdziak's contribution.

In his contribution entitled "Physical Intentionality and the Thomistic Theory of Formal Objects", Michał Głowala analyses in detail the similarities and differences between the ontology of intentionality and the ontology of powers (or dispositions). Intentional states, such as acts of love or hate, are directed at their objects: this directedness belongs to their very essence. An important feature of objects of intentional states is that they need not exist: one can, after all, easily imagine an impossible object – even though it does not exist, it can still be an object of imagination. Such properties have come to be referred to by contemporary analytic philosophers as "marks of intentionality". On the other hand, Głowala considers dispositions and their manifestations. An exemplary instance of a disposition would be the solubility of table salt or sugar in water. It turns out that dispositions are also characterized by a certain directedness towards their manifestation, along with the fact that the manifestations of some dispositions need not themselves exist: after all, salt can be dissolved in water, but does not have to be.

For these reasons, some philosophers claim that powers and their manifestations are also characterized by such "marks of intentionality".

Głowala shows that the ontology of intentionality and the ontology of powers are indeed similar, but not by virtue of the fact that they share "marks of intentionality" as such: rather, it is just that these ontologies both exhibit "marks of having a formal object". Głowala presents a study of formal and material objects in the manner of the Scholastics, arguing that such a framework, pitched at a general level, furnishes the proper basis for a unifying conception of intentionality and dispositionality. Roughly speaking, the formal object of x is the way in which anything must be related to x (or given to x) in order to be an object of x. For example, if we look at a wheel from the side, we see it as if it were oval: this feature, its being oval, is the formal object of the perception – it is a characteristic of an act of capturing the wheel in experience, not the wheel itself. Głowala warns against hasty analogies with contemporary ontological notions, arguing, for example, that a formal object is not a type whose material objects are tokens. He also argues that it would be a mistake to reify formal objects (the oval I see is not the actual wheel I am looking at). Amongst those who are guilty, in his eyes, of committing this mistake, he mentions both Platonists and advocates of Meinongian ontology. The paper ends with an outline of the difference between the mental and the physical, thus touching on one of the most important ontological issues. It turns out that this fundamental difference is based on the stability or changeability of formal objects.

Głowala joins current ontological disputes provoked by analytical philosophers, drawing inspiration from late-scholastic Thomism, including the thought of Paul Soncinas and Cardinal Cajetan. In this way, he makes up for the absence of a historically informed dimension to many of the analyses put forward within the framework of contemporary analytical philosophy, while at the same time complementing historically oriented approaches themselves with his own pursuit of systematic philosophical analysis. (One might add here that a historical approach to philosophical issues is probably the one most frequently adopted by Polish philosophers. This may be because it represents the easiest option from the point of view of one's not then being obliged to take responsibility for one's thoughts: after all, one is allowed to just reside safely, and respectably, in the shadows cast by the great thinkers of the past.) The approach exemplified by Głowala here is nowadays known as analytically oriented Thomism. While not as popular in Poland as the more firmly rooted existential form of Thomism, it is nevertheless being ever more actively pursued.

The present volume closes with a multi-author contribution entitled "An Assessment of Contemporary Polish Ontology". Here we consider the state of contemporary ontology in Poland. We point out both the strengths and weaknesses of ontology as practiced in that country in the early part of the 21st century. Among its strengths, we list thirty major achievements on the part of Polish ontologists. Meanwhile, we also seek to address our weaknesses, which we diagnose as possessing many dimensions (including those of an academic, social, or organizational kind, etc.), proposing to remedy these by taking appropriate action aimed at furthering the development of ontology in the future. The authors of this article include not only philosophers, but also representatives of science and engineering – a fact which allows us, we think, to formulate some fairly objective assessments. Such claims are, for sure, not typically encountered in volumes devoted to ontology, so their presence may surprise readers – but perhaps, in our era of global interconnectedness where human beings so quickly become bored, this will be no bad thing! Anyway, we ourselves do believe that such assessments can be helpful when it comes to increasing the self-awareness of a given discipline of knowledge and allowing for its appropriate positioning within the complex network of contemporary academic knowledge and institutions. My own personal hope is that this paper will inspire ontologists from outside of Poland to set out their own aspirations and assessments in similar terms, thus helping to build, and further define the identity of, an international ontological community.

In 2016, Warsaw University of Technology hosted a conference entitled "Polish Contemporary Ontology", during which most of the authors contributing to this volume presented their papers. The conference was organized by the International Center for Formal Ontology, which itself was established in 2015 as part of the Faculty of Administration and Social Sciences of that same university. The Center brings together philosophers and scientists interested in the use of formal tools in ontology – in particular logic and contemporary mathematics, including probability theory, category theory and topology. As can be seen from the contributions to this volume, opening oneself up to a broad-based engagement with mathematical structures and methods need not entail any narrowing of scope in respect of one's field of research. At the Center we address issues that are currently popular in contemporary analytical metaphysics, but also those more commonly associated with what might be described as more of a non-analytical paradigm of ontological research, such as includes phenomenological and even hermeneutics-based approaches.

Acknowledgment: It only remains for me to thank all of the reviewers and commentators involved in the preparation of this book, be they anonymous or not. In particular, I would like to express my sincere thanks to Piotr Błaszczyk, Andrea Bottani, Ben Caplan, Christopher Daly, Samuel Fletcher, Bob Hale, Christian Kanzian, Wojciech Krysztofiak, Uwe Meixner, Thomas Mormann, Kevin Mulligan, Roman Murawski, Thomas Müller, Marek Nasieniewski, Kristopher McDaniel, Jeff

Mitscherling, Francesco Orilia, Elisa Paganini, Alexander Pruss, Robert van Rooij and Anna Wójtowicz. I would also like to extend my thanks to Zbigniew Bonikowski for typesetting the first versions of the volume in LaTeX. Moreover, I wish to express my sincere thanks to the anonymous reviewer of the book as a whole. His or her critical remarks have allowed us to significantly improve the quality of the entire volume. Furthermore, special thanks are due to Mirosław Szatkowski for his kind support in the initial stages of preparing the book, and to Zbigniew Król, Marek Kuś, Józef Lubacz, Tomasz Bigaj and Christian Kanzian, for all their support and kindness. Without their generous encouragement, I would never have completed the long process of editing it. Last but not least, I would like to thank Carl Humphries for the proofreading of some of the texts, including this introduction. I sincerely hope that after reading the texts contained in this volume, the reader will be left with the impression that Polish ontology is doing well, and that it is set to do even better in the future.

The publication of this book was financially supported by the Faculty of Administration and Social Sciences of the Warsaw University of Technology.

Bibliography

Leitgeb, H. (2013). Scientific Philosophy, Mathematical Philosophy, and All That. Metaphilosophy, 44: 267-275. doi:10.1111/meta.12029.

Perzanowski, J. (1988). Ontologie i ontologiki. Studia filozoficzne, 6-7 (271-272), 87-99.

Simons, P. (2015). "Stanisław Leśniewski", The Stanford Encyclopedia of Philosophy (Winter 2015 Edition), Edward N. Zalta (ed.), URL = <https://plato.stanford.edu/archives/win2015/entries/lesniewski/>.

Twardowski, K. (1904). Przemówienie na otwarciu Polskiego Towarzystwa Filozoficznego we Lwowie (Address on the occasion of the inauguration of the Polish Philosophical Society in Lvov). Przegląd Filozoficzny 7, 293-241.

Woleński, J. (2015). "Lvov-Warsaw School", The Stanford Encyclopedia of Philosophy (Spring 2019 Edition), Edward N. Zalta (ed.), URL = <https://plato.stanford.edu/archives/spr2019/entries/lvov-warsaw/>.

Tomasz Bigaj
On Essential Structures and Symmetries

Abstract: The main goal of the paper is to revisit the concept of a symmetry for relational structures in light of the ontological position dubbed "essentialist structuralism". It is argued that the standard definition of the notion of a symmetry commits us to the existence of non-qualitative, haecceitistic differences between possible worlds. An alternative notion of symmetry is developed, based on the distinction between essential and contingent structures. It is claimed that this new concept is better suited for the doctrine of structuralism, and moreover it offers a new perspective on some well-known problems in the foundation of physical theories, such as the problem of permutation invariance in quantum mechanics and diffeomorphism invariance in general relativity.

Keywords: symmetries, relational structures, essentialism, structuralism, haecceitism, permutation invariance, space-time.

Symmetries of Relational Structures

In recent years the ontological doctrine of structuralism has gained widespread popularity. Structuralism, also known under the moniker Ontic Structural Realism, can be traced back to the observation that physical objects do not exist in isolation, but instead constitute greater wholes via links of mutual interconnections. These wholes are identified as relational structures in which objects participate and which are the proper subjects of scientific investigations. Structural realists insist that structures are ontologically more fundamental than their participating objects. This vague claim can receive many non-equivalent interpretations, ranging from radical eliminativism (there are no objects, only structures) to the relatively mild statement that the criteria of identity and distinctness for objects should somehow involve the qualitative relations they participate in.[1] In my (Bigaj, 2014) I proposed yet another interpretation of structuralism, which focuses on the following much-discussed problem in the metaphysics of modality: how to identify individual objects in possible scenarios (across possible worlds). I sug-

---

[1] The literature on different variants of Ontic Structural Realism is vast and growing. Instead of giving here a semi-complete list of relevant publications, I will refer the reader to the recently updated overview (Ladyman, 2014) and the works quoted therein.

Tomasz Bigaj, Institute of Philosophy, University of Warsaw, Poland.

https://doi.org/10.1515/9783110669411-002

gested there that we should combine the essentialist approach to this question with the structuralist-motivated assumption that relational structures are ontologically prior to the objects participating in them. The result of this fusion is a position I call "essentialist structuralism", which states roughly that the identification of objects in alternative possible worlds should be done with the help of selected qualitative structures dubbed "essential".[2] In this paper I will further probe this idea by formalizing criteria of transworld identity in terms of essential structures and their isomorphism- or homomorphism-based similarities. The starting point of my discussions will be the concept of a symmetry of a relational structure. I will argue that the textbook characterization of this concept contains elements that are difficult to accept by (essentialist) structuralists, and that therefore they may be compelled to look for alternative formalizations. Such an alternative approach will be developed and applied to some well-known cases of symmetries in fundamental physical theories.

Broadly speaking, a symmetry of an object is a transformation that leaves this object invariant. This general characterization already reveals a hidden tension within the concept of symmetry, as it combines two apparently inconsistent elements. On the one hand, a non-trivial symmetry must change something; otherwise it would be the uninteresting identity transformation. But on the other hand, in order to be a symmetry and not just any transformation, its action on the object in question must leave it exactly as it was before. This tension can be resolved in various ways. One approach, common in mathematical applications of the concept of symmetry, is to embed the object whose symmetries we analyze in a broader "environment". In this case a symmetry may be a non-identity when applied to the entire environment, and yet it may transform the object of interest identically onto itself.[3] As an example illustrating such a situation let us consider the case of a square on a two-dimensional plane with Cartesian coordinates, whose vertices are located at points $(1, 1), (1, -1), (-1, 1), (-1, -1)$. Taking into account the rotation of the entire plane about the point of origin by 90 degrees clockwise, we can see that this transformation is clearly non-trivial, as it moves all points of the plane except the point $(0, 0)$. And yet when applied to the square (understood as the set of all points whose both coordinates are within the interval $[-1, 1]$), the rotation results in the same object.

---

[2] To my knowledge, this interpretation of structural realism has never been subject to extensive studies. Notable exceptions that should be mentioned here are (Gołosz, 2005) and (Glick, 2015).
[3] This broad approach to symmetries is adopted e.g. in the recent discussion on the empirical consequences of physical symmetries in (Greaves & Wallace, 2014).

Another option is to insist that a symmetry leaves an object invariant only in certain respects, while altering it with respect to other, irrelevant features.[4] This construal of the concept of symmetry is typical in applications to physical systems. Rotating a cube whose sides are painted in different colours can be considered a symmetry if we forget about colours and focus instead on the orientation of the sides themselves with respect to the rest of the universe. However, this rotation creates a physical change in some relational properties of the cube (different colours of the sides will be facing different directions), which we nevertheless choose to ignore.

As stated earlier, we will be interested exclusively in the concept of symmetries of a relational structure, where by a relational structure we understand an n-tuple consisting of a non-empty set D of objects and a set of relations defined on it: $\Re = \langle D, R_1, \ldots, R_n \rangle$. Symmetries of relational structures are transformations that involve the elements of their domains. Speaking figuratively, a symmetry "shuffles" the elements of the domain in such a way that the entire relational structure remains the same. This general depiction suggests that the change brought about by a symmetry transformation will be a change exclusively with respect to the arrangement of the non-qualitative elements of a structure, while the qualitative characteristics should remain the same in order to preserve the entire relational structure.[5] However, things are slightly more complicated than that. Consider the following simple example of a relational structure involving three geometric points a, b, c and the three relations $R3, R4, R5$ of being spatially separated by, respectively, 3 units, 4 units and 5 units, as depicted in Fig. 1. Suppose that we permute the points labeled a and b in the original structure. The resulting structure, as presented in Fig. 2, differs from the original one only with respect to the labels associated with particular points. Qualitatively, the structure looks the same, and therefore it may be tempting to conclude that the permutation π_{ab} is a symmetry of the considered structure. And yet this conclusion is incorrect. It is easy to observe

---

4 Among the advocates of this approach are Ladyman and Presnell (2016).
5 Recently Møller-Nielsen (2015) has argued that symmetries in physics can connect models that differ qualitatively. However, his concept of symmetry is based on the assumption, common in the context of philosophy of physics, that a symmetry of a given theory connects two solutions of the equations of this theory. If the formulation of the theory contains some "surplus" theoretical elements, such as the concept of absolute velocity in Galilean spacetime or absolute simultaneity in Minkowski spacetime, it may happen that different solutions may differ with respect to some qualitative characteristics. For instance, applying a velocity boost to a particular model of Newtonian gravitational theory yields a new model which is also a solution to the equations of the theory, and yet differs from the previous model with respect to its absolute velocity. However, the difference in absolute velocity is assumed to have no physical meaning, and as such is merely an artefact of the mathematical description. See (Møller-Nielsen, 2015, p. 190).

that the triangular structure from Fig. 1 admits no non-trivial (i.e. distinct from the identity) symmetries.

To see why this is the case, let us recall that according to the standard, model-theoretic definition a symmetry (also known as an automorphism) of a relational structure is a bijective mapping $\pi: D \to D$ of the domain onto itself such that for any relation R_i of the structure, $R_i(x_1, \ldots, x_n)$ iff $R_i(\pi(x_1), \ldots, \pi(x_n))$.[6] In the case of the permutation π_{ab} it is clear that it does not satisfy the above condition, since the relation $R3$ of being separated by three units holds between a and c but it does not hold between points b and c (thus $R3(a, c)$ but $\neg R3(\pi_{ab}(a), \pi_{ab}(c))$).[7] But this means that in order to assess whether a particular transformation is a symmetry we have to compare the transformed and untransformed structures with respect to something more than their purely qualitative similarity. We have to track the identity of bare, non-qualitative elements of the domain, so to speak. In particular, we have to know that the point in the upper left corner in Fig. 1 is the very same point that now occupies the upper right corner in Fig. 2. Only when the points in the transformed structure occupy the same qualitative places as their "counterparts" in the original structure, can we say that the considered transformation is a symmetry.

However, the necessity of identifying the bare elements of a structure pre- and post-transformation is bad news for the structuralist, because it suggests that objects possess their individualities independently of the relations they participate in. This observation can be expounded further using the metaphysical concept of

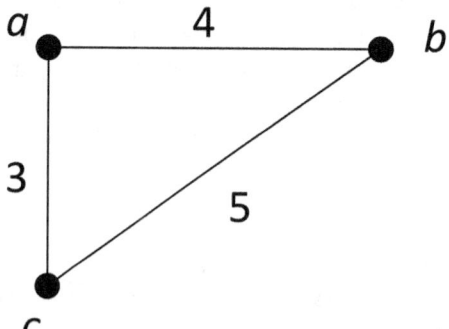

Fig. 1: A simple non-symmetric geometric structure.

---

[6] For a full definition of a model-theoretical symmetry (automorphism) of a structure see e.g. (Hodges, 1997, p. 5).

[7] The permutation π_{ab} would be a symmetry if the element c stood in the same relation to both a and b (for instance if the triangle was isosceles).

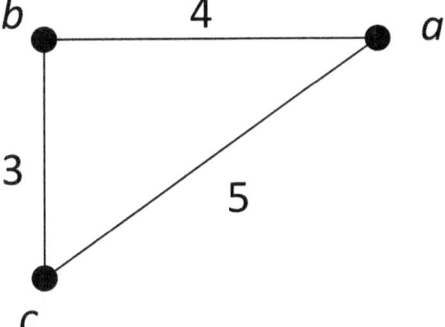

Fig. 2: The same structure after the permutation π_{ab}.

identity across possible worlds. The result of a particular transformation applied to the considered structure can be interpreted as representing a possible state of the same underlying reality (in short: a possible world). The post-transformation state of affairs consists of the same objects as the pre-transformation one, and the relation connecting the "same" objects is precisely identity across possible worlds. However, in order to distinguish transformations that are symmetries from transformations that are not, we have to treat as distinct possible worlds that differ only with respect to which objects are identical with which. This gives rise to what is called in the literature a "haecceitistic difference".[8] Two possible worlds display a haecceitistic difference if they are qualitatively identical and yet differ with respect to what their objects represent *de re*. Considering the structures depicted in Fig. 1 and 2 we can notice that the haecceitistic difference between them consists (partly) in the fact that the vertex located at the right angle of the triangle represents two distinct objects in each universe: in one case it represents the object labeled a, and in the other the object labeled b. But this difference is associated with no difference in the structural properties of both systems. Thus we have to admit that the numerical distinctness between the two vertices occupying two distinct possible worlds cannot be grounded in any structural differences between these worlds, and thus is a primitive fact. This is not acceptable for philosophers of the structuralist stripe.

---

[8] The concept of a haecceitistic difference between possible worlds goes back to (Lewis, 1986, especially p. 221). Of course the concept of a haecceity is a much older one, dating back to Duns Scotus. See also (Skow, 2008, 2011), (Fara, 2009) and (Bigaj, 2015a) for further discussions of different variants of haecceitism.

One conceivable response to this "metaphysical" argument against the standard model-theoretical concept of symmetry is that the standard definition need not be interpreted in terms of distinct possible structures (pre- and post-transformation) and identifications of objects across structures. Read literally, the model-theoretical definition of a symmetry of a relational structure refers only to "actual" objects in the domain and various mappings connecting these objects. My reply is that we are always free to take the standard definition literally; however, the broad intuition of how symmetries are used in mathematics and physics undeniably involves the notion of transforming an object. If we insist, as I think we should, on incorporating the narrow model-theoretical concept of a symmetry into this broader intuition, we have to face the interpretational question of how to make sense of the invariance of structure under such a transformation.

The main goal of this paper is to sketch an alternative conception of a symmetry of relational structures that would be fully compatible with the structuralist, anti-haecceitistic point of view. The starting point of this proposal is the requirement that the only way to identify objects in alternative scenarios be through their qualitative features. Thus we cannot simply point at a particular actual thing, e.g. this tree, and then pick any possible object (for instance a star) and pronounce that it is the very same object as that tree in some counterfactual situation. It is commonly assumed that objects possess certain properties essentially, which means that without these properties an object cannot retain its identity.[9] On top of that assumption the structuralist would like to add that the essential characteristic of a group of objects should include relations in which these objects stand to each other. These assumptions lead naturally to the introduction of the notion of an *essential structure*. Roughly speaking, a relational structure $\langle D, R_1, \ldots, R_n \rangle$ is essential if any possible structure that is meant to contain *the exact same* objects as $\langle D, R_1, \ldots, R_n \rangle$ has to stand in a particular "similarity" relation to it.[10] In the next section we will look closely at what choices we have for the requisite similarity relation between structures.

---

[9] For an extensive analysis of the concept of essential properties see e.g. (Mackie, 2006).

[10] We are not going to offer any formal definition of an essential structure besides this vague description. The main reason for this is that any such definition would have to rely on the concept of transworld identity, and we don't have an independent method of formally characterizing this concept without invoking the notion of essential properties or relations in turn. But I believe that no direct definition of essential and accidental structures is needed. Relational structures are essential not in virtue of some formal features they possess, but by stipulation. Identifying some properties or relations as essential is part of the modal characteristic of the entities that exemplify these properties and relations. We can choose different properties and relations as essential, and as a result end up with differently delineated entities.

Structuralist Identifications of Objects Across Possible Worlds

Let us select a distinguished set of relations $R_1, R_2, \ldots, R_n$ defined on some domain D whose instantiations are assumed to be essential, i.e. such that objects participating in these instantiations cannot cease to do so on pain of losing their identities. Suppose that we consider a certain actual object x. This object generates a particular structure $\mathfrak{R}_x = \langle D_x, R_1, \ldots, R_n \rangle$ in the following sense: $\mathfrak{R}_x$ is the smallest structure whose domain D_x contains x and all objects directly or indirectly connected to x via relations $R_1, \ldots, R_n$. More precisely, we will define structure $\mathfrak{R}_x$ as the result of restricting relations $R_1, \ldots, R_n$ to the subset $D_x \subseteq D$ defined as follows:
(1) $x \in D_x$,
(2) For all $y \in D_x$ and $z \in D$, if $R_i(\ldots y \ldots z \ldots)$, then $z \in D_x$,
(3) D_x is the smallest set satisfying (1)–(2).

The structure $\mathfrak{R}_x$ is the smallest structure containing x whose elements are fully identifiable in all possible worlds by their participation in the selected essential relations.

It can be easily verified that the structure generated by x does not depend on which representative object from the domain D_x we select; therefore we can drop the subscript x in $\mathfrak{R}_x$. The proof of this fact can be sketched as follows. Let $y \in D_x$. We can first prove that $D_y \subseteq D_x$. Clearly D_x satisfies conditions (1) and (2) with x in condition (1) replaced by y. Because D_y is by definition the smallest such set, the inclusion $D_y \subseteq D_x$ follows. Repeating the same reasoning with respect to x we can prove that $D_x \subseteq D_y$, hence both sets are identical. This means that once we have identified the structure $\mathfrak{R}_x$ we can equally well refer to it as $\mathfrak{R}$.

Structure $\mathfrak{R}$ encompasses all the necessary (and sufficient) information about the identity of the objects in its domain, so to speak. But in order to decipher this information we have to explicitly formulate a condition under which a possible structure $\mathfrak{R}' = \langle D', R'_1, \ldots, R'_n \rangle$ can be said to contain the same objects as $\mathfrak{R}$.[11] We will consider the following three options.

---

11 We have to keep in mind that while relations $R'_1, \ldots, R'_n$ may differ *extensionally* from their counterparts $R_1, \ldots, R_n$ in the actual world, they are supposed to be the *very same* relations, speaking intentionally. Thus we are committed to *quidditism* – the idea of the primitive identity of properties and relations across possible worlds.

1. Isomorphism. Suppose that we accept the following condition: for a (possible) structure $\mathfrak{R}'$ to consist of the very same objects that occur in $\mathfrak{R}$, structures $\mathfrak{R}$ and $\mathfrak{R}'$ have to be isomorphic. Two relational structures are considered isomorphic if there is a bijection f between their domains such that $R_i(x_1, \ldots, x_k)$ iff $R'_i(f(x_1), \ldots, f(x_k))$ for each R_i in the structure. This assumption seems intuitive, as the existence of an isomorphism between two structures means essentially that they are one and the same structure, barring the differences in the primitive individualities of their constitutive objects, which we consider irrelevant. However, it may be argued that the condition of identity across possible worlds based on isomorphism is too strong. It may happen, for instance, that the original essential structure contains two qualitatively indistinguishable objects. In that case it may be argued that we should allow for the possibility that in an alternative scenario there is only one possible object that represents *de re* both actual entities. The admissibility of such a situation is explicitly endorsed by David Lewis, who famously claimed that the counterpart relation between objects in alternative possible worlds need not be one-to-one (see Lewis, 1968, 1986).[12] An illustration of such a case is depicted in Fig. 3. Two objects in the actual world are assumed to have one and the same counterpart in a possible scenario, and yet the actual structure and the possible structure are clearly not isomorphic, as their domains are not even equinumerous.

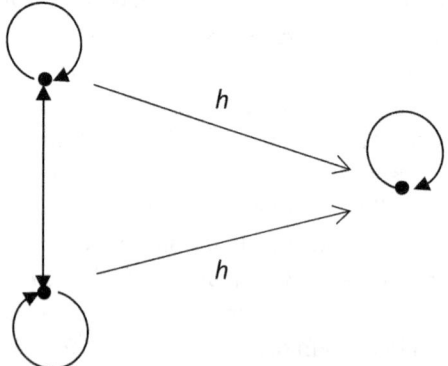

Fig. 3: Two homomorphic but not isomorphic structures. It can be claimed that the object on the right is a counterpart of both objects on the left.

---

12 Thus Lewis insists that the counterpart relation is not identity. In spite of that we will continue talking informally about identity (and identification) across possible worlds, keeping in mind that these expressions do not denote a relation that has the formal features of numerical identity.

2. Homomorphism. A concept weaker than isomorphism is that of a homomorphism. A homomorphism between structure $\Re$ and structure $\Re'$ is a map $h: D \to D'$ that need not be a bijection and that nevertheless preserves the relations of $\Re$ in the sense of satisfying the implications $R_i(x_1, \ldots, x_k) \Rightarrow R'_i(h(x_1), \ldots, h(x_k))$ for each R_i in the structure. The function mapping two objects in the first structure to the single object in the second structure in Fig. 3 is a homomorphism. Yet there may be arguments that the existence of a homomorphism between structures $\Re$ and $\Re'$ is too weak for securing the claim that the objects in $\Re'$ are the exact same entities as the objects in $\Re$. First of all, h does not have to be a surjection, which means that some objects in domain D' may not represent any objects in D. This can be easily remedied by considering a structure whose relations are restricted to the subdomain $h(D)$. But, more importantly, the existence of a homomorphism from $\Re$ to $\Re'$ does not guarantee that there will be a homomorphism in the opposite direction, from $\Re'$ to $\Re$. That is, some relations in structure $\Re'$ may turn out not to be interpretable within $\Re$. And this spells trouble for those who want to develop the concept of identity across possible worlds based on homomorphism. For now we have to accept that objects in $\Re'$ represent *de re* objects from $\Re$, but not vice versa. Or, to put it slightly differently, structure $\Re'$, represents one way the objects from $\Re$ might be, but $\Re$ does not represent a way the objects from $\Re'$ might be. This consequence seems to be in conflict with Lewis's stipulation that the terms "actual world" and "possible world" should be treated as indexicals. If an object that is actual from "our" perspective is represented *de re* by some possible object in another world, from the perspective of this other world it is our object that is possible and that should perform the *de re* representation.

3. The final proposal is to adopt a condition that falls somewhere between the too restrictive isomorphism and too relaxed homomorphism. It can be suggested that the required identity between objects occupying essential structures $\Re$ and $\Re'$ holds iff there is a surjective map $f: D \to D'$ such that $R_i(x_1, \ldots, x_k)$ iff $R'_i(f(x_1), \ldots, f(x_k))$ for each R_i in the structure. This condition is clearly weaker than the existence of a full isomorphism, since f in this case does not have to be one-to-one. Thus we can allow for situations in which two actual objects in $\Re$ are represented *de re* by one and the same possible object in $\Re'$. And yet it can be easily verified that the existence of a "reversed" homomorphism from D' to D is guaranteed. Even though f may not have an inverse, we can produce a homomorphism $h: D' \to D$ by first restricting f to a particular subset of D that does not contain two elements x and y such that $f(x) = f(y)$, and then taking its inverse $h = f^{-1}$. This inverse clearly satisfies the condition $R'_i(x_1, \ldots, x_k) \Rightarrow R_i(h(x_1), \ldots, h(x_k))$, thus it is a homomorphism. Moreover, in the same way we can produce alternative homomorphisms from D' into D so that the sum of all images of D' in these homomorphisms

will give us back the entire set D. Thus it can be claimed that each object in D represents *de re* some object in D'.

Proceeding slightly more formally, we can describe the proposed structural interpretation of the relation of transworld identity (counterpart relation) as follows. Starting with a selected object x whose counterparts in a given possible world we seek, we first identify its essential structure $\mathfrak{R}_x$ by following the above procedure with respect to the essential relations $R_1, \ldots, R_n$ that x enters into. An object x' in a given possible world is a counterpart of x, iff x' is an element of the domain D' of a structure $\mathfrak{R}' = \langle D', R'_1, \ldots, R'_n \rangle$, and there is a surjective function $f: D \to D'$ such that $R_i(x_1, \ldots, x_k)$ iff $R'_i(f(x_1), \ldots, f(x_k))$ for all R_i in $\mathfrak{R}_x$. Now we can verify that by the same definition x will be a counterpart of x' (and thus the counterpart relation is symmetric). In order to do that we have to take any function f^* which is a restriction of f to a subdomain D^* of D such that $f^*: D^* \to D'$ is one-to-one, and $x \in D^*$. Taking the inverse f^{*-1} we can immediately see that it is a homomorphism (in fact, an isomorphism) between $\mathfrak{R}'$ and the restriction of structure $\mathfrak{R}_x$ to the domain D^*. And clearly $f^{*-1}(x') = x$, which completes the proof.

While proposal 3 is the one that we are willing to ultimately accept as a formal explication of the structuralist concept of identity across possible worlds, we may note that in the context of the symmetries of relational structures the concept of transworld identity based on isomorphism is sufficient, since permutations of elements of relational structures are bijections. Thus in subsequent discussions we will follow the simpler concept 1. However, we should remember that generally the requirement of isomorphy is too strong for the identification of the counterparts of the elements of a given structure. There is no reason why an actual object participating in a given essential structure that contains N elements cannot have a counterpart in a possible world which consists of fewer than N objects.

A New Concept of Symmetry

Now we are in a position to make a distinction between two types of relational structures that objects can participate in. First, there are essential structures which, as explained above, equip the objects with their qualitative identities and thus enable us to identify actual objects in counterfactual, possible scenarios. But in addition to that, objects can participate in numerous relations that hold only accidentally. These relations constitute contingent structures that may vary from possible world to possible world. In particular, one type of change that a contingent structure can undergo is with respect to the arrangement of objects participating in it. If a rearrangement of objects produces the same qualitative structure, we call this transformation a symmetry. But we have to keep in mind that merely rearranging bare objects with no qualitative features is not allowed as a legitimate procedure. (Or, to put it differently, the result of such a "transformation" is always identical to the initial state, as we effectively rearrange the labels of objects, not objects themselves.) Instead, we have to use the essential structure in order to track the identity of rearranged objects. Thus we have to come up with a new definition of a transformation of a given contingent structure that would replace the ordinary concept of a permutation (bijective mapping) of its domain. In this new approach, a transformation of a given structure will be characterized "globally" as "acting" on the entire structure rather than on its elements. Thus it may be advisable to define a transformation not as a mapping but in terms of its "outcome" instead. A particular transformation of a given structure will be just a new structure that stands in a certain formal relation to the original one, as described below.

Let $\mathfrak{R} = \langle D, R_1, \ldots, R_n \rangle$ be designated as a contingent structure, and $\mathfrak{S} = \langle D, S_1, \ldots, S_m \rangle$ as an essential one.

Definition 1. *A transformation of structure $\mathfrak{R}$ with respect to $\mathfrak{S}$ is any structure $\mathfrak{R} \oplus \mathfrak{S}' = \langle D, R_1, \ldots, R_n, S_1', \ldots, S_m' \rangle$ such that $\mathfrak{S}$ and $\mathfrak{S}'$ are isomorphic. We will symbolize a transformation of $\mathfrak{R}$ with respect to $\mathfrak{S}$ as $\phi(\mathfrak{R} \oplus \mathfrak{S}) = \mathfrak{R} \oplus \mathfrak{S}'$, where $\mathfrak{R} \oplus \mathfrak{S} = \langle D, R_1, \ldots, R_n, S_1, \ldots, S_m \rangle$. We will call a transformation of $\mathfrak{R}$ with respect to $\mathfrak{S}$ "trivial" (resp. "identity"), if $\mathfrak{S}' = \mathfrak{S}$.*

The broad idea behind the proposed definition is that the rearrangement of objects in the original structure is performed not by permuting the domain D but by "moving" the entire essential structure $\mathfrak{S}$ with respect to the contingent structure $\mathfrak{R}$. The requirement that $\mathfrak{S}$ and $\mathfrak{S}'$ be isomorphic amounts to the assumption that both structures contain the very same objects. However, the same objects can now occupy different places in the contingent structure $\mathfrak{R}$.

We may now ask the question of how the new notion of a transformation of a structure relates to the old one, based on the concept of a permutation of individual objects. It turns out that their relation is a close one, albeit for obvious reasons these two concepts do not coincide. First of all, let us observe that each bijective mapping f of the domain D generates a unique corresponding transformation $\phi_f(\mathfrak{R} \oplus \mathfrak{S})$ of structure $\mathfrak{R}$ with respect to $\mathfrak{S}$ in the sense of Def. 1. We simply select $\phi_f(\mathfrak{R} \oplus \mathfrak{S})$ as identical with $\mathfrak{R} \oplus \mathfrak{S}'$, where the transformed essential structure $\mathfrak{S}'$ is defined as follows: $S_i'(x_1, \ldots, x_k)$ iff $S_i(f^{-1}(x_1), \ldots, f^{-1}(x_k))$ (thus $\mathfrak{S}'$ can be alternatively symbolized as $f(\mathfrak{S})$). However, a particular structure-level transformation ϕ does not select a unique object-level mapping f. This is so because there can be more than one isomorphism connecting essential structures $\mathfrak{S}$ and $\mathfrak{S}'$. This happens when structure $\mathfrak{S}$ admits non-trivial automorphisms. It can be easily verified that when f is an isomorphism between $\mathfrak{S}$ and $\mathfrak{S}'$ and g is an automorphism of $\mathfrak{S}$, then $f \circ g$ is another isomorphism connecting the two structures $\mathfrak{S}$ and $\mathfrak{S}'$. More precisely, we can prove the following fact:

Fact 1. *The following two conditions are equivalent:*
(1) *For every transformation $\phi(\mathfrak{R} \oplus \mathfrak{S}) = \mathfrak{R} \oplus \mathfrak{S}'$ there is a unique isomorphism f between structures $\mathfrak{S}$ and $\mathfrak{S}'$;*
(2) *Structure $\mathfrak{S}$ has no non-trivial (i.e. different from the identity) automorphisms.*

Proof.
(1) → (2). Suppose that g is an automorphism of $\mathfrak{S}$ different from the identity. Let f be an isomorphism between $\mathfrak{S}$ and $\mathfrak{S}'$ (its existence is guaranteed by Def. 1). Clearly $f \circ g$ is another isomorphism connecting $\mathfrak{S}$ and $\mathfrak{S}'$, and $f \circ g$ is distinct from f. Thus (1) is not satisfied.
(2) → (1). Suppose that there are two distinct isomorphisms f and f' connecting $\mathfrak{S}$ and $\mathfrak{S}'$. Now we can easily confirm that the composition $f^{-1} \circ f$ is an automorphism of $\mathfrak{S}$, and that it is different from the identity. Hence (2) is violated. □

The existence of non-trivial automorphisms of the essential structure $\mathfrak{S}$ can be interpreted as indicating that $\mathfrak{S}$ does not uniquely identify all its elements. Thus we can conclude from Fact 1 that if $\mathfrak{S}$ uniquely identifies all of its elements, there is a one-to-one correspondence between the set of all transformations of a given structure $\mathfrak{R}$ with respect to $\mathfrak{S}$ and the set of all permutations of the domain D. However, in the case when $\mathfrak{S}$ fails to discern some of its distinct elements, the set of transformations of $\mathfrak{R}$ with respect to $\mathfrak{S}$ will be smaller than the set of all permutations. Some distinct permutations will correspond to one and the same transformation of $\mathfrak{R}$ with respect to $\mathfrak{S}$ in the sense of Def. 1.

One clear advantage of Def. 1 over the standard approach to the concept of symmetry is that now we are able to distinguish symmetry from non-symmetry transformations on the basis of the qualitative difference between the initial and the final state of the transformation, with no need to invoke the suspicious notion of primitive identity (haecceity). The definition of a symmetry is straightforward.

Definition 2. *A transformation $\phi(\mathfrak{R} \oplus \mathfrak{S}) = \mathfrak{R} \oplus \mathfrak{S}'$ is a symmetry of $\mathfrak{R}$ with respect to $\mathfrak{S}$ (in short: an $\mathfrak{S}$-symmetry) iff structures $\mathfrak{R} \oplus \mathfrak{S}$ and $\mathfrak{R} \oplus \mathfrak{S}'$ are isomorphic.*

Putting together definitions 1 and 2 we can see that a transformation which is not an $\mathfrak{S}$-symmetry relates two combined structures $\mathfrak{R} \oplus \mathfrak{S}$ and $\mathfrak{R} \oplus \mathfrak{S}'$ that are not isomorphic, even though the substructures $\mathfrak{S}$ and $\mathfrak{S}'$ are isomorphic. As the notion of isomorphism is clearly acceptable to a structuralist, the new concept of a symmetry should not be objectionable either.

In the last step we will contrast the concept of symmetry as expressed in Def. 2 with the standard interpretation mentioned in section 1. Again, these two notions are related, but it may be claimed that Def. 2 provides us with a broader conception of what a symmetry is. First, let us note that there is a close correspondence (albeit not one-to-one) between model-theoretic symmetries and $\mathfrak{S}$-symmetries of a given structure $\mathfrak{R}$. Each mapping $f: D \to D$ which is a model-theoretic symmetry of $\mathfrak{R}$ induces an $\mathfrak{S}$-symmetry of $\mathfrak{R}$, as seen in the following fact:

Fact 2. *If f is a model-theoretic symmetry of $\mathfrak{R}$, then $\phi_f(\mathfrak{R} \oplus \mathfrak{S}) = \mathfrak{R} \oplus f(\mathfrak{S})$ is an $\mathfrak{S}$-symmetry of $\mathfrak{R}$.*

Proof of this fact is straightforward. By definition, the structure $\mathfrak{R} \oplus f(\mathfrak{S})$ has the form $\langle D, R_1, \ldots, R_n, S'_1, \ldots, S'_m \rangle$, where $S'_i(x_1, \ldots, x_k)$ iff $S_i(f^{-1}(x_1), \ldots, f^{-1}(x_k))$. By assumption f is an automorphism of $\mathfrak{R}$, hence $f(\mathfrak{R}) = \mathfrak{R}$. Thus f is an isomorphism connecting structures $\mathfrak{R} \oplus \mathfrak{S}$ and $\mathfrak{R} \oplus f(\mathfrak{S})$. This means that the transformation given by $\mathfrak{R} \oplus f(\mathfrak{S})$ is an $\mathfrak{S}$-symmetry of $\mathfrak{R}$.

Analogously, and rather obviously, we can prove that all model-theoretic symmetries of the essential structure $\mathfrak{S}$ induce $\mathfrak{S}$-symmetries of $\mathfrak{R}$ too, as seen in the following fact:

Fact 3. *If f is a model-theoretic symmetry of $\mathfrak{S}$, then $\phi_f(\mathfrak{R} \oplus \mathfrak{S}) = \mathfrak{R} \oplus f(\mathfrak{S})$ is an $\mathfrak{S}$-symmetry of $\mathfrak{R}$.*

Fact 3 follows immediately from the assumption that $f(\mathfrak{S}) = \mathfrak{S}$ (in this case the corresponding symmetry ϕ_f is just the identity). Facts 2 and 3 can be illustrated with the help of a simple example. Let us consider a domain consisting of three objects x_1, x_2, x_3, and let the contingent structure include just one binary relation R, while the essential structure $\mathfrak{S}$ contains two properties (i.e. one-argument relations) S_1 and S_2. Both structures are depicted in Fig. 4a, where arrows represent relation R, and properties S_1 and S_2 are assigned to appropriate objects as indicated.

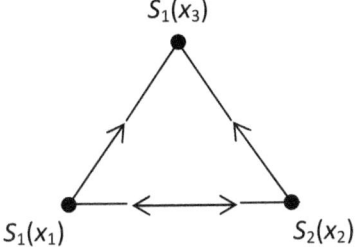

Fig. 4a: Essential and accidental structures.

Now we can consider two bijections f_1 and f_2 of the domain, defined as follows: $f_1(x_1) = x_2, f_1(x_2) = x_1, f_1(x_3) = x_3, f_2(x_1) = x_3, f_2(x_3) = x_1, f_2(x_2) = x_2$. It can be immediately verified that f_1 is a symmetry of $\mathfrak{R}$ but not $\mathfrak{S}$, while f_2 is a symmetry of $\mathfrak{S}$ but not $\mathfrak{R}$. Results of corresponding $\mathfrak{S}$-transformations ϕ_{f_1} and ϕ_{f_2} of $\mathfrak{R}$ are depicted on Figs. 4b and 4c. From the diagrams we can clearly see that the structure ϕ_{f_1} is isomorphic with the original one, while ϕ_{f_2} is just identical with it (in the strong, model-theoretical sense – but see a comment in footnote 13).

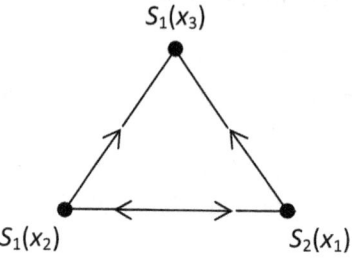

Fig. 4b: Result of the application of transformation $\phi_{f_1} = \mathfrak{R} \oplus f_1(\mathfrak{S})$.

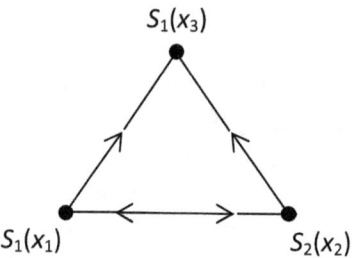

Fig. 4c: Result of the application of transformation $\phi_{f_2} = \mathfrak{R} \oplus f_2(\mathfrak{S})$.

Moving on, we should note that there may be two distinct model-theoretic symmetries of $\mathfrak{R}$ f and f' such that they induce the very same $\mathfrak{S}$-symmetry of $\mathfrak{R}$. This possibility is excluded only when structure $\mathfrak{S}$ uniquely identifies its elements, i.e. does not admit any non-trivial automorphisms. Again, this can be stated in the form of the following fact, analogous to Fact 1.

Fact 4. *If structure $\mathfrak{S}$ does not admit any non-trivial automorphisms, and f and f' are model-theoretic symmetries of $\mathfrak{R}$ such that $\mathfrak{R} \oplus f(\mathfrak{S}) = \mathfrak{R} \oplus f'(\mathfrak{S})$, then $f = f'$.*

Proof. Suppose that $\mathfrak{R} \oplus f(\mathfrak{S}) = \mathfrak{R} \oplus f'(\mathfrak{S})$ but $f \neq f'$. Since both f and f' are isomorphisms between structures $\mathfrak{R} \oplus \mathfrak{S}$ and $\mathfrak{R} \oplus f(\mathfrak{S})$ (as implied by Fact 2), the composition $f'^{-1} \circ f$ is an automorphism of $\mathfrak{R} \oplus \mathfrak{S}$ and thus of $\mathfrak{S}$. But $f'^{-1} \circ f$ is different from the identity, hence $\mathfrak{S}$ admits non-trivial automorphisms. □

We can conclude that a necessary and sufficient condition for the one-to-one correspondence between $\mathfrak{S}$-symmetries of $\mathfrak{R}$ and symmetries of $\mathfrak{R}$ in the standard, model-theoretic sense, is that $\mathfrak{S}$ uniquely identifies all its elements ($\mathfrak{S}$ does not admit any non-trivial automorphisms). Thus as long as this condition is satisfied, the structurally acceptable concept of a symmetry coincides with the ordinary notion.[13] However, when $\mathfrak{S}$ admits non-trivial automorphisms, the number of symmetries of a given structure $\mathfrak{R}$ with respect to $\mathfrak{S}$ will be reduced in comparison to the number of ordinary symmetries. In the extreme case when $\mathfrak{S}$ does not distinguish any of its elements (i.e. when all permutations are its automorphisms), the symmetries of $\mathfrak{R}$ with respect to $\mathfrak{S}$ (as a matter of fact, all transformations of $\mathfrak{R}$ with respect to $\mathfrak{S}$) reduce to the identity.

We can close this section with an interesting side question: how do Def. 1 and 2 behave in the case when the essential and contingent structures are identical? In other words, what are the symmetries of a structure $\mathfrak{R}$ with respect to itself? It can be easily verified that in this case all permutation-based symmetries of $\mathfrak{R}$ will be represented by the trivial identity transformation of $\mathfrak{R}$ with respect to $\mathfrak{R}$. That is, if f and g are two distinct symmetries of $\mathfrak{R}$, then $\phi_f : \mathfrak{R} \oplus \mathfrak{R} \to \mathfrak{R} \oplus f(\mathfrak{R})$ and $\phi_g : \mathfrak{R} \oplus \mathfrak{R} \to \mathfrak{R} \oplus g(\mathfrak{R})$, but because $f(\mathfrak{R}) = g(\mathfrak{R}) = \mathfrak{R}$, $\phi_f = \phi_g$. This means that in the structural approach defended above we cannot properly analyze the set of symmetries for the essential structure $\mathfrak{S}$, unless we do this with respect to a different underlying essential structure whose symmetries have to be identified with respect to yet another essential structure, and so forth. Obviously, the threat of an infinite regress looms large here.

---

[13] We have to keep in mind, however, that the individuation (and hence counting) of various structure-level symmetries is not possible in a language acceptable for the essential structuralist. In order to distinguish between two $\mathfrak{S}$-symmetries ϕ_f and ϕ_g we have to discern two structures $\mathfrak{R} \oplus f(\mathfrak{S})$ and $\mathfrak{R} \oplus g(\mathfrak{S})$ which by assumption are isomorphic with $\mathfrak{R} \oplus \mathfrak{S}$, and therefore are isomorphic with each other. And two isomorphic structures can differ from each other only with respect to the arrangement of bare individuals. From the structuralist perspective all symmetries of a given structure are *de facto* identities.

Possible Applications in Physics

The concept of symmetry plays a crucial role in the foundational analysis of physical theories. It is often claimed that symmetries of a given physical theory provide important information regarding the metaphysical consequences of this theory. They namely separate parts of theoretical description that represent genuinely real features of the world from those that do not (cf. Saunders, 2003; Møller-Nielsen, 2015, p. 180). Structures that remain invariant under a given theory's symmetries are considered real, while those that vary with symmetry transformations are treated as mere artefacts of the theory (so-called "surplus" structures). Below we will present a few typical examples of symmetries in physical theories, and we will show how the structuralist-friendly concept of symmetry developed in the previous section can be helpful in analyzing the meaning of these examples. Due to the restrictions of space the subsequent analysis will be rather sketchy at places, but I hope that I will be able to fill in the gaps on a different occasion.

Symmetries encountered in the context of physical theories can be discrete or continuous.[14] The prime example of a discrete symmetry is the permutation symmetry of the quantum-mechanical states of many particles of the same type (e.g. electrons, photons). If $\psi(1, 2, \ldots, n)$ represents the physical state of n indistinguishable particles, then the transformed mathematical object $\psi(\pi(1), \pi(2), \ldots, \pi(n))$, where π is any permutation of the set $\{1, 2, \ldots, n\}$, should represent the same physical state. Since in the standard quantum-mechanical formalism physical states are represented, up to an inessential phase factor, by normalized vectors in an appropriate Hilbert space, the requirement of permutation symmetry leads to the selection of two types of quantum objects: bosons, for which any permutation applied to the initial vector ψ gives back the same vector ψ, and fermions, whose states are antisymmetric (meaning that odd permutations change the sign of the vector, while even permutations leave the vector unchanged).[15] The restriction of the physical states available for bosons and fermions to, respectively, symmetric and antisymmetric ones, has a profound impact on the physics of quantum particles. In particular, the statistical behaviours of groups of bosons and fermions differ significantly from each other and from the behaviour of classical particles. Bosons obey so-called Bose-Einstein statistics, which implies that the probability of finding all particles in the same quantum state is slightly higher than the probability

---

[14] See (Brading & Castellani, 2003) for an excellent survey of various philosophical problems surrounding both types of symmetries.

[15] For an introduction to the quantum theory of many particles of the same type the reader can consult any textbook, e.g. (Sakurai & Napolitano, 2011, chapter 7).

predicted by Maxwell-Boltzmann statistics for classical particles. Fermions, on the other hand, are governed by Fermi-Dirac statistics, which assigns zero probability to the states in which distinct particles have the same quantum properties (the Pauli exclusion principle), and thus Maxwell-Boltzmann statistics is violated as well.

In order to apply the new concept of symmetry to the case of quantum particles of the same type, we have to first identify essential structures determining the identity of these particles. It seems quite natural to assume that the essential structure for a given type of particle should consist of their so-called state-independent properties, i.e. properties that do not change over time. These properties, which are typically used to identify and categorize particles, include such measurable quantities as rest mass, electric charge, and total spin. The key fact to note is that all particles belonging to the same category possess precisely the same state-independent properties. Thus the essential structure $\mathfrak{S}$ consisting of particles of a certain type (e.g. electrons) and their state-independent properties is maximally symmetric in the sense that each permutation of the underlying domain is an automorphism of structure $\mathfrak{S}$. This means, as we have already noted in the previous section, that all transformations of any contingent structure $\mathfrak{R}$ with respect to $\mathfrak{S}$ (in the sense of Def. 1) collapse into the identity. Consequently, when we take as $\mathfrak{R}$ the structure consisting of the domain D of particles of the same type and their total state ψ: $\mathfrak{R} = \langle D, \psi \rangle$, all transformations of $\mathfrak{R}$, with respect to the essential structure $\mathfrak{S}$ are symmetries in the sense of Def. 2.

This result may seem paradoxical. As all available transformations of structure $\mathfrak{R}$ preserve it, it looks as though there is no need to make any restrictions on the admissible states of particles of the same type. All structures of the form $\langle D, \psi \rangle$ remain invariant under any "shift" of the essential structure $\mathfrak{S}$, since $\mathfrak{S}$ is perfectly symmetric. And this is true regardless of what particular form the state vector ψ has. The conclusion that can be drawn from this example is that the restriction of the state subspaces accessible for bosons and fermions to, respectively, symmetric and antisymmetric sections of the whole Hilbert space, is not a simple consequence of permutation invariance. After all, permutation invariance is automatically guaranteed once we accept that particles of the same type possess the same essential properties. Exchanging two objects with the same essential properties leads to the joint state identical with the original one, regardless of what contingent properties or relations these objects instantiate (cf. Bigaj, 2015b). The quantum-mechanical postulate demanding that the states of bosons (fermions) of the same type be symmetric (antisymmetric) – and, consequently, the Pauli exclusion principle applied to fermions – should be seen as an extra addition to the usual permutation invariance requirement, specific only to quantum mechanics and confirmed independently in numerous experiments.

The most typical examples of continuous symmetries are symmetries of spatiotemporal structures. In this survey we will discuss three well-known types of spatiotemporal transformations: the Leibniz shift, the Galilean boost, and the diffeomorphisms of general relativity (in particular the famous "hole transformation"). A Leibniz shift is a transformation that translates uniformly all the matter in the universe in an arbitrary spatial direction by a fixed distance. This transformation is typically considered in the context of Newtonian mechanics with the notion of absolute space. A model of this theory can be given in the form of the following structure: $\mathfrak{M} = \langle M, t_a, h^{ab}, \nabla_a, \sigma^a, \Phi \rangle$, where M is a set of spatiotemporal points (equipped with the basic topological features, and thus referred to as a "differentiable manifold"), t_a is the temporal metric, h^{ab} is the Euclidean metric defined on the 3-dimensional instantaneous spaces, ∇_a is an affine connection, σ^a encodes the notion of being the "same" point of space over time, and Φ represents the distribution of matter in space-time.

Strictly speaking, the structure identified above is not a relational structure in the sense introduced in section 1, as the characteristics of the domain are not relations but functions assigning mathematical objects to elements of the domain. But it is straightforward to generalize all the requisite notions of symmetries and the like so that they apply to structures involving functions as well as relations. This can be done as follows. The definition of an isomorphism requires a separate clause covering functions: if ϕ is an n-argument function on elements of the domain in one structure, and ϕ' its "counterpart" in another structure, then for a function f to be an isomorphism the following has to hold: $\phi(x_1, \ldots, x_n) = \phi'(f(x_1), \ldots, f(x_n))$. Then Defs 1 and 2 of transformations and symmetries remain unchanged, except now all descriptions of structures (essential and contingent) may contain function symbols as well as relation symbols. For a more extensive discussion of this and other spatiotemporal structures encountered in physics see (Møller-Nielsen, 2015, p. 189ff), (Pooley, 2013), and (Friedman, 1983, ch. 3).

In order to give a structuralist interpretation of the Leibniz shift we have to again identify essential and contingent structures in this case. Not surprisingly, we will assume that the essential structure determining the identity of points in M consists of all spatial and temporal relations, leaving the distribution of matter as the only contingent feature of the world. The essential spatiotemporal structure $\mathfrak{S} = \langle M, t_a, h^{ab}, \nabla_a, \sigma^a \rangle$ is highly symmetric (spacetime in Newtonian mechanics is obviously flat), and translations of points in instantaneous spaces are among its automorphisms. That is, if f is a transformation of M that uniformly translates all points along the hypersurfaces of simultaneity, then clearly points x and $f(x)$ will have the same temporal characteristic expressed in one-form t_a, as well as the same vector σ^a orthogonal to the hypersurface of simultaneity. Moreover, the Euclidean distance between simultaneous points x and y, as encoded in the tensor

h^{ab}, will be identical to the distance between $f(x)$ and $f(y)$, which proves that f is a model-theoretical automorphism of structure $\mathfrak{S}$. Consequently, the corresponding structure-level transformation is the identity (this follows directly from Fact 3), and therefore evidently is a symmetry of the contingent structure $\mathfrak{R} = \langle M, \Phi \rangle$. The structural approach to symmetry confirms the well-known fact that the Leibniz shift is among the symmetries of Newtonian mechanics.

The situation changes when we move to the case of a transformation that applies a uniform velocity to the whole universe (the so-called Galilean boost). In that case, unless $\Phi(x)$ has a highly symmetric form (if, for instance, Φ is a constant function throughout the whole spacetime), the transformation will lead to a new structure that is not isomorphic with the original one. To illustrate this, let us use a simple example of a universe with only one, stationary and spatially unextended object. In the original structure all the spatiotemporal points x such that $\Phi_0(x) \neq 0$ by assumption satisfy the predicate of coinciding spatially (having the same spatial location). But when we apply a boost to the object, thus obtaining a new "density function" Φ'_0, its spatial locations at different times will no longer coincide, and thus no isomorphism can exist between structures $\langle M, t_a, h^{ab}, \nabla_a, \sigma^a, \Phi \rangle$ and $\langle M, t_a, h^{ab}, \nabla_a, \sigma^a, \Phi' \rangle$. On the other hand, if we abandon the notion of absolute space encoded in the sigma-field σ^a, as in Galilean spacetime, a velocity boost will become a symmetry.

So far the new concept of symmetry leads to the same conclusions regarding the status of the considered spatiotemporal transformations as the standard analysis. However, the case of a hole transformation is slightly different. A model of GR can be presented in the form of a triple $\langle M, g_{ab}, T_{ab} \rangle$, where g_{ab} is the metric tensor, and T_{ab} the stress-energy tensor, representing the distribution of matter in the universe. Applying a particular diffeomorphism d of the manifold M, we can create a new model $\langle M, d^*g_{ab}, d^*T_{ab} \rangle$ in the usual way, by "dragging along" the mathematical objects g_{ab} and T_{ab} following the point-level transformation d (i.e. the new object d^*g_{ab} is assigned to a point x iff the old object g_{ab} is assigned to the point $d^{-1}(x)$). Thanks to the feature of GR known as general covariance, this new model will be a solution to the equations of the theory, hence diffeomorphisms are considered symmetries of GR. The widely discussed consequence of this fact is that GR seems to be affected by a peculiar sort of indeterminism. Considering a compact region H of spacetime, we can define a so-called hole diffeomorphism by stipulating that d is the identity outside of H, and smoothly transforms into a non-identity inside H.[16] This means that the transformed model $\langle M, d^*g_{ab}, d^*T_{ab} \rangle$ will assign the same mathematical objects to points outside of H but will "shuffle"

---

16 For an introduction to the hole argument see (Norton, 2015).

these objects within H. Hence the assignment of objects g_{ab} and T_{ab} to the outside of H does not uniquely determine an appropriate assignment inside H.

The hole argument is typically considered a challenge to the substantivalist who believes in the objective and independent existence of spatiotemporal points (see a discussion in Pooley, 2013). A popular strategy of defence for the substantivalist is to adopt the stance of sophisticated substantivalism which rejects the concept of "bare points" and instead insists that points possess their metrical features essentially.[17] This assumption offers two possible ways of rejecting the hole argument. We can either claim that the transformed mathematical model $\langle M, d^*g_{ab}, d^*T_{ab}\rangle$ does not represent any physical reality, since it is metaphysically impossible for a given point to possess a value of the metric tensor g_{ab} different from the actual one (cf. Maudlin, 1988, 1990). Or, we can maintain that the model $\langle M, d^*g_{ab}, d^*T_{ab}\rangle$ represents the same reality as $\langle M, g_{ab}, T_{ab}\rangle$ when we reidentify the points in both models according to the metric properties they possess (cf. Butterfield, 1989; Brighouse, 1994; Bartels, 1996; Pooley, 2006).

The new approach to the concept of symmetries advocated in this paper adds more strength to the essentialism-based defense of substantivalism. Again, we will start by making the distinction between the essential structure encompassed in the metric tensor g_{ab}, and the contingent structure containing the physical stress-energy tensor T_{ab}. The key point to observe is that in the proposed approach the structure-level transformation from $\langle M, g_{ab}, T_{ab}\rangle$ to $\langle M, d^*g_{ab}, d^*T_{ab}\rangle$ does not even qualify as a legitimate transformation of the contingent structure $\langle M, T_{ab}\rangle$ with respect to the essential structure $\langle M, g_{ab}\rangle$. As we can recall, Def. 1 implies that in our particular case a transformation of the original joint structure $\langle M, g_{ab}, T_{ab}\rangle$ is a structure of the form $\langle M, g'_{ab}, T_{ab}\rangle$ such that the structures $\langle M, g_{ab}\rangle$ and $\langle M, g'_{ab}\rangle$ are isomorphic. This condition is satisfied if we apply a particular diffeomorphism d to the metric structure only, and not to the metric and physical structures simultaneously; thus the outcome of the structure-level transformation should be $\langle M, d^*g_{ab}, T_{ab}\rangle$ rather than $\langle M, d^*g_{ab}, d^*T_{ab}\rangle$. Speaking figuratively, we are allowed to move the essential metric structure with respect to

---

[17] It may be objected here that the metric tensor is an unlikely candidate for an essential property of points in the context of general relativity, since metrical properties of spacetime in this theory vary concurrently with the contingent distribution of matter in the universe. An extensive response to this worry can be found in (Maudlin, 1990, pp. 548–551). Here I will not repeat Maudlin's elaborate arguments in detail, noting only that GR does not imply that metrical properties of spatiotemporal regions literally vary in time, since they themselves constitute what time is. As for the link between the essential (metrical) and contingent (physical) structures, Maudlin observes that the regularity expressed in Einstein's field equation is itself metaphysically contingent, hence it is metaphysically (but not nomically) possible that a given region of space-time could have a different distribution of matter from what it has in actuality.

the contingent, physical structure, but not to move both. Technically, the transformation $\langle M, g_{ab}, T_{ab}\rangle \to \langle M, d^*g_{ab}, d^*T_{ab}\rangle$ could be subsumed under the concept described in Def. 1 if we reinterpreted the image structure as consisting of the domain d^*M, whose points $d(x)$ are identified with points x in M. But in that case the result would be the trivial identity, and both models would refer to the same physical reality.

In contrast to that, the structure-level transformations of the form $\langle M, g_{ab}, T_{ab}\rangle \to \langle M, d^*g_{ab}, T_{ab}\rangle$ are definitely non-trivial. But it is easy to observe that, barring cases of highly symmetric distributions of matter, these transformations will usually not be symmetries of the contingent physical structure $\langle M, T_{ab}\rangle$. Generally, points with the same values of the metric tensor in both structures may receive different values of the stress-energy tensor, so the two metric-physical structures will not be isomorphic. If it were possible to prove generally that some hole diffeomorphisms d would always produce models $\langle M, g_{ab}, T_{ab}\rangle$ and $\langle M, d^*g_{ab}, T_{ab}\rangle$ that are isomorphic, this indeed would create a great challenge to the substantivalist. For in that case literally the same points (in the much more substantive sense of the word "same" than "having the same haecceity") inside the hole could have different physical properties in spite of the fact that the outside would remain the same. But the general covariance of GR clearly does not guarantee that.

Conclusion

The proposed reconstruction of the standard model-theoretic concept of a symmetry in terms of essential structures offers several advantages. It does not rely on the haecceitistic, non-qualitative identification of objects in alternative structures, and therefore appears to be compatible with the position of essentialist structuralism. It highlights the often neglected fact that the existence and number of symmetries for a given structure should depend on the details of how we identify the individual elements of the domain. The suggested, structuralist-friendly approach to symmetries provides a new and intriguing perspective on the role of the concept of symmetry in physics. Among its potentially fruitful applications is a novel way of rejecting the hole argument against substantivalism in GR, as well as a non-standard analysis of the relation between permutation invariance and symmetrization postulate in the quantum theory of many particles.

Acknowledgment: I am deeply grateful to Thomas Müller and Sam Fletcher for their extensive critical comments to an earlier version of this paper. The work on the paper was supported by the Marie Curie International Outgoing Grant FP7-PEOPLE-2012-IOF-328285.

Bibliography

Bartels, A. (1996). Modern essentialism and the problem of individuation of spacetime points. *Erkenntnis, 45*, 25–43.
Bigaj, T. (2014). In defense of an essentialist approach to ontic structural realism. *Methode: Analytic Perspectives, 3*(4), 1–24.
Bigaj, T. (2015a). Essentialism and modern physics. In T. Bigaj, & C. Wüthrich (Eds.), *Poznań Studies in the Philosophy of the Sciences and the Humanities: Vol. 104. Metaphysics in contemporary physics* (pp. 145–178). Leiden–Boston: Brill–Rodopi.
Bigaj, T. (2015b). Exchanging quantum particles. *Philosophia Scientiae, 19*(1), 185–198.
Brading, K., & Castellani, E. (Eds.). (2003). *Symmetries in physics: Philosophical reflections*. Cambridge: Cambridge University Press.
Brighouse, C. (1994). Spacetime and holes. In D. Hull, M. Forbes, & R. M. Burian (Eds.), *PSA 1994, Vol. 1* (pp. 117–125). East Lansing: Philosophy of Science Association.
Butterfield, J. (1989). The hole truth. *British Journal for the Philosophy of Science, 40*, 1–28.
Fara, D. G. (2009). Dear haecceitism. *Erkenntnis, 70*, 285–297.
Friedman, M. (1983). *Foundations of space-time theories. Relativistic physics and philosophy of science*. Princeton: Princeton University Press.
Glick, D. (2015). Minimal structural essentialism. In A. Guay, & T. Pradeau (Eds.), *Individuals across the sciences* (pp. 1–28). Oxford: Oxford University Press.
Gołosz, J. (2005). Structural essentialism and determinism. *Erkenntnis, 63*, 73–100.
Greaves, H., & Wallace, D. (2014). Empirical consequences of symmetries. *British Journal for the Philosophy of Science, 65*, 59–89.
Hodges, W. (1997). *A shorter model theory*. Cambridge: Cambridge University Press.
Ladyman, J. (2014). Structural realism. In E. Zalta (Ed.), *The Stanford Encyclopedia of Philosophy (Spring 2014 edition)*. Retrieved from http://plato.stanford.edu/archives/spr2014/entries/structural-realism
Ladyman, J., & Presnell, S. (2016). *Representation and symmetry in physics*. (Manuscript).
Lewis, D. (1968). Counterpart theory and quantified modal logic. *The Journal of Philosophy, 65*, 113–126.
Lewis, D. (1986). *On the plurality of worlds*. Oxford: Blackwell.
Mackie, P. (2006). *How things might have been. Individuals, kinds, and essential properties*. Oxford: Clarendon Press.
Maudlin, T. (1988). The essence of space-time. *Proceedings of the Biennial Meeting of the Philosophy of Science Association, Vol. 2*, 82–91.
Maudlin, T. (1990). Substances and space-time: what Aristotle would have said to Einstein. *Studies in History and Philosophy of Science, 21*, 531–561.
Møller-Nielsen, T. (2015). Symmetry and qualitativity. In T. Bigaj, & C. Wüthrich (Eds.), *Poznań Studies in the Philosophy of the Sciences and the Humanities: Vol. 104. Metaphysics in contemporary physics* (pp. 179–214). Leiden–Boston: Brill–Rodopi.

Norton, J. D. (2015). The hole argument. In E. Zalta (Ed.), *The Stanford Encyclopedia of Philosophy (Fall 2015 edition)*. Retrieved from http://plato.stanford.edu/archives/fall2015/entries/spacetime-holearg

Pooley, O. (2006). Points, particles, and structural realism. In D. Rickles, S. French, & J. Saatsi (Eds.), *The structural foundations of quantum gravity* (pp. 83–120). Oxford: Oxford University Press.

Pooley, O. (2013). Substantivalist and relationalist approaches to spacetime. In R. Batterman (Ed.), *The Oxford handbook of philosophy of physics* (pp. 522–586). Oxford: Oxford University Press.

Sakurai, J. J., & Napolitano, J. (2011). *Modern quantum mechanics* (2nd ed.). Addison-Wesley.

Saunders, S. (2003). Physics and Leibniz's principles. In K. Brading, & E. Castellani (Eds.), *Symmetries in physics: Philosophical reflections* (pp. 289–307). Cambridge: Cambridge University Press.

Skow, B. (2008). Haecceitism, anti-haecceitism and possible worlds. *Philosophical Quarterly, 58*, 98–107.

Skow, B. (2011). More on haecceitism and possible worlds. *Analytic Philosophy, 52*, 267–269.

Mariusz Grygianiec
Prospects for an Animalistically Oriented Simple View

Abstract: Animalism is a metaphysical doctrine according to which human persons are biological organisms of the species *Homo sapiens*. The Simple View of personal identity claims that there are no non-circular informative metaphysical criteria of personal identity in the form of both necessary and sufficient conditions. In the philosophical literature, both views are presented almost exclusively as competing. The main goal of the paper is to find out whether it is at all possible to reconcile some variant of animalism with the idea that there are no criteria of personal identity. In other words, the text attempts to determine what conditions must be met for someone to be both an animalist and an adherent of the Simple View. The first part of the paper offers a detailed reconstruction of the Simple View and various forms of animalism. In the second part, three different strategies for an integration of the Simple View and animalism are presented and briefly discussed. Although these strategies are put forward very tentatively, some points in favour of the third strategy are provided. According to this approach, "person" is a semantically complex, referentially permanent, and open-ended phase sortal term, expressing both metaphysical and normative (forensic) aspects of meaning, and, consequently, its character calls into question the possibility of providing appropriate criteria of personal identity.

Keywords: the Simple View, animalism, person, organism, animal, human being, personal identity, identity criteria, natural kind, reconciliation.

Introduction

Animalism is a metaphysical doctrine according to which human persons are biological organisms. This view has a few non-equivalent interpretations. The weakest of them states that we are just animals. This version is also the most trivial and intellectually the least interesting. A stronger version of the doctrine says that each of us is identical with some organism of the species *Homo sapiens* (in other words: each of us is a human person, and each human person is identical to a human organism). The strongest variants make use of modal terms, claiming, for

---

Mariusz Grygianiec, 1. Lehrstuhl für Philosophie mit Schwerpunkt analytische Philosophie und Wissenschaftstheorie, Institut für Philosophie, Universität Augsburg, Deutschland; 2. Institute of Philosophy, University of Warsaw, Poland.

https://doi.org/10.1515/9783110669411-003

instance, that human persons are essentially (i.e. of necessity or fundamentally) human biological organisms. A certain form of animalism is sometimes accepted within Thomism; it is claimed, following Aristotle, that human persons are rational animals: human rationality is grounded in our animal nature in the sense that this kind of rationality would be impossible without the existence of the power of sensation, which is determined by our animal nature. In its basic articulation, it is also important to note that animalism does not imply any responses to the problem of the persistence of persons through time; at least it does not itself constitute such a response.

The Simple View of personal identity claims, roughly speaking, that there are no non-circular informative metaphysical criteria of personal identity in the form of both necessary and sufficient conditions. The necessary and sufficient conditions are interpreted here as *truth-conditions* for statements about identity. It is sometimes added that these conditions are simply certain ontological factors that determine or ground personal identity through time – they are something that personal identity, in fact, consists in or depends on. Since the Simple View denies the existence of these types of conditions, it rejects, *ipso facto*, the possibility of reducing the relation of identity to some allegedly more basic facts. At the same time, it is noteworthy that the existence of truth-conditions is not questioned here: it is only claimed that these conditions are facts of identity. The consequence of this is the rejection of the existence of criteria which are both informative and non-circular: possible criteria of this type can only be trivial, circular, or uninformative. At best, they can be treated as purely evidential criteria that provide relevant conditions of assertability for identity. These may simply be statements that say what constitutes evidence for identity, i.e. what constitutes a good reason for saying of an object x that it is or is not identical with an object y.[1]

In the philosophical literature, animalism and the Simple View are presented almost exclusively as competing. Meanwhile, it seems, at least *prima facie*, that there is some kind of logical space for a possible combination of both positions. First, animalism is not a view that determines what personal identity consists in. Admittedly, animalism is a metaphysical doctrine of the human person, but it leaves the question of personal identity through time basically open. On the other hand, the Simple View primarily concerns personal identity, putting, as it were, the question of the metaphysical nature of a person on the sidelines. Second, within both views the meanings of the terms *person, human person, human, human being, persistence* and *identity* are understood in different ways, allowing

---

[1] Incidentally, it should be emphasised that the evidence for a given proposition very rarely coincides with its truth-conditions. See Chisholm (1976, p. 112).

for various reconstructions of the relevant claims. In consequence, both views may only seemingly be in opposition: they may, for example, associate different semantic roles with the concept of *person*. Third, a great deal of misunderstanding has been generated, I think, around the concepts of *persistence*, *identity* and *person*; as a consequence, these misunderstandings are responsible for apparent conflicts of intellectual opinion that either do not really exist or are less pronounced than is commonly believed.

These three circumstances constitute a stimulus for me to find out whether it is at all possible to reconcile some variant of animalism with the Simple View sketched out above. In other words, I would like to ponder which preliminary conditions, if any, have to be fulfilled in order to be both an animalist and an adherent of the Simple View.[2]

Basic Tenets of the Simple View and of Animalism

The philosophical debate on the issue of persistence through time, in particular of the persistence of persons, is mainly concerned with the question of what persistence consists in.[3] The very fact that objects persist through time is decidedly not the subject of the debate. Philosophers are more interested in finding out *how* this happens, i.e. what it takes for an object to persist from one time to another. In response to this question they usually formulate criteria of diachronic identity for objects of the examined kinds, assuming that the persistence of objects is to be reconstructed and analysed in terms of identity, and that identity itself consists in, or is reducible to, some further facts.

---

2 In this paper, I am not going to argue separately for the Simple View nor for animalism.
3 According to the usual approach to persistence (endurantism), ordinary objects persist by enduring – that is, by being *wholly present* at each of the times at which they exist. According to the doctrine of temporal parts (perdurantism), things persist not by enduring, but by perduring – that is, by being only *partially present* at each of the times at which they exist. The doctrine of temporal parts is sometimes developed in such a way that things persist not by having temporal parts at each of the times at which they exist (*the worm view*), but rather by having different temporal counterparts at each of those times (*the stage view*). For more on the classical endurantist approach, see Oderberg (1993), Lowe (1998); for *the worm view*, see Heller (1991); for *the stage view*, see Sider (2001) and Hawley (2001); for alternative solutions, see, e.g. Brower (2010), Simons (2008), Piwowarczyk (2010) and Costa and Giordani (2013, 2016). The collection Haslanger and Kurtz (2006) is an excellent resource of many essential texts. A very useful overview of the whole debate can be found, e.g. in Effingham (2012), or, alternatively, in Haslanger (2003). See also Wasserman (2004, 2016).

It is common to draw two distinctions: one between identity conditions and criteria of identity, and another between criteria in the metaphysical sense and in the evidential sense. Identity conditions are certain ontological factors that determine or ground identity; they are (intended to be) something that identity, in fact, consists in. In other words, identity conditions as I mentioned above, can be taken to be the truth-conditions of statements about identity. By contrast, identity criteria are verbalisations (articulations), in an appropriate form, of the mentioned conditions. In light of this distinction, it would be possible, for instance, for something to possess some conditions of identity, while nevertheless, it would be not possible to spell out the relevant identity criteria. In turn, the difference between criteria in the metaphysical sense and criteria in the evidential sense lies in the fact that, while metaphysical criteria articulate identity conditions, criteria in the evidential sense supply us with indicators of the assertability of identity – means by which we may recognise identity without characterising its nature.[4] One possible consequence of that difference is the existence of evidential criteria in the absence of the relevant metaphysical criteria. It is generally expected that the identity criteria should be informative and non-circular, i.e. that they articulate non-trivial information concerning the objects in question, without applying, in an overt manner or tacitly, the notion of *identity* to those objects. Sometimes, if the criteria are closely connected with natural kinds, an additional requirement of minimality is added: the criteria should only refer to objects of the given kind and not to objects that belong to this kind and to other kinds as well.

In the current philosophical literature, identity criteria are standardly formulated, as mentioned above, in the form of necessary and sufficient conditions.[5] The criteria of identity, formulated as necessary and sufficient conditions, are additionally divided into so-called one-level and two-level criteria. The form of a one-level criterion is as follows (see Lowe, 2013, pp. 14, 71):

(K1) $\forall x \forall y \, \{K(x) \land K(y) \to [x = y \equiv R_K(x, y)]\}$

(in other words: for every two objects x and y belonging to kind K: x is identical with y if and only if x and y stand in the relation R_K to one another; the relation R_K must be an equivalence relation).

---

[4] A very useful description of different interpretations of identity criteria can be found in Fine (2016) and Horsten (2010).

[5] Specifying necessary and sufficient conditions differs from specifying conditions that are only necessary or only sufficient. Perhaps we should not *a priori* rule out the following situation: although might be possible with respect to certain objects to formulate necessary identity conditions *and* to formulate sufficient conditions, it may still prove impossible to provide conditions that are both necessary and sufficient.

A two-level criterion can take the form of the following general formula (see e.g. Williamson, 2013, p. 146; Lowe, 2012, p. 142):

(K2) $\forall x \forall y \, \{K(x) \wedge K(y) \rightarrow [f_k(x) = f_k(y) \equiv R_K(x, y)]\}$

(in other words: for every two objects x and y belonging to kind K: the K-function of x is identical with the K-function of y if and only if objects x and y are R_K-interrelated).

The difference between (K1) and (K2) boils down to the fact that in (K1) the criterial equivalence relation R_K between x and y constitutes an identity condition for the objects x and y, whereas in (K2) this relation does not serve as a criterial relation for the identity of x and y, but for the identity holding between certain other things, appropriately related to x and y (namely, as *values* of a certain function, with x and y as *arguments*). Therefore, a two-level criterion specifies the identity condition of certain objects, which are not necessarily of kind K, in terms of an equivalence relation (R_K) holding between objects of kind K.

The following formulas are standardly invoked as paradigmatic examples of identity criteria: (i) the axiom of extensionality for sets – $\forall x \forall y \, \{set(x) \wedge set(y) \rightarrow [x = y \equiv \forall z \, (z \in x \equiv z \in y)]\}$; (ii) the Fregean criterion of identity for directions – $\forall x \forall y \, \{line(x) \wedge line(y) \rightarrow [(direction\ of\ x = direction\ of\ y) \equiv x$ is parallel to $y]\}$; (iii) Davidson's criterion of identity for events – $\forall x \forall y \{event(x) \wedge event(y) \rightarrow \langle x = y \equiv \forall z \{[cause(x, z) \equiv cause(y, z)] \wedge [cause(z, x) \equiv cause(z, y)]\}\rangle\}$ (see Lowe, 1998, pp. 620–621).[6]

It should be noted that schemata (K1) and (K2) are schemata of statements that provide synchronic (or atemporal) criteria of identity. What follows are – partially specified – schemata of statements that provide diachronic criteria of identity:

(K1*) $\forall x \forall y \forall t_1 \forall t_2$ (if x is a person and y is a person, then x at t_1 is the same person as y at t_2 if and only if x at t_1 is related *via* the relation R to y at t_2),

(K2*) $\forall x \forall y \forall t_1 \forall t_2$ (if x is a human being and y is a human being, then the person of x at t_1 is the same person as the person of y at t_2 if and only if x at t_1 is related *via* the relation R' to y at t_2).[7]

Regardless of which of the above formulas is ultimately preferred, the condition of personal identity is, in each case, constituted by the criterial relation R or R'. It should be noted from the outset that relations R and R' have formal properties very similar to those of identity: they are reflexive, symmetric and transitive, and

---

[6] Whether the last one is non-circular is in fact debatable due to the fact that causes and effects are usually taken to be events.
[7] Cf. Lowe (2012, p. 151).

therefore they are equivalence relations. At the same time, however, they are not, in contrast to identity, minimal equivalence relations.[8] This simple difference seems to be very important because it may serve as an additional basis for questioning the metaphysical interpretation of identity criteria. Although each criterial relation is an equivalence relation, it cannot be a minimal one: the relation of identity is included in any other equivalence relation, but not *vice versa*. Hence identity cannot be coextensive with any potential criterial equivalence relation.[9]

In proposing various formulations of identity criteria, metaphysicians usually declare that the criteria in question:

a) are not criteria only in the evidential sense (i.e. purely heuristic or cognitive conditions of the assertability of identity claims);
b) are not definitions of the relation of identity itself;
c) are not definitions of kind-relative relations of identity (i.e. definitions of identity-relations relative to given natural kinds);
d) are not definitions of natural kinds that are invoked in the criteria (see Lowe, 1998, p. 45; Williamson, 2013, pp. 144–145).

At the same time, philosophers often indicate that the criteria are semantic or metaphysical in character since they determine *truth-conditions* for statements about identity. As above, such criteria are meant to provide conditions in virtue of which objects that satisfy them are identical: satisfying those conditions makes the objects the same. The significance of identity criteria is additionally highlighted by their role in cognitive individuation: we can single out and count objects, *inter alia*, by means of such criteria; in the case of their absence, it can be doubted whether, for instance, a potential counting of objects is correct (if we were unable to single out the counted objects, then it could turn out that we counted an object twice or two different objects once). Some even claim that identity criteria ultimately determine whether something should conclusively be classified as an object; if no relevant criteria for certain entities can be formulated, then those entities are either

---

[8] For more on both the logical requirements for identity criteria and the formal constraints for relation R, see Carrara and Gaio (2011, pp. 227–233). Cf. also Horsten (2010).

[9] These matters are not so simple, however. First, for any given criterion of personal identity the criterial relation seems to entail numerical identity. Second, for certain kinds of objects, say sets, their numerical identity seems to come down to the criterial relation of co-extensiveness. Therefore, following Quine, we can agree that under specific ontological conditions (e.g. within an appropriately parsimonious ontology) some equivalence relations can function as the relation of identity. Still, although a given criterial relation can be locally coextensive with identity, it cannot be globally coextensive. Furthermore, even if identity were coextensive with the criterial relation R, this would not automatically mean that this relation simply is identity. I thank Uwe Meixner for pointing out this complication to me.

called *quasi*-objects or are regarded as fundamental and primitive elements of reality (see e.g. Lowe, 1994). In any case, it is a common trend to interpret identity conditions expressed by identity criteria in the form of metaphysical-cum-semantic principles as *truth-conditions* for statements about identity.

In contemporary analytic metaphysics, the question of which relation R provides conditions of personal identity has led to three general answers: (a) the reductionist-physicalist solution, according to which personal identity consists in (or depends on, or holds in virtue of) the occurrence of certain physical relations (e.g. spatio-temporal continuity, causal continuity, continuity of life processes, identity of the body or identity of the brain); (b) the reductionist-psychological solution for which personal identity is grounded in an appropriate mental or spiritual relationship (e.g. continuity of memory or *quasi*-memory, causal connectedness between mental states, or identity of the soul); (c) the non-reductive account, according to which personal identity is a primitive, brute, irreducible relation, consisting in nothing other than itself.

Interestingly, among advocates of the Simple View there is a lack of unanimity regarding the interpretation of the fundamental conviction which is intended to be expressed by the main thesis. Indeed, this view is usually articulated by one of the following claims:
(1) personal identity is (or consists in) a simple, irreducible fact;
(2) personal identity is grounded in something that is itself not further reducible;
(3) there are no conditions of personal identity;
(4) there are, except for identity itself, no other truth-conditions for statements about personal identity;
(5) it is not possible to formulate any appropriate criteria of personal identity.

However, there are reasons[10] to interpret almost all variants of the Simple View as committed to the following claim:
(SV) There are no non-circular informative metaphysical criteria of personal identity through time, i.e. criteria specifying *truth-conditions* for statements about personal identity.[11]

---

10 For more on these reasons, see e.g. Grygianiec (2016c).
11 Arguments in favour of this thesis can be found, for instance, in Jubien (1996, pp. 343–356; 2009, pp. 46–54) and Merricks (1998, pp. 106–124). See also Lowe (2009, pp. 23, 120, 133–140). Supporters of the Simple View include Chisholm (1986, pp. 73–77; 1989, pp. 124–127), Baker (2012, pp. 188–189), Nida-Rümelin (2012, pp. 157–176) and Madell (2015). For an independent analysis of this view, see Noonan (2011, 2012) and Olson (2012).

The main point in favour of this interpretation is an observation that, although this formulation is not entirely accurate,[12] it depicts the main idea of the Simple View fairly well, and it seems to be common to almost all its variations. It is worth noting that the above statements (1)–(4) do not mention identity criteria. Nevertheless, even an advocate of (1) would be ready to accept, in my opinion, the claim that the pertinent criteria cannot be formulated. More or less the same applies to the other versions. On a specific understanding of the concept of identity criteria, position (1), (3), (4) and (5) entail (SV). The only exception here is claim (2) which does not rule out the possibility of formulating non-circular informative criteria of identity. However, due to the fact that it is – at least explicitly – rarely represented, it can be omitted here for the sake of clarity of presentation.

It is noteworthy that the Simple View, as has already been hinted above, denies neither the existence of identity criteria in the evidential sense nor the existence of identity criteria in the metaphysical sense – as long as such criteria are formulated in a trivial, circular or uninformative manner. The interesting defectiveness of identity criteria in the metaphysical sense does not consist in any formal mistakes or in violation of logical rules – rather, the point is that interestingly defective metaphysical criteria do not permit us to express anything more than is already covered by the notion of identity itself. After all, if the relation of numerical identity is completely primitive and unanalysable, then, as a matter of fact, no other relations can constitute it.[13] Such relations may, of course, accompany identity and be testimonial components of identity criteria in the evidential sense. The Simple View by no means denies this. What is questioned, however, is an interpretation, according to which the relation of identity is grounded in – or reduced to – some other, more basic facts.

---

[12] Unfortunately, thesis (SV) does not fully encompass all the positions that can be called non-reductive as there are variations of the doctrine, described here as reductive ("complex"), whose representatives refer to them as non-reductive. An example of this is Swinburne's position. According to him, the identity of a person is grounded in the identity of that person's soul (a soul being an immaterial thinking substance), while the latter is no longer reducible to any other elements. One could say, following Olson, that the Simple View is the thesis that personal identity is irreducible, or is grounded in something that is itself not further reducible. Having recalled this complication here, I just want to express the fact that the systematising decision I made above is not entirely accurate. For Olson's proposal, see Olson (2012, p. 56). For Swinburne's views, omitted in this text, see e.g. Swinburne (1997, pp. 145–173; 2013, pp. 141–173).

[13] See Salmon (2005, pp. 153–154). For some arguments in favour of methodological scepticism with regard to identity criteria see Grygianiec (2016a, pp. 5–7).

Animalism is a metaphysical theory about human nature, claiming that, to put it crudely, we are animals.[14] Such a formulation of animalism appears to be a trivial assertion that does not even deserve to be called a philosophical theory. Indeed, it looks similar to intellectually uninteresting trivialities like "Snails are animals", "Dogs are mammals" or "Cars are vehicles". It may seem hard to believe that such a banal and trivial doctrine is the subject of an intense philosophical debate. Nevertheless, animalism is considered to be philosophically interesting, which is sufficiently evidenced by the increasing number of professional publications (see e.g. Olson, 2007; Snowdon, 2014a; Blatti and Snowdon, 2016) concerned with it.

The claim that "we are animals" is the conclusion notoriously repeated by one of the most prominent and zealous proponents of animalism, Eric T. Olson. The main trouble with this claim, I think, lies in the fact that none of its component expressions has a transparently philosophical meaning, and consequently, that the assertion in question does not have, at least *prima facie*, any specific philosophical significance. Neither the personal pronoun "we" nor the term "animal" expresses, without additional interpretation, philosophical content. Earlier, in the course of writing his first book, *The Human Animal: Personal Identity without Psychology* (Olson, 1997, p. 24), Olson was still willing to express the fundamental thesis of animalism differently, namely as the following belief:

(A1) Each human person is identical with a human organism, i.e. with an instance of the primate species *Homo sapiens*.

This thesis looks philosophically more interesting. It should be noted that in animalism there is, firstly, a limitation of the term "person" to "human person" (thus animalism avoids any assertions about non-human persons, such as, e.g., God, angels or Martians). Secondly, by making use of quantification and the notion of identity, a more precise formulation is achieved. Thirdly, the thesis makes use of the term "organism" which is, from an ontological point of view, perhaps preferable to the term "animal". Although in one of his most recent texts Olson tried to prove that the identity form of the main thesis of animalism is not appropriate (see Olson, 2015), it is this formulation that is commonly used in outlining the doctrine in question. This is confirmed by other advocates of animalism, such as Paul F. Snowdon (Snowdon, 2014b):

(A2) For each x: if x is a human person, then x is identical with a human animal.

---

14 The main arguments for this doctrine are: (i) *The Thinking Animal Argument* (Olson, 2003); (ii) *The Animal Ancestors Argument* (Blatti, 2012); (iii) *The Association Argument* (Bailey, 2015); (iv) *The Argument from Causal Powers* (Licon, 2012). For a concise discussion on these arguments, see Grygianiec (2016b).

Even more precisely, this thesis can be expressed as follows:
(A3) ∀x {x is a human person → ∃y [y is a human organism ∧ x = y]}.

It is worth emphasising once again that animalism does not claim that each and every person is an organism – it is a thesis exclusively about human persons. Nor does animalism claim that each human organism is a human person (see Mackie, 1999, pp. 230–233). The question of whether the terms "person", "human person" or "organism" denote natural kinds, whether they are substantial concepts, is also not settled here. In the first period of his intellectual career, Olson was willing to admit that the term "human organism" was a substantial concept, whereas he regarded the term "person" to be a phase sortal (Olson, 1997, p. 121). Other advocates of animalism are inclined to propose alternative formulations of the main thesis. For example, Stephan Blatti (2014) has put forward the following formula:
(A4) □∀x [x is a human person → x is a human animal].

By contrast, Christopher Belshaw (2011, p. 401) prefers the following formulation:
(A5) We are essentially animals,

which should probably be rendered more accurately as:
(A6) ∀x [x is a human person → □(x is a human animal)].[15]

A slightly different proposal has been set forth by Harold W. Noonan, an indefatigable critic of animalism. According to him, animalism amounts to the conjunction of two claims (see Noonan, 1998, p. 304):[16]
(A7) (i) All persons that at some time coincide with animals are animals; (ii) It is a necessary truth that all persons that at some time coincide with animals are animals.[17]

In addition to the formulations listed above, one can find further elaborated versions or completely new proposals. An example of the first is Jens Johansson's variant (2007, p. 205):

---

[15] An interpretation according to which we are necessarily both persons and animals is opted for by Sharpe (2015). The analysis offered by Melin (2011) goes in a similar direction.
[16] Noonan's use of the notion of *coincidence* appears to be due to the desire to get a convenient starting point for a polemic against animalism in the context of Olson's fundamental argument.
[17] Since (i) follows logically from (ii), animalism, according to Noonan, seems to amount to (ii).

(A8)   All or nearly all human persons are identical with animals.[18]

An example of the second is Patrick Toner's Thomistic proposal (2011, p. 79):
(A9)   (i) Human persons are all and only those things that belong to a kind whose instances naturally have rationality and sensation; (ii) To belong to this kind entails being identical with an animal.[19]

To my mind, Belshaw, rather than Olson, is right on the following matter. If we were willing to admit that all animals are necessarily animals, then the thesis of animalism would entail formula (A6). Accordingly, one could propose a less trivial version of animalism as follows:
(A10) $\forall x \{x \text{ is a human person} \rightarrow \exists y [x = y \land \Box(y \text{ is a human organism})]\}$.

There are two arguments supporting (A10): first, it is the most accurate formulation, omitting personal pronouns and making use of the terms "person" and "organism", which have some evident metaphysical connotations; second, it involves a strong modal notion and thus locates its content beyond the sphere of trivial assertions. The identity version *ab initio* eliminates interpretations of animalism according to which a human person is constituted by an organism, supervenes on it, or emerges from it. However, it has to be kept in mind that (A10) does not reflect the views of all animalists; it may reasonably be suspected, for instance, that Olson, the main defender of animalism, would not agree with this formulation of the doctrine.

Indeed, it is impossible to explore all logical possibilities related to the central thesis of animalism. These possibilities concern not only its logical form (quantification, identity, appropriate modal expressions), but also an entire range of non-coextensive predicates, such as *person* and *human person* on the one hand, and *animal*, *human animal*, *organism*, *human organism*, etc. on the other. Interestingly enough, as has been brought up earlier, in one of his recent papers, Olson explicitly rejects almost all versions of animalism quoted here as over-interpretations and insists on so-called weak animalism ("we are animals"; see Olson, 2015). He expresses doubts concerning the formulations referring to modal concepts, to

---

18 Johansson seems to allow here for the possibility that some human persons might not be animals.
19 In Toner's proposal, the influence of the Aristotelian definition of *human being* is easily identifiable. Undoubtedly, the very idea lying behind this proposal, namely the conviction that the capacity for sensation implies, in the case of human person, the property of being an animal and is, as a matter of fact, the precondition for our rationality, is a very interesting (and by no means trivial) thought. Of course, having sensation, taken quite generally, does not logically entail having a body.

metaphysical nature or natural kinds, or even to the very notion of identity. In my opinion, weak animalism is rather inadequate from a metaphysical point of view: perhaps Olson's strategy brings some clarity to the whole debate, but the thesis of weak animalism is by no means able to avoid a charge of triviality. It does not seem to be a significant philosophical claim; it does not contain any substantial metaphysical content.[20]

On the surface, it seems that animalism is a physicalist doctrine: since organisms, according to most animalists, are entirely physical objects and do not have any non-physical constitutive parts, animalism should, it seems, be interpreted as a specific local version of physicalism. However, matters are not so simple. Firstly, there are problems in determining what physicalism (or materialism) actually claims.[21] Secondly, there is some ambiguity as to what exactly the relation is between the concept of *body* and the concept of *organism* (e.g. see Olson, 2007, pp. 25–26). And thirdly, there is a difficulty as to whether the life of an organism can be fully explained in physicalist terms. If physicalism were true and if we had a complete physicalist description of the universe at our disposal (thus, if animal organisms, as purely physical bodies and life processes – together with their origins, were fully explained within physicalist terminology), then animalism would certainly be some version of physicalism. But otherwise the matter is not so clear. Thus, although most animalists officially declare themselves to be physicalists, it is difficult to see that physicalism would automatically follow from the doctrine of animalism without any additional assumptions (for an opposing interpretation, see Bailey, 2014). Indirect evidence in favour of such a diagnosis is present in those versions of animalism that originated from the Thomist tradition (e.g. see Toner,

---

20 Uwe Meixner has indicated to me that the potential triviality of weak animalism depends on what one means by "animal", and on whether one is ready to add to the thesis "we are animals" a further assertion: "and nothing more".

21 Problems with a precise formulation of physicalism are connected, *inter alia*, with Hempel's dilemma. In short, the point is that it is not very clear which type of physics is spoken of when the predicate *physical* is defined: is it contemporary physics, in terms of the current content of our best scientific theories, or is it a future "ideal" or "completed" physics? In the first case – since physical theories sometimes happen to be false – the thesis of physicalism might prove to be false as well. In the second case, since it is not currently known which scientific theories future physics will include, it is also not known at present what physicalism really claims (thus, physicalism seems to be an unacceptably indeterminate and vague claim; apart from that, if futurism (i.e. the second option) were right, then physicalism might prove to be a completely trivial doctrine). I am not in a position to discuss these alternatives in detail here. In my opinion, the best reconstruction of these issues is presented in Stoljar (2010, pp. 93–108). For more on various reactions to the dilemma, see e.g. Crane and Mellor (1990), Melnyk (1997; 2003, pp. 11–20), Crook and Gillett (2001), Montero and Papineau (2005), Wilson (2006) and Ney (2008).

2011, 2014; Oderberg, 2005; Hershenov, 2008, 2011). It is more than obvious that no adherent of Thomism can accept the physicalist doctrine. The fact that adherents of Thomism accept animalism allows the supposition that animalism, at least in its basic formulation, is not *ipso facto* a form of physicalism. Needless to say, an animal in the hylomorphist – Aristotelian, Thomistic – tradition is not just a physical organism.

Some animalists claim that physicalism (materialism) itself does not deliver any specific description of the metaphysical nature of a human (person), and they are inclined to say that the responsibility for this task rests precisely with animalism (see, e.g. Bailey, 2014, p. 475). Therefore, animalism would be a certain natural supplementation of the physicalist doctrine. It seems that this point of view is rather typical for most present-day animalists. In such an interpretation, of course, animalism would imply physicalism with respect to human persons, but not *vice versa*.

Advocates of animalism usually also discuss the issues of persistence and personal identity over time, mostly arguing against psychological approaches, which seek identity criteria in suitable psychological relations holding between the mental states of persons. On the other hand, the animalist thesis itself, and especially the thesis favoured by Olson, does not imply any specific position on the issues of identity and persistence (see Olson, 2007, pp. 7, 17–18). The main reason for this is, as I have suggested above, that animalism is a doctrine that primarily offers to answer the question of what we are, metaphysically speaking, the question of what our basic metaphysical nature is, and not the question of what it takes for us to persist identically through time.

It should, then, be agreed that the trivial version of animalism does not imply any solution to the question of personal identity. However, if we opted for a stronger, non-trivial version of animalism, the matter would seem much more complicated. If one claimed, for instance, that human persons are necessarily animals, i.e. that it is true of every human person that it is necessarily an animal, and thus that it is true of every human person that it is not possible for it to exist without being an animal, then one would consequently have to concede that human persons have the persistence conditions that are typical for animals. Such an approach is far from trivial and immediately inserts animalism into the debate on personal identity (e.g. see Melin, 2011). A further-reaching detailed analysis of persistence conditions would provide a clear animalistic response to the question of personal identity and would unambiguously locate animalism in opposition to the psychological approach.[22]

Nevertheless, the analysis of persistence conditions just mentioned is not uniform. According to some animalists (e.g. van Inwagen, 1990, pp. 142–168), these conditions comprise the continuity of the life processes of an organism. Others are prone to appeal to the causal and spatio-temporal continuity of the body of an organism. Still others invoke the life-apt functional organisation of the animal body as a factor responsible for the persistence of an organism (incidentally, it is supposed that the internal organisation of a living and growing organism is maintained even after the end of its life processes – see Mackie, 1999, p. 237). Looking for appropriate persistence conditions, one could even – assuming the weak (or trivial) version of animalism – appeal to some mental or spiritual elements. However, animalists formulate these conditions mostly in terms of either the continuity of metabolic life processes (the continuation of life-sustaining functions) or the preservation of the organism's complex internal structure.[23] It is important to observe once again that the weak thesis of animalism says nothing about the persistence conditions of organisms; they may instead follow from a more detailed and stronger version of animalism. These stronger versions seem to constitute the antitheses of both the psychological approach and the Simple View of personal identity.

If there were such a thing as standard animalism, i.e. a set of convictions typical for an animalist, then it might be presented as follows:

(1) Each human person is identical with a human animal (i.e. an organism of the species *Homo sapiens*);
(2) Human persons are governed by the persistence conditions typical for human animals (which follows logically from (1));
(3) Persistence conditions for human persons are specified either in terms of the continuity of the life processes of organisms or in terms of the preservation of the organism's internal biological structure, allowing for a continuation of life processes;
(4) Human persons are material objects (which follows logically from (1));

---

[22] It may be worth mentioning that animalism, in restricting the notion of person to the notion of human person (a restriction, I think, it has to adopt), is in a disadvantageous polemic position in the debate relative to psychological approaches, since the latter do not need, at least *prima facie*, to adopt that restriction.

[23] For an alternative approach, see Hoffman and Rosenkrantz (1997, pp. 80–90, 128–134) and Rosenkrantz (2012).

(5) Most, if not all, causal powers, mental states and other properties of each human person are properties of a human animal identical with it (which follows logically from (1)).[24]

Possible Reconciliation Strategies

It might reasonably be asked, at the outset, who would wish for a reconciliation between animalism and the Simple View? Well, it might be someone who, on the one hand, is inclined to call into question the existence of identity criteria in the metaphysical interpretation and who, on the other hand, has clear animalist predilections. It is somewhat surprising that in the philosophical literature there is practically no representation of such a combination of beliefs. An exception perhaps, are Trenton Merricks' views (see Merricks, 2001), which approximate, in a sense, the above-mentioned combination: Merricks is an advocate of the Simple View and an adherent of a biological position on material composition (according to him, only elementary particles and organisms exist, the latter being the only composite objects). Although Merricks himself has never clearly acknowledged being an advocate of animalism in the form discussed here, animalism seems to follow from his position (what else can human persons be, if all composite objects are organisms, but human organisms?).

The mere juxtaposition of animalism and the Simple View shows, at first glance, that there is no contradiction between them: an animalist can also be an advocate of the Simple View. However, this is only true of the weakest version of animalism, which makes no claim about the nature of persistence through time. That version of animalism does not regard the terms "animal", "organism" and "human organism" as natural kind terms but rather as general classificatory terms, which have no identity or persistence criteria associated with them (it is standardly assumed that such criteria are associated with natural kind terms or with substantival general terms).[25] A drawback of the *simple* combination of the Simple View and animalism is the fact that it is intellectually trivial: first, it would not automatically (and in some cases – *ex definitione*) offer any specific description of persistence through time; second, it would not have at its disposal any interesting conception of human nature – merely saying that a human being is an animal. Although we can take

---

[24] Since (2), (4) and (5) follow logically from (1), it seems that (1) and (3) would entirely suffice to characterise the average animalist.
[25] For more on natural kind terms and their role in metaphysics, see e.g. Koslicki (2008).

note of the *simple* combination as the first potential reconciliation option, it should be rather obvious that it is the least satisfying one.

In search of a better option for reconciliation we should, to my mind, take a closer look at the very concept of *person*. The fact that the term "person" functions very ambiguously in public discourse is incontestable. It is rather obvious that, normally, philosophical terms are not (or only to some extent) respected in common linguistic practices. Their use is usually – implicitly or explicitly – purely ideological and rhetorical in character. In order to restore, as it were, the philosophical character of the term "person", we need to appeal to certain regulative procedures which could help us determine its meaning and role.

The first step in this direction is making a decision on whether the term "person" is or is not a natural kind term. Possible answers to this question equip us *a priori* with two different paths of reconstruction: (i) the term "person" is a natural kind term which has a criterion of identity associated with it;[26] (ii) the mentioned term is not a natural kind term. Within the first option, an advocate of (SV) would in fact, following Lowe (2009, pp. 138–140; 2012, pp. 152–153), have to concede that although natural kind terms typically have identity conditions associated with them, in the case of "person" these conditions cannot be adequately specified. There may be either metaphysical reasons for such an impossibility (for instance, persons might constitute a fundamental ontological category of objects and we might not be in a position to identify any more basic category in which the alleged criteria could ultimately be grounded) or purely methodological reasons (for instance, certain formal properties of the relation of identity might exclude any possible reductive analysis).[27] In the case of the second option, an advocate of (SV) may concede that the term "person" is an expression which does not denote any natural kind, while still insisting that it denotes an artificial or functional kind, such as "spy", "shepherd dog" or "cut flower". These expressions are used in order to identify objects only in some periods of their existence: after all, one can continue to exist without being a spy – and someone might exist prenatally or in a persistent vegetative state without being a person. In such a case, the advocate of (SV) may claim, on the one hand, that each of us (i.e. subjects of experience, without prejudging their being human persons) is identical to a member of the primate species *Homo sapiens* (where the term "Homo sapiens" would be a natural

---

[26] As a matter of fact, I am willing to admit that identity criteria are associated rather with well-defined ontological categories than with natural kinds. For instance, the term "set" is the clearest instance of a term that has a criterion of identity associated with it and is by no means a natural kind term. I owe the change of my own view in this respect to Uwe Meixner.

[27] Of course, if the formal properties of the relation of identity excluded any possible reductive analysis, then it would not matter which kind, natural or not, we are talking about.

kind term with certain identity conditions associated with it) and, on the other hand, that human organisms are sometimes persons, but since the term "person" is only a *phase sortal*, it is not *ex definitione* possible to formulate the relevant criteria of identity.[28] The above option is closely related to the first reconciliation option outlined earlier. There is, however, one significant difference: the potential criteria of identity and persistence would be associated here primarily with the natural kind term "Homo sapiens" (and not with the term "person", although *secondarily* they are also associated with the latter term). It should perhaps be noted in passing that an interpretation according to which the term "person" encodes essentially normative (forensic) and prescriptive aspects of meaning might elegantly be adopted as a part of this option.

There is another possible strategy for reconciling both views, founded on a different conceptual analysis of the term "person". Namely, it can be assumed that "person" is a semantically hybrid term, i.e. neither a pure natural kind term nor a typical phase sortal.[29] According to this option, the term "person" would express some ontological as well as normative aspects of meaning. Furthermore, it would be interpreted as an open-ended term, i.e. a term which absorbs, so to speak, different aspects of meaning depending on a given context of use. What would support such a hybrid interpretation? Note that the term "person" is, after all, a classificatory term and is standardly applied, in accordance with common practice, in the individuation and re-identification of objects. Nonetheless, in this case identification practices (e.g. a certificate of a taxpayer's identification number assignment, a naturalization, a judicial declaration of death, authorship of works) semantically exceed the frame typical for phase sortal terms. It might be supposed that these practices are indirectly and implicitly based on semantic functions of the term "human being", but for some reasons (perhaps pragmatic, legal or moral) the mentioned procedures blatantly prefer the term "person" here. Indeed, it is rather difficult to imagine, for instance, that the parents of a newborn baby would treat their child like a conglomerate of human cells or merely as human animal, due to a lack of the relevant criteria of personhood. In defiance of this lack, parents,

---

[28] In fact, all these matters depend on how we are identified and who we are. If we were identified as human persons from the very outset, then the advocate of (SV) would be claiming that each of *us* – each human person – is identical to a primate of the species *Homo sapiens*. But therewith he would be admitting that there are, nevertheless, identity conditions for *us*, the human persons: the same identity conditions as for the members of the primate species *Homo sapiens*. Therefore, it would be possible, after all, to formulate appropriate identity criteria: they would simple be the same identity criteria as for primates of the species *Homo sapiens*. Given the aim of reconciliation, this sort of move would obviously undermine the endeavor.

[29] A similar, though not identical, interpretation is favoured by Kanzian (2012).

in fact, *ab initio* treat their baby as a person – as someone who can in principle feel, be in pain, be happy, laugh, reason, think about itself and about others and the like.[30] These circumstances indicate two basic things. First, the term "person" stands in a close semantic relationship to the term "human being". Second, the scope of use of both terms is different: the term "human being" is preferred in purely descriptive and reporting contexts, whereas the term "person" is favoured in prescriptive (i.e. normative and forensic) contexts. Both contexts of use require further consideration.

Starting from the second context, I am inclined to believe that among the whole class of phase sortals we could single out a specific subclass of them. This subclass contains terms which behave in a special way: being phase sortals, they nevertheless possess a peculiar permanence of reference. Such terms as "citizen", "Pole", "taxpayer" and so on, behave in a similar fashion as the term "person" (with a bit of tolerance, we can admit that we are always some kind of citizen, that someone who is a Pole in terms of nationality will remain so forever, that we are constantly taxpayers). Interestingly, these terms do not only describe something but their use in specific situations is an indication of allegiance, respect/disrespect, or legal obligation. They overtly express normative and forensic meaning-aspects, while what they refer to seems, at least at first glance, much more permanent than in the case of most phase sortals. This permanence of reference should by no means be treated as a so-called rigid designation; it is rather questionable whether the mentioned terms could be interpreted as rigid designators in Kripke's sense (even in the case of the term "person" there is no consensus concerning this). Nevertheless, there is a significant difference between phase sortals with (more or less) permanent reference and such terms as, for instance, "student", "shoemaker" or "mascot". Under the proposed interpretation, it is therefore assumed that the term "person" is a referentially permanent phase sortal. Moreover, it seems to be an open-ended term which takes on different meaning-aspects depending on the context of use. Therefore, it would be difficult to classify it as a precise well-defined ontological (categorial) term. I am prepared to affirm at this juncture that this fact may constitute one of the principal reasons why it is not possible to specify any appropriate criteria of personal identity.

As for the semantic relationship between the terms "person" and "human being", matters are somewhat more complicated. Despite the pragmatic differences between these notions, briefly indicated above, it should be noted that the term "person" borrows, so to speak, some semantic functions from the term "human being". This does not, I think, happen completely accidentally. The foundation for

---

30 Chappell (2011) argues persuasively against personhood criterialism.

this connection can be sought in different factors, depending on the preferred view. One view is that the foundation in question consists of an ontic element. Although I allow for alternative solutions in this regard, I will not conceal that I strongly prefer this view. According to it, the semantic relationship between the terms "person" and "human being" is founded on the fact that being a human being makes it possible (for an organism) to be a human person; being a human being is, generally speaking, a sufficient condition of being a human person.[31] In other words, owing to the fact that there are some potentialities rooted in human nature,[32] each human being is – naturally, automatically, and permanently – a person. This interpretation is, obviously, Aristotelian in spirit: there is an implicit reference to the concepts of entelechy and potency, both anchored in human nature.

What are the reasons for preferring this particular interpretation of the reconciliation of animalism and the Simple View,[33] and not the other? Firstly, this interpretation elegantly explains the above-mentioned semantic relationship between the notions of *human being* and *person* (it explains, among other things, the relative isomorphy in individuation and re-identification of persons and human beings). Secondly, it fairly convincingly explains the aforementioned permanence of reference of the term "person". Thirdly, an appeal to ontic potentiality as the metaphysical basis for personal attributes simplifies attempts to specify criteria of personhood (i.e. indicators of the necessary and sufficient conditions for being a

---

[31] I want to leave open the question of whether it is also a necessary condition. At first glance, it seems quite obvious that being a human being is a necessary condition of being a human person too. On the other hand, it cannot be such a condition in the case of person *simpliciter* (i.e. human as well as non-human persons). Moreover, as suggested by Johansson (2007, p. 204), the possibility of the existence of completely disembodied human persons should not be ruled out a *priori*.

[32] See Vetter (2015) for contemporary attempts at reconstructing and making precise the notion of *potentiality*, attempts which fit the interpretation favoured in my paper. The potentialities I am speaking about here, whatever else they can be, are the metaphysical basis of all personal attributes typically ascribed to human persons.

[33] It seems that if being a human being is a sufficient condition for being a human person (but not necessarily a necessary one), then the main thesis of the discussed option should be, as suggested to me by Uwe Meixner, the claim that *every human being is a human person*. Now, animalism claims that *human persons* are *animals of the species Homo sapiens*. It looks as though, contrary to the earlier assertion, I have to let being a human being be also necessary for being a human person on pain of contradiction with animalists intuitions and, therefore, extensionally identify being a human being with being a human person. As regards this complication, I am inclined to reply that although it is true that all human persons are human animals and every human being is a human person, it does not automatically follow from this that every human animal is a human person nor that every human person is a human being. Even if these assertion were true, one could, on the option being discussed, remain neutral on this. As noted earlier, I do not want to argue separately for any of the above identifications.

person).³⁴ Now, as regards some formulations of these criteria, there is sometimes an objection raised concerning their adequacy for purely pragmatic reasons. It is said, for instance, that even after having formulated criteria of this type, one still would not be able to decide whether something (or someone) – in order to qualify as a person – has to meet all those criteria at once, and whether it/he/she has to satisfy them to the highest degree or not (it is considered unclear, for instance, whether, in the case of not fulfilling one of the criteria or fulfilling some of them only to a certain extent, the final verdict would still be positive). The suggested interpretation does not arouse this type of concern; accordingly, any object whose inner nature metaphysically allows for the instantiation of personal attributes is a person. Needless to say, in determining what a person is, it is necessary to have recourse to those attributes (consciousness, rationality, being a moral subject, etc.). However, according to this interpretation, insofar as these attributes are made to play the role of sufficient criteria for personhood, they can hardly be expected to guarantee that whatever satisfies them will be a person. Rather, that something is a person is ensured without any criteria by the ontic potentiality permanently embedded in human nature (i.e. by the in-principle possibility of instantiating those attributes).³⁵ So, human beings remain persons even in a permanently vegetative state, and they are persons long before they develop any capacity for rational thought. I believe that this view of the matter not only explains the legitimacy of the semantic dependence of the term "person" upon the term "human being", but also that it helps to explain the former term's permanence of reference.

According to the above position, a reconciliation of animalism and the Simple View could run as follows – animalism offers a metaphysical description of the nature of human persons, implying that "Homo sapiens" can be interpreted as a natural kind term, providing the appropriate evidential criteria of identity and persistence for human organisms. On the other hand, the Simple View of personal identity still remains in force. The specification of identity criteria in the metaphysical sense for persons is precluded due to the fact that the term "person" is neither a natural kind nor a well-defined categorical notion and, as an open-ended term, does not have any specific (metaphysical) identity criteria associated with it. This assertion can be additionally supported by the supplementary metaphysical and

---

34 Fourthly, grasping the foundation of the mentioned semantic relationship in terms of a sufficient (but not necessary) condition does not *a priori* exclude the possibility of the existence of other, non-human persons.

35 One could argue at this point that "the ontic potentiality embedded in human nature" would be, nevertheless, a sufficient and necessary criterion of personhood. I am prepared to agree, but, at the same time, I would be inclined to treat this criterion as a purely metaphysical one (i.e. as non-evidential in any sense).

methodological reasons mentioned earlier. Although being a person and being a human being are not exactly the same in a metaphysical sense, they stand in such close relation that they are, on the empirical level (*via* identification and re-identification), indistinguishable. The difference that remains is primarily based on the normative character and open-endness of the notion *person*,[36] while the permanent relationship of both concepts is grounded in the potentiality of human nature, i.e. in the modal factor that makes it possible for a human being to instantiate specific personal attributes.

Conclusion

According to the reconstruction presented above, there are at least three strategies for integrating the Simple View and animalism. The first one refers to a weak version of animalism and refrains from treating "human animal" (*resp.* "human organism") as a natural kind term. According to this strategy, it is not possible to formulate appropriate (metaphysical) identity criteria for human beings, and it is also not possible to formulate them for persons (as the term "person", too, is neither a natural kind term nor a well-defined categorical term). It seems that this strategy cannot avoid the charge of triviality. The second strategy, a position with dualist overtones, uses both as natural kind terms. Although this interpretation does not usually call into question the generally assumed relationship between identity criteria and natural kind terms, it rejects, in the case of the term "person", the possibility of specifying such criteria: either because of the ontological simplicity of persons (a metaphysical reason) or because of some properties of the relation of identity (a logical reason).[37] With this strategy, a stronger version of animalism can easily be accepted, e.g. one in line with (A3). The third approach interprets the term "person" as a semantically complex, referentially permanent, and open-ended phase sortal term, expressing both metaphysical and normative (forensic) meaning-aspects. This strategy, though it generally allows for the specification of identity criteria in the case of the terms "human being" and "human organism", calls into question the possibility of also providing them for the term "person".

---

[36] Its normative and open-ended character seems to make the term "person" both anomalous with regard to any attempt at defining it and, consequently, anomalous in relation to a possible specification of identity criteria.

[37] One can argue, for instance, that the relation of identity is completely primitive, unanalysable, perfectly general, admitting of no degrees, always holding (or not) of necessity and "kind-transcendent" – regardless of whether it relates to numbers, persons, sentences, or ships.

This is due both to the semantic complexity of the term in question and its open-endedness. A stronger version of animalism can also be accepted in this case. Among the options presented, the third strategy seems to be the philosophically most interesting; it allows, on the one hand, for different reconstructions of the relationship between the term "person" and "human being", and, on the other hand, it seems ontologically much safer than the second strategy. Nor does it smack of triviality, as is the case with the first strategy. But, of course, it should be further elaborated and go well beyond what has provisionally been presented here.

Acknowledgment: This paper has been prepared within a project that has received funding from the European Union's Horizon 2020 Research and Innovation Programme under Marie Skłodowska-Curie grant agreement No 650216 (*The Ontology of Personal Identity*). I would like to express my gratitude to my supervisor at the University of Augsburg, Prof. Uwe Meixner, who persuaded me to write this text and who offered many critical comments on my work.

Bibliography

Bailey, A. M. (2014). The elimination argument. *Philosophical Studies, 168*, 475–482.
Bailey, A. M. (2015). Animalism. *Philosophy Compass, 10/12*, 867–883.
Baker, L. R. (2012). Personal identity: A not-so-simple simple view. In G. Gasser, & M. Stefan (Eds.), *Personal identity: Complex or simple?* (pp. 179–191). Cambridge: Cambridge University Press.
Belshaw, C. (2011). Animals, identity, and persistence. *Australasian Journal of Philosophy, 89*, 401–419.
Blatti, S. (2012). A new argument for animalism. *Analysis, 72*, 685–690.
Blatti, S. (2014). Animalism. In E. Zalta (Ed.), *The Stanford Encyclopedia of Philosophy (Summer 2014 edition)*. Retrieved from http://plato.stanford.edu/archives/sum2014/entries/animalism/
Blatti, S., & Snowdon, P. F. (Eds.) (2016). *Essays on animalism: Persons, animals, and identity*. Oxford: Oxford University Press.
Brower, J. E. (2010). Aristotelian endurantism. A new solution to the problem of temporary intrinsics. *Mind, 119*(476), 883–905.
Carrara, M., Gaio, S. (2011). Towards a formal account of identity criteria. In M. Trobok, N. Miščević, & B. Žarnić (Eds.), *Between logic and reality. Modeling inference, action and understanding* (pp. 227–242). Dordrecht: Springer.
Chappell, T. (2011). On the very idea of criteria for personhood. *The Southern Journal of Philosophy, 49*(1), 1–27.
Chisholm, R. M. (1976). *Person and object. A metaphysical study*. La Salle, IL: Open Court.
Chisholm, R. M. (1986). Self-profile. In R. J. Bogdan (Ed.), *Roderick M. Chisholm* (pp. 3–77). Dordrecht: Reidel.
Chisholm, R. M. (1989). *On metaphysics*. Minneapolis, MN: University of Minnesota Press.

Costa, D., & Giordani, A. (2013). From times to worlds and back again: A transcendentist theory of persistence. *Thought: A Journal of Philosophy, 2*(1), 210–220.
Costa, D., & Giordani, A. (2016). In defence of transcendentism. *Acta Analytica, 31*(2), 225–234.
Crane, T., & Mellor, D. H. (1990). There is no question of physicalism. *Mind, 99*, 185–206.
Crook, S., & Gillett, C. (2001). Why physics alone cannot define the 'physical': Materialism, metaphysics, and the formulation of physicalism. *Canadian Journal of Philosophy, 31*, 333—360.
Effingham, N. (2012). Endurance and perdurance. In N. A. Manson, & R. W. Barnard (Eds.), *The continuum companion to metaphysics* (pp. 170–197). London: Bloomsbury.
Fine, K. (2016). Identity criteria and ground. *Philosophical Studies, 173*(1), 1–19.
Grygianiec, M. (2016a). Criteria of identity and two modes of persistence. *Filozofia Nauki, 2*(2016), 1–13.
Grygianiec, M. (2016b). O banalności animalizmu. *Przegląd Filozoficzny – Nowa Seria, 3*(99), 295–309.
Grygianiec, M. (2016c). Criteria of personal identity. Reasons why there might not be any. In A. Brożek, A. Chybińska, M. Grygianiec, & M. Tkaczyk (Eds.), *Myśli o języku, nauce i wartościach. Seria druga. Profesorowi Jackowi Juliuszowi Jadackiemu w 70-tą rocznicę urodzin*. Warszawa: Wydawnictwo Naukowe Semper, 197–213.
Haslanger, S. (2003). Persistence through time. In M. J. Loux, & D. W. Zimmerman (Eds.), *The Oxford Handbook of Metaphysics* (pp. 315–354). Oxford: Oxford University Press.
Haslanger, S., & Kurtz, R. M. (Eds.) (2006). *Persistence. Contemporary readings*. Cambridge, MA: Bradford Books–MIT Press.
Hawley, K. (2001). *How things persist*. Oxford: Oxford University Press.
Heller, M. (1991). *The ontology of physical objects. Four-dimensional hunks of matter*. Cambridge: Cambridge University Press.
Hershenov, D. B. (2008). A hylomorphic account of thought experiments concerning personal identity. *American Catholic Philosophical Quarterly, 82*, 481–502.
Hershenov, D. B. (2011). Soulless organisms?: Hylomorphism vs. animalism. *American Catholic Philosophical Quarterly, 85*, 465–482.
Hoffman, J., & Rosenkrantz, G. S. (1997). *Substance: Its nature and existence*. London: Routledge.
Horsten, L. (2010). Impredicative identity criteria. *Philosophy and Phenomenological Research, 80*(2), 411–439.
Johansson, J. (2007). What is animalism? *Ratio, 20*, 194–205.
Jubien, M. (1996). The myth of identity conditions. *Philosophical Perspectives, 10*, 343–356.
Jubien, M. (2009). *Possibility*. Oxford: Oxford University Press.
Kanzian, Ch. (2012). Is "person" a sortal term? In G. Gasser, & M. Stefan (Eds.), *Personal identity: Complex or simple?* (pp. 192-205). Cambridge: Cambridge University Press.
Koslicki, K. (2008). Natural kinds and natural kind terms. *Philosophy Compass, 3*(4), 789–802.
Licon, J. A. (2012). Another argument for animalism: The argument from causal powers. *Prolegomena, 11*, 169–180.
Lowe, E. J. (1994). Primitive substances. *Philosophy and Phenomenological Research, 54*(3), 531–552.
Lowe, E. J. (1997). Objects and criteria of identity. In B. Hale, C. Wright (Eds.), *A companion to the philosophy of language* (pp. 613–633). Oxford: Blackwell.
Lowe, E. J. (1998). *The possibility of metaphysics. Substance, identity, and time*. Oxford: Clarendon.

Lowe, E. J. (2009). *More kinds of being. A further study of individuation, identity, and the logic of sortal terms.* Oxford: Wiley-Blackwell.
Lowe, E. J. (2012). The probable simplicity of personal identity. In G. Gasser, & M. Stefan (Eds.), *Personal identity: Complex or simple?* (pp. 137–155). Cambridge: Cambridge University Press.
Lowe, E. J. (2013). *Forms of thought: A study in philosophical logic.* Cambridge: Cambridge University Press.
Mackie, D. (1999). Personal identity and dead people. *Philosophical Studies, 95*, 219–242.
Madell, G. (2015). *The essence of the self: In defense of the simple view of personal identity.* London: Routledge.
Melin, R. (2011). Animalism and person as a basic sort. *Argument, 1*, 69–85.
Melnyk, A. (1997). How to keep the 'physical' in physicalism. *The Journal of Philosophy, 94*, 622–637.
Melnyk, A. (2003). *A physicalist manifesto. Thoroughly modern materialism.* Cambridge: Cambridge University Press.
Merricks, T. (1998). There are no criteria of identity over time. *Noûs, 32*(1), 106–124.
Merricks, T. (2001). *Objects and persons.* Oxford: Clarendon Press.
Montero, B., & Papineau, D. (2005). A defense of the via negativa argument for physicalism. *Analysis, 65*, 233–237.
Ney, A. (2008). Physicalism as an attitude. *Philosophical Studies, 138*, 1–15.
Nida-Rümelin, M. (2012). The non-descriptive individual nature of conscious beings. In G. Gasser & M. Stefan (Eds.), *Personal identity: Complex or simple?* (pp. 157–176). Cambridge: Cambridge University Press.
Noonan, H. W. (1998). Animalism versus lockeanism: A current controversy. *The Philosophical Quarterly, 48*, 302–318.
Noonan, H. W. (2011). The complex and simple views of personal identity. *Analysis, 71*(1), 72–77.
Noonan, H. W. (2012). Personal identity and its perplexities. In G. Gasser, & M. Stefan (Eds.), *Personal identity: Complex or simple?* (pp. 82–101). Cambridge: Cambridge University Press.
Oderberg, D. S. (1993). *The metaphysics of identity over time.* Basingstoke: Palgrave Macmillan.
Oderberg, D. S. (2005). Hylemorphic Dualism. *Social Philosophy and Policy, 22*, 70–99.
Olson, E. T. (1997). *The human animal: Personal identity without psychology.* New York: Oxford University Press.
Olson, E. T. (2003). An argument for animalism. In R. Martin, & J. Barresi (Eds.), *Personal identity* (pp. 318–334). Oxford: Basil Blackwell.
Olson, E. T. (2007). *What are we? A study in personal ontology.* Oxford: Oxford University Press.
Olson, E. T. (2012). In search of the simple view. In G. Gasser, & M. Stefan (Eds.), *Personal identity: Complex or simple?* (pp. 44–62). Cambridge: Cambridge University Press.
Olson, E. T. (2015). What does it mean to say that we are animals? *Journal of Consciousness Studies, 22*, 84–107.
Piwowarczyk, M. (2010). Paradoks tożsamości w czasie. Uwagi metaontologiczne. *Studia Philosophica Wratislaviensia, 5*(2), 137–152.
Rosenkrantz, G. S. (2012). Animate beings: Their nature and identity. *Ratio, 25*(4), 442–462.
Salmon, N. U. (2005). *Metaphysics, mathematics, and meaning. Philosophical papers I.* Oxford: Oxford University Press.
Sharpe, K. W. (2015). Animalism and person essentialism. *Metaphysica, 16*, 53–72.

Sider, T. (2001). *Four-dimensionalism. An ontology of persistence and time*. Oxford: Oxford University Press.
Simons, P. (2008). The thread of persistence. In Ch. Kanzian (Ed.), *Persistence* (pp. 165–183). Frankfurt: Ontos.
Snowdon, P. F. (2014a). *Persons, animals, ourselves*. Oxford: Oxford University Press.
Snowdon, P. F. (2014b). Animalism and the lives of human animals. *The Southern Journal of Philosophy, 52*, 171–184.
Stoljar, D. (2010). *Physicalism*. London: Routledge.
Swinburne, R. (1997). *The evolution of the soul*. Oxford: Oxford University Press.
Swinburne, R. (2013). *Mind, brain, and free will*. Oxford: Oxford University Press.
Toner, P. (2011). Hylemorphic animalism. *Philosophical Studies, 155*, 65–81.
Toner, P. (2014). Hylemorphism, remnant persons and personhood. *Canadian Journal of Philosophy, 44*, 76–96.
van Inwagen, P. (1990). *Material beings*. Ithaca: Cornell University Press.
Vetter, B. (2015). *Potentiality: From dispositions to modality*. Oxford: Oxford University Press.
Wasserman, R. (2004). Framing the Debate over Persistence. *Metaphysica, 5*, 67–80.
Wasserman, R. (2016). Theories of persistence. *Philosophical Studies, 173*(1), 243–250.
Williamson, T. (2013). *Identity and discrimination*. Oxford: Blackwell.
Wilson, J. (2006). On characterizing the physical. *Philosophical Studies, 131*, 61–99.

Filip Kobiela
How Long Does the Present Last? The Problem of Fissuration in Roman Ingarden's Ontology

Abstract: In the philosophy of time the standard view on the present holds that it has no duration. The classic proponent of this view, St Augustine, claims that the present is the blade of a knife separating the future from the past. Despite its dominant position, this view might be questioned on both phenomenological and ontological grounds. An interesting attempt at accounting for the duration of the present can be found in Roman Ingarden's analyses of temporal being. In his ontology Ingarden discerned two features that characterize present temporal objects – activeness and fissuration. The former outlines a distinctive quality of present temporal objects – their "fullness of being": a complete qualitative determination and efficaciousness. The latter portrays a limitation to activeness – the actual, effective existence of temporal objects is restricted to their present being, it is only a "fissure" between the past and the future. But according to Ingarden this fissure might vary for different objects, which raises a question concerning the duration of the present. In this article, I point to some motifs that led to entertaining the possibility of a certain duration of the present – a non-zero value of the fissure. I also investigate the relation between the duration of the present of objects existing in time and their ontological structure. In the conclusion, I propose an outline of an ontological theory of relativity of the duration of the present inspired by Ingarden's analyses.

Keywords: Ingarden, time, fissuration, specious present, duration, eternity.

Because the following introductory remarks aim mainly at explaining basic problems of Ingarden's ontology, the readers already familiar with these topics might skip over it and proceed to section 1 – "Temporality and existential moments".[1]

---

[1] Daniel von Wachter's article "Roman Ingarden's Ontology: Existential Dependence, Substances, Ideas, and Other Things Empiricists Do Not Like" offers a "Europe-in-seven-days tour through

Note: This text is a revised version of the previously published article *Problem szczelinowości w fenomenologii Romana Ingardena* (Kobiela, 2011, in Polish).

Filip Kobiela, Section of Philosophy and Sociology in Tourism, University of Physical Education in Cracow, Poland.

Perhaps the most important Polish contribution to ontology is Roman Ingarden's *Controversy over the Existence of the World* (henceforth the *Controversy*). The title originates from Ingarden's disagreement with Edmund Husserl's transcendental idealism and suggests that the book is devoted to a particular debate. But, in fact, the *Controversy* presents Ingarden's original account on almost all fundamental ontological problems, and thus it might be called the modern "summa of ontology". The language of Ingarden's considerations might seem a bit exotic, but his analyses are deeply rooted in the continental tradition and have also some features of analytic ontology.[2]

According to Ingarden, ontology is an *a priori* field of study of pure possibilities. Ontology understood this way has some resemblance to mathematics and is set against fields of study that investigate facts such as metaphysics and the natural sciences. Ingarden divides ontology into three domains: existential, formal, and material. Existential-ontological problems concern the modes of being of objects (real, ideal, absolute, etc.); formal-ontological problems concern forms of objects (events, processes, objects enduring in time, etc.) and material-ontological problems concerning the material endowment of objects, their qualitative determination (Ingarden, 2013, 87–89). Existential ontology is a foundation of Ingarden's investigations. Its starts with an analysis of the notion of existence. According to Ingarden, analysis of the notion of existence is possible and reveals several *existential moments* – "elements" of modes of being. Existential moments "can be intuitively discerned and grasped in a mode of being by means of abstraction" (Ingarden, 2013, 108). In the first stage of his investigations (not with regard to the problem of temporality) Ingarden discerns four different pairs of opposing existential moments: 1) autonomy vs. heteronomy; 2) originality vs. derivativeness, 3) self-sufficiency vs. non-self-sufficiency, 4) independence vs. dependence (Ingarden, 2013, 109). Some combinations of these moments are contradictory (e.g. heteronomy and originality) (Ingarden, 2013, 155-156); the non-contradictory combinations of existential moments create a rich set of possible modes of being. The most important modes of being are absolute, real, ideal, and purely intentional.[3]

In the second phase of his investigations (that starts with chapter V of the *Controversy* – "Time and Modes of Being") Ingarden focuses on existential moments that characterise a temporal mode of being. All objects existing in time have

---

Ingarden's ontology" (Wachter, 2005). A detailed introduction to the *Controversy* in the context of Ingarden's polemics with Husserl might be found in Jeff Mitscherling's book "Roman's Ingarden Ontology and Aesthetics" (Mitscherling, 1996).

2 Cf., e.g., (Thomasson, 2017).

3 A good introduction to Ingarden's ontological "combinatorics" might be found in (Simons, 2005) and (Chrudzimski, 2015).

some imperfections – they cannot remain present (transience) and their present is always only a fissure between the two remaining domains of time – future and past. Inspired by some of Bergson's analysis, Ingarden entertains the hypothesis that different real objects have different relations to time – their ontological structure generates different values of fissuration. There are several problems that emerge from these considerations. The most important concerns the objective character of the duration of the present (non-zero fissure). From the phenomenological perspective such a "wide" present is a data of consciousness. But on ontological grounds one may ask if it is only a kind of illusion or is there an objective, real feature of certain objects? In the latter case, another fundamental problem arises – what is the relation between the structure of the object and its relation to time? These problems are rarely discussed but they are of high importance for the phenomenologically oriented ontology of time.

The structure of this article is as follows. In the first four sections, I present the relevant considerations of Ingarden. In section 1, I introduce Ingarden's notion of fissuration against the background of his ontology of temporal being. In sections 2–4, I present a hierarchy of objects (inanimate objects, living individuals, conscious objects, and absolute being) in each case demonstrating the peculiarities of their present. In section 5, I proceed to a more detailed analysis of the problem of the duration of the present – I analyse St. Augustine's theory as well as the theory of a *specious present*. In section 6, I analyse different arguments in favour of the non-zero span of the present – Ingarden's phenomenological "argument from Neon", Karl Popper's empirical argument and Bogdan Ogrodnik's argument linking the duration of the present and the formal complexity of the structure of objects. In this section, I also discuss a similar idea presented in Stanisław Lem's short novel "137 seconds". In the conclusion – section 7 – I present a generalisation of these considerations in the form of the ontological theory of the relativity of the duration of the present.

Temporality and Existential Moments

One of the main goals of Ingarden's ontology is to show the difference between the mode of being of purely intentional objects and the mode of being of real (temporal) objects. The crucial difference between these two modes of beings is that only the temporal mode of being is characterized by *Activeness* or *Aktualität*,[4]

---

[4] Ger. *Aktualität*; Pol. Aktualność. In Arthur Szylewicz's translation of the *Controversy*, the term Activeness has been designated as the equivalent of German *Aktualität*, because the term *Actuality*

an existential moment which, according to Ingarden, is a certain perfection of an object. Activeness, in turn, may be subjected to further analysis, revealing such phenomena as fragility and fissuration. These phenomena are the sub-moments of activeness that reveal different aspects of its limitation. Whereas activeness is a kind of perfection of a temporal object, fragility and fissuration concern its imperfection: fragility determines the boundaries of the object's existence (persistence),[5] whereas fissuration determines the boundaries of its activeness.

The notion of fissuration appears in the *Controversy* in the discussion of the mode of being of the so-called *objects persisting in time* (OPT). Ingarden claims that these objects:

> exist during the *entire* time that they exist, but in accordance with the essence of time any given instance of their active being is always confined to only a *single* present moment beyond the bounds of which they can at no time reach. The activeness of their being spans at any particular time only a single – if we may put it that way – narrow fissure [*Spalte*]. Beyond it in the one direction there is the retroactively derived past being, and in the other the first intimations of the future being. This so to speak "fissure-like [*spaltartige*]" existence is characteristic of every temporally extended being, and of every persistent object in particular. (Ingarden, 2013, p. 274)

Ingarden introduces the notion of fissuration as a further characteristic of the activeness of temporal objects. The notion emphasizes the limitation of the activeness of temporal objects – its "narrowness" in comparison to non-active domains of past and future. Fissuration understood in this way is a derivative of transience, which, according to Ingarden, is based chiefly on

> the *constant* transformation of the being-active of what is present into this puzzling "no-longer-being-in-the-present," whereby it is nonetheless somehow sustained in being in the past, as something bygone [*Vergangenes*]. This transformation – comprising the innermost essence of temporality – is of course nothing accidental, but is essentially bound up with a certain deficiency of the entity existing in this fashion: namely, with its inability to persist in activeness, as it were, without succumbing to passage. (Ingarden, 2013, p. 239)

(a natural candidate for being an equivalent for *Aktualität*) has already been employed in this translation for *Wirklichkeit*, which cannot be identified with *Aktualität*. Szylewicz explains this decision in note 214 (Ingarden, 2013, p. 99). Hovewer, even if the term *Actuality* already plays a different role in the translation, the term *Effectiveness* might in some cases still seem better than *Activeness* as a translation of *Aktualität*.

5 Because the term persistence has been sometimes used to translate *Dauerhaftigkeit*, which is the opposite of fragility, to avoid possible misunderstanding let us note that *persistence* should be understood here as a temporal mode of being, which allows for the object to perish.

Based on the above we can define fissuration as a specific quality of a temporally determined being that renders it unable to last infinitely in the activeness phase, limiting activeness to one present only, which creates a fissure between non-actual being – on the one hand of the future, and on the other of the past. In his further considerations, Ingarden analyses fissuration in two contexts: 1) within the context of the mode of being of the OPT (living individuals overcoming fissuration, and especially the conscious ones) and 2) within the context of the mode of being of an absolute being. He does not examine closely the fissuration of processes and events, although fissuration is a part of their mode of existence.

The Differences Between Living Individuals and Inanimate Objects

Characterising the OPT mode of being, Ingarden observes living individuals, whose being is (up to the time of their death, conditioned by the fragility of their being) crossing the sphere of activeness of the ever new present. As Ingarden says:

> For living beings, however, there emerges against the background of the fissure-like mode of existence an essential modification that somehow enables the living being to transcend the activeness-fissure of any particular present, and that is because for such a being what has happened in the past makes its mark in an essentially different and more meaningful way on the structure [Ausgestaltung] of what "presently" exists than it does for "inanimate" things. (Ingarden, 2013, p. 274)

A significant difference between inanimate objects and living beings is that the former possess remnants from the past in form of a type of a multitude of qualities, which, from the "point of view" of this object are an effect of random impacts. As Ingarden notes: "For a living being meanwhile, what remains of its past makes up *meaningful* unity" (Ingarden, 2013, p. 275). A living being is characterised by "the 'ingenious' *mode of its reaction* to the assaults directed at its being – a mode that is characteristic for it, promotes the preservation of its life, and rebuilds its inner structure (in a way that is to some extent creative)" (Ingarden, 2013, p. 275). The relation between the phases of development of a living being and its defensive actions against the interference of the outside world that could breach its integrity "is expressed synthetically in the living being's active state and constitutes the *inner unity* of not only the total content of its *present* makeup, but also of its *entire temporally spread out [ausgespannten]* being" (Ingarden, 2013, p. 275). This inner unity of a living being (both synchronic and diachronic) increases the intensity of the retroactively derived past being of an individual, "and in this way elicits at least the *semblance of an expansion [Ausweitung] of the activeness-phase* in the direction of the past" (Ingarden, 2013, p. 276). However, "The 'inanimate' thing deteriorates gradually, until some impact obliterates it completely. Thus the 'fissure-character [*Spalthaftigkeit*]' of its activeness is much more radical than in the case of the living being" (Ingarden, 2013, p. 277).

Therefore, although fragility and fissuration – the existential-ontological imperfections of a real being – are not overcome in the case of living beings, the inner structure of these beings, i.e. their formal-ontological qualification, elicits at least a seemingly "wider" fissuration. This last formulation is very significant – it suggests that the whole existence within the time of one being resembles a process consisting of phases, and one of these phases is activeness. The phase of activeness has a certain finite duration, which Ingarden calls fissuration. Assuming, following Ingarden's considerations, that the length of the activeness phase of a given being may be variable, the fissuration might have different values. Since activeness is the same as the present, the value of fissuration is then the period of the duration of the present. A question about the duration of the present is then a question about the value of fissuration.[6]

Conscious Subjects, Fragility and Fissuration

Ingarden, discussing another "borderline" OPT group, states that there is a category of living beings,

> in which the fissure-character of active being appears to be overcome in a quite pronounced measure, and in an especially distinctive manner: the beings that live consciously. [...] through their acts of recollection, retention, protection and expectation they can look out beyond the structure of their current present, and can at least in principle survey the whole course of their lives [...] They do so only "intentionally," but even this merely intentional, presumptive [vermeinende] intuiting and grasping of what exceeds the bounds of the current activeness-phase entails a jutting out above the uninterrupted lapse of time. (Ingarden, 2013, p. 277)

While the living individuals partially overcome fissuration by combining the histories of these individuals with activeness, for the conscious individuals this process of overcoming is two-sided, i.e. expanding the fissuration represents "looking" both ways – into the past as well as into the future. Consciousness opens a possi-

---

[6] The maximum possible value of fissuration is determined by the boundaries of the duration of this object.

bility to additionally strengthen the inner unity of a living being, which, according to Ingarden, somehow reduces the fragility of its being.[7] Concluding his considerations on the specific hierarchy of real objects and their relation to time, Ingarden states that

> All temporally determined entities exist by passing through an ever new activeness-phase and they are unable to overcome the "fissure-character [*Spalthaftigkeit*]" of their existence even in the existentially highest form of conscious living beings. (Ingarden, 2013, pp. 279–280)

However, from the ontological point of view, we cannot exclude the existence of real objects, which are not human, yet overcoming the limitations of existence related to time in a more effective manner than people. Ingarden tends to use this reasoning when considering the mode of being of an absolute being, which can be treated as a continuation of the above-mentioned considerations on the hierarchy of real objects.

Fissuration Within the Context of the Modes of Being of an Absolute Being

Constructing the notion of an absolute being as a being existing in a timeless manner, Ingarden takes two possibilities into consideration: 1) absolute being which is characterized by activeness, durability and *non-fissuration*; 2) paradoxically imperfect absolute being, which is characterized by activeness, durability and *fissuration*.

Ingarden considers the following possibility

> whether the "fissurative character" of the being of what exists actively can be overcome by means of an unrestricted broadening of the span of the present, if we may put it that way. That this span is alterable only in admittedly very modest measure, is something we know from our daily experience – as Bergson has observed. But whether it is possible to broaden these relatively narrow bounds, and effectuate [*aktivieren*] being in such a way that activeness could encompass the collective past and the entire future – that is one of the deepest and most difficult problems of both existential and material ontology. (Ingarden, 2013, p. 290)

This problem is further related to the more general question of

---

[7] It is important to note that this reduction of fragility is an ontological phenomenon, not just *in intentione*, but also *in re*. In case of fragility, Ingarden's view opens a possibility for different interpretations. Taking into account Ingarden's remarks concerning non-fissuration, the ontological interpretation of fragility seems to be consistent with Ingarden's ontology.

> whether time is one and the same for all variants of individual being, or whether the variety of differently structured times are possible which would be characteristic for the various types of individual being – as Bergson argues. (Ingarden, 2013, p. 291)

Presenting Bergson's position, Ingarden states that while the tension of our duration

> may be subject to variation, there are incomparably larger variations of the rhythm of duration for different living beings, *resp.* varied principal types of reality. (Ingarden, 1922, p. 329; my translation – F.K.)

While considerations of fissuration within the context of the OPT mode of being, particularly on living and conscious beings, were mainly conducted based on phenomenological description, and the possibility of an apparent broadening of the span of the present was taken into account, the notion of possible absolute beings fissuration is perceived as a strictly ontological property, characterizing various beings to various extents, determining their degree of overcoming transience. Ingarden is certainly referring here to the theory of Bergson, who identified phenomenal changes in the span of the present and also assigned to different beings – placed on various levels of the hierarchy of real being – different "tensions of duration" (*tension de la durée*), the smallest in inanimate matter, then increasing in living, spiritual individuals and culminating in God. Thus Ingarden considers on the basis of the extrapolation of the phenomenal broadening of the span of the present and projection of this phenomenon onto a being outside the subject – certainly as a possibility only – a hierarchy of real beings, which are characterized by various values of fissuration.

The Problem of the Duration of the Present

The notion of fissuration generates interpretational difficulties. Firstly, they result from some inconsistencies in Ingarden, e.g. fissuration is treated once as a mode of being and another time as a part of a mode of being (existential moment), and yet another time as an existential sub-moment characterizing the moment of activeness. Transience is also sometimes treated as a mode of being by Ingarden. Such difficulties are relatively easy to overcome by consistently describing (in line with the spirit, however not the letter of Ingarden's text) fissuration as a sub-moment of activeness. A more difficult problem is posed by Ingarden's statement that overcoming fissuration is "at least apparent". It renders a possibility for fissuration to be interpreted in a phenomenalistic or ontological way. The notion of fissuration itself is unclear, as it is in fact a metaphor (in the *Controversy* quotation marks are usually put around it). Further considerations then lead to presenting a coherent notion of fissuration, based mainly on Ingarden's ontology, as well as on studies of other philosophers.

Augustine's Theory of Time and the Problem of the Present

The key to explaining the problem of fissuration is to refer to the two fundamentally different ways of experiencing time defined by Ingarden, which generate two basic types of ontological theories of time. According to Ingarden, St. Augustine's theory of time is a theoretical consequence of experiencing time in this way, where the only existing domain of time – the present – is of a punctual nature. St. Augustine's theory of time, which is challenged by Ingarden, consists of two principal theses and a few complementing ones.

The first principal thesis of St. Augustine says that neither the past nor the future exist: "neither things to come nor past are" (Augustine, St., 2008, p. 125), only the present exists. In contemporary philosophy the view is called presentism[8]. Contrary to this theory, Ingarden puts forward an idea that past and future exist as well, however in a mode different from the present.

The second principal thesis of St. Augustine says that the present is devoid of any duration – it is a point or a cross-section, or, in a figurative way, it is the blade of a knife separating the future from the past.

> If an instant of time be conceived, which cannot be divided into the smallest particles of moments, that alone is it, which may be called present. Which yet flies with such speed from future to past, as not to be lengthened out with the least stay. For if it be, it is divided into past and future. The present hath no space. (Augustine, St., 2008, pp. 123–124)

This belief, which may be called the "zero" theory of the present, is countered by Ingarden with the idea that the present is not a point devoid of dimensions, but a specific time quantum. According to Ingarden temporal quanta

> marked-off from each other within the passage of time – without thereby comprising a temporal point or a time-interval. (Ingarden, 2013, p. 231)

The temporal present quantum is a counterpart to the fissure of activeness, and the sizes of the temporal quantum – the present moment – are the sought value of fissuration. Ingarden is a proponent of a non-zero duration of the present, not

---

[8] Compare (Ingram & Tallant, 2018).

ascribing, however, any particular duration to the present, as if it were an interval in time. In Augustine's theory, there is this characteristic tension between the analysis of time as a tripartite structure of past-present-future, where only the present exists yet is devoid of duration, while *an experience* of the present is lasting and has a certain duration.[9]

St. Augustine solves this problem by employing the notion of the function of the mind,[10] and especially the memory: impressions of the mind triggered by the passing presents are kept by the mind, although their origins have passed. As we can keep the traces of the passing presents in our memory, we are able to measure time and understand the words heard, which would have been impossible if we had only perceived a series of non-enduring presents. When we hear an extended – in memory – sound, only its finishing bounds can be considered to be the punctual objective present.

The Theory of the Specious Present

This element of Augustine's theory has been discussed in the late 19th and 20th century under the name of the theory of the specious present. The term *specious present* serves as an explanation of such phenomena as hearing an enduring sound or seeing a falling meteor, which would be impossible if our consciousness of the present had a punctual character. The notion was introduced into philosophy by William James:

> the practically cognized present is no knife-edge, but a saddle-back, with a certain breadth of its own on which we sit perched, and from which we look in two directions of time [...]. We seem to feel the interval of time as a whole, with its two ends embedded in it. (James, 1890, pp. 609–610).

As it is usually assumed that in reality, the ontologically existing objective present is of a punctual character, the duration of the phenomenal present is treated only as a phenomenon, or appearance. We arrive then at the differentiation of the objective present, i.e. a strictly ontological category and the subjective or phenomenal present, thus a category, which is, first of all, epistemological.

---

9 The reality of the span of the present is compatible with experience. Traditional arguments supporting this idea refer to the perception of movement (and the impression of continuity when watching a moving picture in the cinema), understanding words and tunes.
10 The three functions of mind – expectation, consideration and memory – refer respectively to the future, present and past.

The nature of the objective present is an ontological problem, whereas the nature of the subjective present is primarily a psychological problem. Various results of empirical research on the duration of the subjective present lie within a range of a few milliseconds to several seconds. These studies however are controversial, as the notion of the subjective present in itself is not fully clear, likewise the interpretation of the results of these experiments. Let us now consider the relations between Ingarden's theory and the theory of the specious present. For that purpose we need to analyse the troublesome statement by Ingarden referring to the broadening of the span of the present. As this broadening is, according to Ingarden, at least specious, then it is certainly being experienced, and is thereby something subjective, but it can also be something objective, and thereby not merely experienced. By stating that a phenomenally given span of the present is *at least* specious, Ingarden voices his uncertainty regarding this "speciousness", or the merely phenomenal nature of this present. Ingarden tends towards this second possibility, however not all of his statements are unambiguous.

The Problem of Eternity and Absolute Being

Let us then go back to St. Augustine's theory of time. One of the remaining theses is the definition of eternity:

> But the present, should it always be present, and never pass into time past, verily it should not be time, but eternity. (Augustine, St., 2008, p. 123)

This notion is accepted by Ingarden, but he calls this eternity "non-fissuration". Paraphrasing St. Augustine, Ingarden could say: a non fissured activeness is not an activeness of a temporal being, but of a timeless or an absolute being. Let us then consider a divine perspective of perceiving the world in this context – it leads to the reverse of the theory of the specious present, which can be named the theory of the specious non-present (with the combined notion of the past and the future). As *sub specie aeternitatis* the entire history of the world is present in the present, and, indicating a point on the axis of time as the present is only related to the human epistemological perspective and has no basis in the ontic structure of the world. It is then something subjective, a delusion or appearance, but of an entirely different nature than in the case of the theory of the specious present.

In St. Augustine's theory we can locate the origin of two theories:
1. According to the "view from nowhere" (H. Price), regarding some present to be existing and negating the existence of the past and the future is a delusion which I shall call the specious non-present;

2. The view form a certain punctual now, which regards the experienced duration of the present as a delusion, a specious present.

Each perspective is related to some delusion concerning the present. The first theory is called eternalism, the second is called presentism. Ingarden rejects both of them, devising his own theory based on the plurality of the existence of the real being and the notion of fissuration. As this last element is not a fully developed notion, Ingarden's entire theory of time requires additional clarification. Later in this paper I will present three short arguments (one of which is Ingarden's) against the theory of the specious present and at the same time against the punctual interpretation of the present.

Arguments Supporting the Non-Zero Span of the Objective Present

The first group of arguments is focused on determining whether it is possible to pass from this certain time span of the subjective present to the span of the objective present. According to Ingarden, the subjective present, which is non-punctual, is a certain part of the real world, thus there is a certain part of the world where the extended present does exist.

Ingarden's Phenomenological Argument ("Argument From Neon")

The argument might be structured as follows:
1. The lighting up of a sequence of bulbs appropriately ordered in space and time can give the illusion of a continuous movement of one lamp
2. this illusion succeeds only because a *continuous* process plays itself out physiologically and psychologically
3. thus what is past as such, though it has ceased to be strictly active, retains a peculiar way of being (Ingarden, 2013, p. 238).

According to Ingarden, the argument supports a way of experiencing time according to which "what is past and what is future also exist in some way" (Ingarden, 2013, p. 231), and time-instants are peculiar temporal quanta. There are psychological phenomena that are at the same time both real beings and continuous processes, not multiplications of punctual events. These processes include the stream of consciousness as well as the conscious experiences of which that stream consists. It is important in this context that the argument supports the possibility of a non-punctual present, at least within the framework of the concrete time of the knowing subject, and thus a certain real element of the world.

It does, however, seem that in order to resolve the problem of the relation between experienced time and objective time an analysis of acts of consciousness must be carried out. According to Ingarden, each act of consciousness is a certain process, a constituent in a stream of consciousness. Since Ingarden in the end includes the stream of consciousness within the structure of the real world, there appears to arise another problem concerning the possible carrier of the stream of consciousness and all of the separate acts of consciousness. The proponent of the zero present theory might argue that although the content of the act of consciousness relates to some process extended in time, the material carrier of this act of consciousness can be found in the objective present devoid of duration. Notwithstanding the resolution of this problem, a more promising way of defending the non-zero objective present is to refer directly to non-psychological phenomena.

Popper's Empirical Argument

The other group consists of arguments stating that the non-punctual, extended present is mind-independent and occurs in the physical world. These arguments are presented by the philosophy of nature or philosophy of the natural sciences and are based on the interpretation of certain physical phenomena. Karl Popper claims that the present is non-punctual in this respect. In his *Objective Knowledge*, when debating Kant's intuitionism, he claims that the intuition of time can yield to changes and serve as a function of, e.g. language, worldview and relevant theories. According to Popper:

> While particle physics suggests a razor-like unextended instant, a "punctum temporis", which divides the past from the future, and thus a time co-ordinate consisting of (a continuum of) unextended instants, and a world whose "state" may be given for any such unextended instant, the situation in optics is very different. [...] there are temporally extended events (waves possessing frequencies), whose parts co-operate over a considerable distance of time. Thus owing to optics, *there cannot be in physics a state of the world at an instant of time*. [...] what has been called the specious present of psychology is neither specious nor confined to psychology, but is genuine and occurs already in physics. (Popper, 1989, p. 135)

Ogrodnik's Formal-Ontological Account

The third group of arguments indicating the non-zero nature of the present are the ontological arguments. Having identified the value of fissuration with the size of the present temporal quantum we might now focus on the theory of Bogdan Ogrodnik, which is a further analysis and at the same time a modification of Ingarden's ontology. By employing the notion of the temporal quantum, Ogrodnik in fact expands the theory of fissuration, although he does not use this term in his considerations. Ogrodnik develops an expanded ontological theory, allowing for a formal-ontological grounding of various values of fissuration. The course of his argument can be summarized as follows: The past and the future can jointly be named the non-present. For a given object X one might distinguish its absolute and relative non-present. The former refers only to its concrete time, while the relative non-present refers to a hierarchy of concrete times. In a derivatively individual object, one might distinguish the hierarchy of this object's parts belonging to various structural levels and the corresponding hierarchy of concrete times. Each of the concrete times has the same structure, i.e. it consists of quanta. Ogrodnik claims:

> Quantum of the concrete time of the higher order contains quanta (at least one) of the concrete time of the lower order. (Ogrodnik, 1995, p. 113)

The consequence of the above consists of:

> Taking into account the hierarchical structure of the real world, as a result we obtain a whole series of present and non-present "immersed" in each other – until we reach the present of the object of the highest order (provided that such an object exists at all), which is the Universe. (Ogrodnik, 1995, p. 114)

The quantum of the present of the object of the higher order is larger than the quantum of the present of the object of the lower order. In other words, the value of fissuration of the real object depends on the complexity of its structure, and its place in the hierarchy of real beings. The non-fissuration being can be found at the top of this hierarchy.

> Existence of the relative non-present of concrete time of the object of a lower order is guaranteed by the existence of the "quantum" – the present moment of concrete time, which belongs to the object of a higher order. [...] The non-present of a given real object is "placed" between the present of this object and the present of the object of a higher order. (Ogrodnik, 1995, p. 113).

Arguably, in the case of an object of a higher order, its material constitution can determine the value of the fissuration of its concrete time. By adapting these theses Ogrodnik modifies the theory of Ingarden which claims that the past and the future are non-active beings. According to Ogrodnik, the non-present is non-active only within a given real object, but it maintains its activeness within the present of the objects of the higher orders, containing this object. The theory of Ogrodnik then, clearly indicates the primacy of the objects of the higher order. The modification only applies to the mode of being of the non-present, as Ogrodnik accepts Ingarden's account of the mode of being of the present.

A Notion of Fissuration in Lem

Stanisław Lem's short story *One Hundred and Thirty-Seven Seconds* portrays a theory of the present that resembles Ingarden's theory, and is possibly inspired by it. The main character in the story, an agency journalist serendipitously discovers that the computer he is working on renders a factual report of events, even though it is disconnected from the teleprinter transmitting current information. As a result of simple experiments it turns out that despite being disconnected from the teleprinter, which transmitted the initial part of the message, the computer comes to a standstill for a short time, and then renders – without errors – the remaining part of the message for the exact duration of 137 seconds. Within this period of time it knows everything about this event, however a second later it knows nothing. For instance, it provides exact information on the result of casting dice, but only if they are cast no later than 137 seconds from posing the "question" about the result of the throw. As it turns out, the situation only occurs when the computer is connected to the federal IT network. Within these 137 seconds, the computer exploits several dozen percent of the whole network's potential. The first idea when attempting to explain this phenomenon is that the computer uses the performance power of the network to predict the results of experiments. It is then something of a Laplace's demon. However, it turns out that it is a false hypothesis, and the phenomenon is correctly explained by the physicist named Hart.

> He said that the computer cannot indeed predict the future, but that we're in some specific way restricted in perceiving the universe. In his words: "If one imagines time as a straight line, stretched from the past into the future, our consciousness is like a wheel rolling across that line and touching it consistently only in one point; we call that point the present, and that present immediately becomes a past moment making room for the next one. Studies by psychologists demonstrated that what we take for a present moment, devoid of temporal extent, is indeed slightly prolonged and covers a bit less than half a second. Perhaps it is possible that the interface with that line might be a little bit wider; that it's possible to remain

in contact with a longer section at the same time, and that the maximum dimension of that temporal section is exactly one hundred and thirty-seven seconds." If that's indeed true, says Hart, it means that the entirety of our physics is still anthropocentric, since it makes assumptions which are unimportant outside of human sensors and consciousness. [...] It might be the result of the fact that the concept of the present is not only as relative as Einstein's theory proclaims, depending on the location of the observers, but it also depends on the scale of the phenomena in the same "place." The computer resides simply in *its own* physical present, and that present is more broad in time than ours. [...] this has important philosophical consequences since it means that if free will exists, it only happens beyond the limit of one hundred and thirty-seven seconds, even though we'd never know this from introspection alone. Within those one hundred and thirty-seven seconds our brain behaves similarly to our body which is inert and cannot suddenly change direction – for this you need time to allow a force to skew the path – and something like that happens in every human head. [...] the bigger a brain (or a brain-like system), the wider its contact with time, or the so-called "present," whereas the atoms don't properly touch it at all, only dance around it, so to say. In one word, the present is something like a triangle: point-like, near-zero where electrons and atoms reside, and widest around big bodies gifted with consciousness. (Lem, 2015)

Philosophical Commentary on the Conception of Lem

Similarly to Ingarden, Lem rejects the theory of the specious present, treating the subjectively experienced present as something real, however, he perhaps validates the theory of the objective punctuality of the present in reference to the objects of microphysics. The human experience of the present is treated as one of the possible experiences that are dependent on the structure of the subject experiencing it. Like Ingarden, Lem treats fissuration as a certain human limitation.[11] Lem, similarly to Ingarden, accepts various possible values of concrete beings' fissuration based on the complexity of their structure. The hierarchy of fissuration presented by Lem in the form of a pyramid also corresponds to the intuitions of Ingarden (as well as Ogrodnik). Lem assumes some finite value as the maximum value of fissuration, thus rejecting the hypothetical non-fissuration being. A particularly important thing is that Lem, and likewise Ingarden (and Ogrodnik), does not connect the phenomenon of extending the duration of the present to the act of predicting facilitated by the deterministic structure of being, but treats it as an essential part of a given being, based on its innate structure. The concrete objects compared to the wheels of various diameters rolling across the line of time, where the wheel

---

[11] The number 137 – the maximum value of fissurations in Lem's story, is an allusion to both Pythagorean philosophy and modern physics, in which 1/137 is an approximate value of a dimensionless constant called *fine structure constant*.

diameters correspond to the levels in the hierarchy of the objects, present an interesting analogy. The contact with the line of time, or the present of the respective concrete objects (their fissuration), is directly proportional to the sizes of the diameters of the rolling wheels, i.e. the complexity of their structure. By expanding the diameters of the rolling wheel to infinity, i.e. considering an infinite being, we achieve full contact with the line of time, an eternal being - characterized by non-fissuration. This leads to determine God as a sphere, whose centre is everywhere and whose circumference is nowhere.[12] It is also very pertinent to show a relativity of the present understood in a sense different from Einstein's sense of relativity. Human activity occurs within his concrete present, the fissuration of which is "immersed" in a broader fissure of the activeness of an object higher in the hierarchy. It does not exclude the activity of man – however, according to Lem – it limits his freedom, and to be more precise – retards the activity of man by the maximum value of fissuration, if it is initiated by a free impulse of will.

The problem of freedom existing within the framework of the present of the object of a higher order is the counterpart to the theological problem: does divine foreknowledge limit human freedom? To formulate it in our manner, the problem appears as follows: does the value of the divine fissuration (resp. divine non-fissuration) of activeness, which triggers the destiny of the world, from the divine point of view, within the framework of one present, exclude the freedom of human action? And in more general terms, the freedom of action of all the systems of the lower order? Let us note that this formulation is rather stronger than the traditional problem of divine foreknowledge, as it considers not only the knowledge, but also the activeness of the divine subject.

Conclusion: The Spectrum of Possible Values of Fissuration and the Theory of the Relativity of the Duration of the Present

Having presented various arguments supporting the non-zero span of the present, i.e. arguments supporting the ontological interpretation of fissuration, I propose a generalization of the notion of fissuration originating in Ingarden's theory. The following values of the duration of the present of concrete time, i.e. the values of fissuration, are possible in a purely ontological sense:

---

[12] This "geometrical" description of God might be incompatible with the idea of God's simplicity, hovewer, it is a natural extension of Lem's analogy.

1. Punctual (radical) fissuration: the present is a point devoid of dimensions, bears no duration – the value of fissuration = 0 (Augustine). This possibility only allows for phenomenal duration of the present (the theory of the specious present).
2. The interim case: a fissure with a finite, but a non-zero value, with a certain finite "duration". The fissure consists of a certain quantum (Bergson, Ingarden, Popper, Lem).
3. Non-fissuration: there are no limits to the activeness of being; it embraces the entire being (the value of the "fissure" = ∞). The distinction between present and non-present bear only specious character in this possibility (the theory of the specious non-present).

A solution combining all these possibilities would come true in a theory of the relativity of the present boundaries' span, assuming the multitude of perspectives, conditioned by various levels of structural complexity of objects – their position within the hierarchy of real objects and its material constitution. This last theory might perhaps be reduced to one of the previous options of 1 to 3. If we were to assume, as Ogrodnik and Lem do, that the objective present lasts as long as a quantum of the time of the object situated on the top of the real world hierarchy, then the boundaries of the span of the present of the objects of the lower order turn out to be only their cognitive limitation, thus they do not bear an objective character. We then see that it is only a generalization of the theory of specious *non-present*. We may, however, assume an opposite hypothesis, according to which the boundaries of the objective present are determined by the size of the quantum of the time of objects residing on the very bottom of the real world hierarchy. In this case, the length of the duration of the present of the higher-order objects will prove to be subjective. This second possibility illustrates the generalized theory of the specious *present*.

Acknowledgment: I would like to express my gratitude for Professor Jeff Mitscherling for his valuable comments and linguistic counsel. I would also like to thank Professor Andrea Bottani for his inspiring remarks.

Bibliography

Augustine, St. (2008). *Confessions* (E. B. Pusey, Trans.). Rockville, Maryland: Arc Manor.
Chrudzimski, A. (2015). *Ingarden on Modes of Being*, In D. Seron, S. Richard & B. Leclercq (eds.), *Objects and Pseudo-Objects: Ontological Deserts and Jungles From Brentano to Carnap*. Boston/Berlin: De Gruyter, pp. 199-222.

Ingarden, R. (1922). *Intuition und Intellekt bei Henri Bergson*. [*Intuition and Intellect in Henri Bergson*]. Jahrbuch für Philosophie und Phänemonologische Forschung, vol. 5, pp. 285-461.
Ingarden, R. (2013). *Controversy over the existence of the world* (A. Szylewicz, Trans.). Frankfurt am Main: Peter Lang Edition.
Ingram, D. & Tallant, J. (2018). Presentism. In E. N. Zalta (ed.), The Stanford Encyclopedia of Philosophy (Spring 2018 Edition), URL = <https://plato.stanford.edu/archives/spr2018/entries/presentism/>.
James, W. (1890). *Principles of Psychology*. New York: Henry Holt.
Kobiela, F. (2011). Problem szczelinowości w fenomenologii Romana Ingardena. [The problem of fissuration in Roman Ingarden's phenomenology] In A. Węgrzecki (Ed.), W kręgu myśli Romana Ingardena [In the circle of thoughts of Roman Ingarden] (pp. 90-113). Kraków: Wydawnictwo WAM.
Lem, S. (2015). One hundred and thirty-seven seconds (M. Wichary Trans.). Retrieved from https://medium.com/@mwichary/one-hundred-and-thirty-seven-seconds-2a0a3dfbc59e#.17hi9mnf0
Mitscherling, J. (1996). *Roman's Ingarden Ontology and Aesthetics*. Ottawa: University of Ottawa Press.
Ogrodnik, B. (1995). *Ontologia czasu konkretnego* [The ontology of concrete time]. Katowice: Wydawnictwo US.
Popper, K.R. (1989). *Objective knowledge. An evolutionary approach*. Oxford: Clarendon Press.
Simons, P. (2005). *Ingarden and the Ontology of Dependence*, In A. Chrudzimski (ed.), *Existence, Culture, and Persons: The Ontology of Roman Ingarden*. Frankfurt a. M.: Ontos Verlag, pp. 39-53.
Thomasson, A. (2017). Roman Ingarden. In E. N. Zalta (ed.), The Stanford Encyclopedia of Philosophy (Fall 2017 Edition), URL = <https://plato.stanford.edu/archives/fall2017/entries/ingarden/>.
von Wachter, D. (2005). *Ingarden's Ontology: Existential Dependence, Substances, Ideas, and Other Things Empiricists Do Not Like*, In A. Chrudzimski (ed.), *Existence, Culture, and Persons: The Ontology of Roman Ingarden*. Frankfurt a. M.: Ontos Verlag, pp. 55-82.

Zbigniew Król* and Józef Lubacz
The Subject's Forms of Knowledge and the Question of Being

Abstract: This paper considers the diverse forms of knowledge which constitute a subject's conviction that something exists. It is argued that the classical approaches to knowledge within the phenomenological and analytical traditions in fact bypass the problem of existence and thus fail to provide a basis for characterizing the full complexity of cognition. The paper suggests that the problem of attaining knowledge of the existence of something involves, in particular, distinguishing between different forms and acts of consciousness, and also between explicit and implicit components of cognition. Special attention is paid to the following notions: assertion, acceptance, belief, and trust.

Keywords: ontology of knowledge, forms and acts of consciousness, phenomenological and hermeneutical analysis of knowledge, analytical approaches to knowledge, being, existence, belief, trust.

> *Do we in our time have an answer to the question of what we really mean by the word 'being'? Not at all. [...] So first of all we must reawaken an understanding for the meaning of this question.*[1]
> M. Heidegger

Introduction

One of the many aspects of Heidegger's call to reawaken an understanding of the term "being" consists in understanding questions such as these: How does a subject acquire knowledge that something exists, or on what grounds do they believe or trust it to possess genuine being? Obviously, the question concerning the subject's epistemic access to the existence of something remains fundamental

---

[1] M. Heidegger, *Being and Time*. Text taken from the translation of *Sein und Zeit* made by Joan Stambaugh, State University of New York Press, 1996, p. xix.

*Corresponding author: **Zbigniew Król**, International Center for Formal Ontology, Faculty of Administration and Social Sciences, Warsaw University of Technology, Poland.
Józef Lubacz, International Center for Formal Ontology & Faculty of Electronics and Information Technology, Warsaw University of Technology, Poland.

https://doi.org/10.1515/9783110669411-005

in philosophy. We shall consider it here by applying what basically amounts to a phenomenologically and hermeneutically oriented conceptualization, albeit one with an added analytical flavour.

In interpreting the above questions, it first must be realized that anything that is conscious can be a subject of knowledge, but not everything that is conscious is knowledge. For instance, fear is a conscious phenomenon, but only reflection on fear can result in knowledge. In this work we limit our considerations to such forms of consciousness as can constitute a subject's knowledge about the existence of something. This is a special case of a more general problem which we do not consider directly: namely, how awareness of existence is constituted in a subject's consciousness.

In classical Husserlian phenomenology the existence of something is supposed to be "bracketed" (epoché): i.e. the above questions are *de facto* not essential. This also pertains, although for other reasons, to analytically oriented philosophy. Of course such eliminatory approaches do not mean that these questions disappear; there is no way to skirt around them because our conviction or trust in the existence of different kinds of thing is a fundamental existential fact.

In the ensuing discussion we shall be focusing on those forms of consciousness which unveil the existence of something in the form of knowledge. This requires a clarification of the notion of a subject's knowledge, and this, in turn, leads to a consideration of both explicit and implicit forms of knowledge. One of our conclusions will be that the latter are essential to a subject's consciousness of existence.

Acquiring knowledge, including knowledge of existence, is a process that requires conscious activity; products of the mind that are not conscious are not considered knowledge. It is natural to ask what makes a product of conscious activity count as knowledge, and what distinguishes such knowledge from other products of conscious activity. Within the analytical tradition of thought such questions are usually formulated in terms of expressions of the form "S knows ... ", where S stands for a knowing subject. The meaning of this expression can be specified directly or indirectly. An example of direct specification is the Justified True Belief (JTB) formulation of necessary and sufficient conditions for considering a subject to have genuine knowledge. The JTB analysis of knowledge has a long tradition and is associated with many well-known problems. Some believe that the JTB formulation was given by Plato in *Theaetetus*; it is, however, disputable whether it was treated there as a definition of knowledge, since in the same text we find a remark to the effect that justification of belief requires knowledge, so the JTB formulation cannot be a proper definition. In spite of this, JTB has become one of the most influential approaches to the analysis of knowledge. Somewhat paradoxically, the so-called Gettier problems (Gettier, 1963) with the JTB definition provided a new impetus to this approach to the analysis of knowledge. Attempts at "repairing" the JTB definition or somehow bypassing the Gettier problems have proved inconclusive. We ourselves believe that such efforts cannot be successful, in particular due to an inherent inconsistency within the JTB definition, in that it introduces an objective condition, "p is true", with respect to the intrinsically subjective notion "S knows that p". Our further analysis also indicates other problematic features of the JTB definition, thanks to which, we claim, it can hardly be considered to refer comprehensively to real-life aspects of cognition and knowledge: it is restricted to predicative knowledge, does not encompass the cognition of false predications, relies on a one-sided interpretation of the notion of belief, and fails to address such matters as the notion of trust. Similar objections may be formulated with respect to analyses of knowledge based on epistemic modal logics, such as concentrate on the interpretation of a sentence of the form "S knows ..." in terms of its logical properties and correlates (cf. Hintikka's (1962) classical work). A discussion of such problems may be found in, for example, Halpern (1986) and Janowicz (2014).

Our considerations are inspired by a phenomenological type of analysis pertaining to subjective forms and components of knowledge. We shall nevertheless substantially modify and extend the classical phenomenological approach of Husserl, in order to overcome its limitations when applied to a comprehensive, full-blooded analysis of knowledge. Our approach employs an extended phenomenological conceptual framework that at the same time manifests, so to say, a hermeneutical and analytical attitude. This is feasible, given that the traditional analytical and logical apparatus has its own phenomenological and hermeneutical counterpart: for instance, with respect to such terms as "meaning", "sense", "assertion", "concept" and "belief".[2] What we hope to achieve with this approach is to show that the two seemingly distant or even conflicting approaches to the interpretation of knowledge – phenomenological and analytic – can enrich each other conceptually, and hopefully lead to a unified hermeneutical "theory of the knowing subject".

The distinguishing feature of the analysis presented here is that it exposes the importance of trust in acquiring knowledge about the existence of something. Trust is therefore set against belief, which in traditional analyses plays the fundamen-

---

2 Cf. for instance Husserl (1970a). Consider also the following comment by Poli and Simons (1996): "The idea of formal ontology arose around the turn of the present century in the work of Edmund Husserl. It coincides in many respects with what is nowadays sometimes called 'analytic metaphysics' or with attempts to use formal methods to solve classical philosophical problems relating to the notions of being, object, state of affairs, existence, property, relations, universal, particular, substance, accident, part, boundary, measure, causality, and so on."

tal role. In particular, we shall separate out the semantically and pragmatically oriented components of belief, along with the role of trust in reflecting pragmatic aspects of knowledge, particularly in its pre-propositional forms.

One of the conclusions stemming from our considerations is that it is plausible to assume that a subject can genuinely know something that is false. We thus depart from the traditional, fundamental assumption of analyses of the knowing subject, which typically gives rise to atemporal, cumulative and non-realist theories, especially where scientific knowledge is concerned. In this context it is worth noting that any product of cognition (e.g. a proposition or a scientific theory), to become some subject's knowledge, must involve the subject's conscious activity (i.e. their grasping, interpreting something, etc.). Thus in principle, objective and subjective knowledge cannot be considered separate and independent notions.

Our analysis begins with the introduction of basic elements of the conceptual framework applied to the analysis of explicit forms of knowledge, before further extending it to include conceptual components of implicit forms of knowledge and, finally, applying it to the question of the subject's knowledge of the existence of something.

Knowledge and Consciousness

As knowledge is a phenomenon that involves consciousness, it is natural to refer to the concepts established within phenomenology that pertain to forms and acts of consciousness. Our main goal is, however, to expose some essential limitations of the phenomenological approach to analysing knowledge. These considerations constitute a background for some features and forms of knowledge which, we believe, have not so far been brought into focus where the traditional conceptual analysis of cognition and knowledge are concerned.

According to the classical distinctions put forward by phenomenologists, forms of consciousness may be divided into *acts of consciousness* and *non-active* (non-act-like) *forms of consciousness*. Acts of consciousness are forms which contain an *intention*-involving component (i.e. "*moment*"): every act refers to something ("is about something") that is presented in the act – the *content* of the act. A frame of mind or a mood are examples of non-act-like forms of consciousness; such forms may interact with acts of consciousness.[3] A subject's knowledge is associated with the content of an act, but only with a specific kind of content: namely, the con-

---

[3] In the following, we use a rather schematic and simplified presentation of the main ideas of phenomenology, as well as a somewhat modified terminology. In particular, we neglect some

tent that is given directly (in phenomenological terms, given in an instant or in a source-like way), meaning that it is "purified" of any indirect components such as attitudes, convictions, presuppositions, prejudices, superstitions, etc. Such content is referred to as the *pure phenomenon*; acts consisting of such content are referred to as *transcendental* or *pure* conscious acts. So in other words, according to the classical phenomenological conception a subject acquires genuine knowledge only in a pure act of consciousness. Moreover, knowledge acquired in this manner is assumed to be intrinsically true. Such a conceptualization in phenomenological terms does not refer to the question of whether the content of an act corresponds to some external (transcendent) object or whether the content exists. In the concluding section we shall revisit this issue, considering it in the broader context of the relation of existence to essence.

In real-life situations, the content of acts of consciousness usually includes both direct and indirect components. Purifying the content of indirect components is one of the principal postulates of the phenomenological methodology where the acquisition of true knowledge is concerned. Whatever one's opinion concerning how realistic such a methodology might be, because pure acts of consciousness are rare in real life, the phenomenological approach to knowledge is limited in scope and leaves aside, as will be explicated below, many historically disclosed knowledge-related and ontology-related questions which are important to a full-blooded conceptualization of knowledge. We may thus conclude that the classical phenomenological approach requires significant modification and extension, and set out to propose some possibilities for this in what follows.

Apart from intention-involving ones, acts of consciousness may also consist of other types of component. It is especially important to distinguish the component of *assertion* from that of *acceptance*. For the sake of concision, we shall refer to an act of consciousness consisting of an assertion component as an *act of assertion* or just an *assertion*, and to an act of consciousness associated with an acceptance component as an *act of acceptance* or just an *acceptance*. Acts of assertion may be considered to belong to the field of semantics, acts of acceptance to that of pragmatics.[4]

An assertion may consist of many different intellectual operations and/or "partial assertions" (sub-components), such as the predicating of a property of something, the predicating of something about an object, process or situation or of combinations of something with something else, etc. For example, in the case

---
aspects of the conceptual evolution of Husserl's phenomenology (cf. Husserl (1970a, 1982, 1970b, 1989), and also Ingarden (1964, 2013)).

4 Note that a similar distinction was introduced by R. Ingarden and E. Husserl; cf. Ingarden (1972), Husserl (1970a).

of the cognition of the concept of a "square-like circle", two different objects are combined to form one. Moreover, assertions may be propositional or conceptual. The latter are responsible for the creation of intentional objects (cf. Ingarden (2013), Husserl (1970a)). An assertion may result from a complex intellectual operation which relies on some non-propositional components. A propositional assertion usually attributes a property to an object or specifies a relation between objects.

Acceptance is an attitude on the part of a subject towards something, usually containing (or built out of) one or more a propositional assertions. While assertions that are limited to pure phenomena are legitimate components of acts of consciousness in classical phenomenology, acts of acceptance are not. This is because an act of acceptance is a presupposition, which according to classical phenomenology should be eliminated from acts of consciousness in order to achieve knowledge of "things in themselves". As will be explained below, eliminating acceptance from legitimate components of acts of consciousness also leads to the elimination of belief: i.e. one of the fundamental notions of most analyses pertaining to knowledge.

In considering the notion of belief it is important to carefully distinguish assertion from acceptance. One can make assertions without accepting them, as, for instance, in the following sentence: "the weather tomorrow will be good". Acceptance (or rejection) can pertain to the truth or falsity of an assertion, or to the reality of something (the assertion's reference), so acceptance is, in principle, a psychological, pragmatic attitude (pertaining primarily to assertions). On the other hand, assertions are *sine qua non* conditions of a "purified" understanding of something, in that they potentially enable one to follow the content of a sentence or theory without embracing any presuppositions. Assertions can be entertained in a purely theoretical and hypothetical mode: i.e. without acceptance (or rejection).

Belief and Trust

By invoking the above distinctions we have defined *beliefs* as accepted assertions; in doing so we are departing from classical phenomenology. Assertions can be true or false in a semantic sense, whereas beliefs may be considered to possess a pragmatic character. Acceptance does not imply the truth or falsity of a belief: a subject can accept (or reject) something false. Moreover, some assertive acts, such as conventional definitions, count as true only when freely endorsed – i.e. accepted – by a subject. For instance, if the objects of mathematics are considered fictions, then the truth of mathematics requires their acceptance. In the following, we shall consider cases in which a subject can accept (or reject) not only an assertion, but also the content of an act of consciousness and/or a non-act-like form of awareness; for instance, one can accept a frame of mind or mood.[5]

We may note in passing that an unjustified identification of beliefs with assertions may lead to confusing interpretations. For instance, some believe that the concept of truth is redundant, because to say that "2 + 2 = 4 is true" is just the same as saying simply that "2 + 2 = 4". The former sentence, however, can also express acceptance of the assertion "2 + 2 = 4". If an assertion expresses a real attitude on the part of a subject,[6] then the expression conveying acceptance may be identical to a sentence expressing an assertion.

There are many kinds of belief, among them scientific ones. Scientific beliefs are based not only on acceptance of their content, but also on acceptance of their justification and such requirements and circumstances as consistency and coherence with other beliefs, conformity with experimental data, etc. It is a well-known fact that there can be competing theories and explanations of the same experiments, experimental facts, observations etc., so scientific theories encompass some content that is independent from empirical facts. Practically speaking, we accept the plausibility of some propositions that are not necessarily true, but nevertheless "explain" something. Even objectively false theories and false propositions can be useful: for instance Newtonian mechanics, scientific hypotheses and indirect proofs in mathematics. Such theories and systems of knowledge, though only temporarily true, enable the evolution of scientific knowledge.

We believe that what counts for genuine knowledge cannot be based solely on beliefs, i.e. on the acceptance of assertions, and that it also has to involve *trust*. Trust is a complex notion, which is traditionally considered within sociology and psychology rather than epistemology. Involving trust in the latter indicates that knowledge is not being treated as a kind of game, but rather as a vital phenomenon in the development and well-being of societies, and of science in particular. Losing trust is usually an indication of an approaching crisis, which in turn leads to cultural, societal, and scientific changes that are sometimes revolutionary. There are numerous arguments lending support to the view that social trust, as a kind of shared attitude, constitutes one of the most important factors in the development of science.[7]

---

5 Obviously, a mood can become the object of an act in which the subject of cognition realizes the presence (the "phenomenon") of mood.
6 The same two-sided structure of expressions is well-known in the case of assessments. An assessment, when formulated in a verbal form, may be treated as a sentence: i.e. a statement expressing a report concerning a given fact.
7 We believe that diminishing levels of trust among scholars and scientists is symptomatic of a grave crisis in modern science.

Before embarking upon our own analysis of trust in the context of knowledge, we shall offer one more general remark. On commonsensical interpretations, trust is sometimes considered to be a state of mind that is free of doubt. A more realistic view holds that trust and doubt replace be simultaneous and interdependent attitudes of a subject. Trust may replace doubt completely, and *vice versa*, but trust may also diminish doubt, and *vice versa*. Trust and doubt are thus not mutually exclusive notions in the logical sense. The term "doubt" may be substituted with the term "distrust", understood as lack of trust, only if trust is completely replaced by doubt (in the latter case, trust may be considered a kind of faith). The interplay between trust and doubt is important in science and takes the form of rational criticism, which is the cornerstone of all scientific knowledge.

Recall that a belief (in particular, a propositional belief) was defined as an accepted assertion (in particular, a propositional assertion): i.e. as acceptance of an assertion pertaining to the content of an act of consciousness. With reference to the same conceptual framework, we propose defining trust as *acceptance of the content* of an act of consciousness. In short, trust is acceptance of content (by analogy with belief construed as acceptance of assertion). Thus both belief and trust are attitudes pertaining to the content (or something involved in the content) of an act of consciousness. In the case of a belief, however, the attitude consists in the acceptance of an assertion, whereas in the case of trust it pertains to the content itself (i.e. without additional "intermediate" components of a conscious act such as an assertion). However, if an assertion is considered to be a component of the content of an act, rather than an attitude referring to the content, then trust becomes a more general notion than belief. We shall, in due course, arrive at a specific relation between trust and belief in the context of the implicit components of conscious acts and knowledge considered in the next section.

It is important to note that in contrast to the definition of belief, the definition of trust can be extended so as to pertain also to non-act-like forms of consciousness: i.e. to non-intentional forms of consciousness such as frames of mind or moods, meaning intellectual or personal[8] attitudes or inclinations towards something (not excluding ourselves) not associated with reflection or any kind of justification. Trust, as an attitude, does not reside only in acts of consciousness; such acts are tools by which a subject can express its trusting attitude and convictions. The acceptance (rejection) of some content of a non-act-like state of consciousness may be the source of this state, or may even in some cases be identified with it. It seems plausible to suppose that a non-act-based kind of trust may consist, at least partially, in an innate predisposition (consider very young children, who obviously

---

8 Personal attitudes overstep the realm of the purely intellectual domain of human behaviour.

cannot be suspected of being involved with beliefs or trust of an act-based sort), or result from some form of "background consciousness", experience, or spiritual belief.

The above definitions of belief and trust are enclosed within the mental realm: i.e. they do not involve the existence of objects external to the mind. If such objects are assumed, then the definition of belief and trust will require adjustment: belief will be the acceptance of an assertion pertaining to content concerning an external object, and trust will be the acceptance of content concerning an external object. Note that these adjustments do not refer to any aspects of a subject's interaction with an external object that result in forming the content of the act of consciousness and its assertion and acceptance components. In general, acts of consciousness are formed through complex subjective and intersubjective, time-dependent interactions and activities: i.e. epistemic *processes*. We shall not go more deeply into these, as this would require a separate analysis of such notions as perception, justification and conformity – a set of issues that deserves to be pursued in its own right.

Non-Explicit and Non-Verbal Knowledge

So far we have basically been considering *explicit* forms of acts of consciousness. In a real-life characterization of trust and belief in the context of cognition and knowledge, the implicit components that are constitutive of beliefs and trust (the act's content, assertion and acceptance) play an important role. The case of trust is somewhat less complex in this respect, because it does not have to rely on assertions. In the following, we shall begin by focusing on belief – a more complex issue. Most of the comments and conclusions presented will also apply, however, to trust. In the final part of this section, we shall consider the distinction between, and role of, verbal and non-verbal aspects of cognition and knowledge, and present a few remarks concerning *a priori* knowledge.

Every real-life conscious act has a two-sided structure: besides something that is actually and explicitly given, there are many implicit and "hidden" components that co-constitute the act. Such components co-determine the content of acts.[9] The co-determining components may be made explicit or "visible" to the subject as a result of some additional conscious activity, or may become premises in some future activity that can, for instance, require that one draw certain conclusions

---

[9] To such concepts belong also the so called passive forms of consciousness, as introduced in Husserl (1982), para. 5; compare also some ideas presented in Husserl (1970b) and Husserl (1989).

or make decisions based on multiple premises. These circumstances imply yet another reason for a departure from classical phenomenology, which treats implicit and hidden components of conscious acts as located "outside" the act in question, to be separated from explicit components in a phenomenological reduction.

It is usually difficult to identify and separate implicit and explicit components of conscious acts. In this respect, the use of formalized theories and tools of classical analytical approaches to cognition and knowledge may be useful. Such analysis must also make use of hermeneutic methods, because conscious acts are components of a broader environment determined by a given culture, historical epoch, etc. This sort of a comprehensive environment for conscious acts may be referred to as the *hermeneutic horizon*.

The interplay of explicit and implicit components and factors of conscious acts can be illustrated by means of the following examples. Consider grasping the notion of a square. The explicit content consists mainly of definitional properties of the concept of a square: 4 equal sides, 4 equal angles, flatness or being a polygon. The implicit content of the square in this instance is composed of some co-posited components and purported commitments: e.g., concerning the continuity of the square and its sides, that the figure is symmetric, and also relating to secondary qualities of the explicitly defined properties (properties of "side", "angle", "polygon"), etc. Although a formal interpretation of such notions as "continuity" or "flat surface" might require a sophisticated theory (in some cases competing theories or points of view are available), they are usually readily grasped intuitively, based on some implicit "background" knowledge. The explicit content of an act is open to determination by implicit components which may vary over time. More examples and case studies are presented in (Król, 2015).

Taking another example, one looks at a leaf, and the surface of the leaf is perceived as continuous and homogenously green. We are usually aware that "in reality" the surface is not continuous and homogenous, and that there are implicit components that determine the explicit content, some of which might even be intuitively contradictory. There is no "cognitive dissonance" in such cases, because the subject "automatically" reinterprets the explicit results of perception. A dissonance will only show up if a contradiction appears between simultaneous explicit components of acts, as when one observes the discontinuity exhibited by a spoon immersed in a glass of water.

In the context of implicit acts and components of acts of consciousness, it is especially important to distinguish between assertive acts and beliefs. A subject may be considered to treat an assertion "seriously" if the assertion represents a real attitude towards the object of assertion. Such assertive acts are also beliefs (usually implicit, and sometimes referred to as "convictions"); because of this, assertions are sometimes simply treated as beliefs. A subject of cognition may entertain many assertions without being seriously convinced of their veracity. For instance, one can accept Quantum Mechanics and the General Theory of Relativity (i.e. theories composed of consistent or coherent sets of propositions) knowing that they are false. There are also some which may be called "as if" assertions: hypotheses, conventions, agreements, commitments, etc., in which the assertion (and its subsequent acceptance) may be only tentative, temporary, hypothetical, or purely theoretical. The subject is not always aware of all of the explicit content of an assertion, or the entire content of what is accepted in the context of a given act. Some common, as well as scientific superstitions, fall into this latter category, as of the various kinds of intuitive reasoning based on hidden assumptions.[10] Because of the implicit components of assertive acts, the content of these acts cannot, in fact, be treated as explicitly conscious. It is also important to distinguish between *verbal assertions* and *verbal beliefs*, as well as their non-verbal versions.

Note the following: (a) not everything that is non-verbal is implicit, but everything that is implicit is non-verbal; (b) intuitive knowledge can, in principle, be conceptualized and thus, verbalized; (c) implicit assertions and non-verbal acts may form the context of explicit acts; (d) implicit acts and their components are constitutive elements of so-called intellectual intuition and intuitive knowledge.

Fig. 1: Kepler's star-polyhedron (source: Kepler images at https://commons.wikimedia.org/wiki/Category:Harmonices_Mundi;_Smithsonian_scans)

---

10 Consider, for example, historical evidence in the field of mathematics: the axiom of choice was used as a hidden assumption in many proofs by its opponents (Borel, Lebesgue) not only *before* any axioms were explicitly stated, but *after* they had been explicitly formulated by Zermelo in 1904. Cf. Zermelo (1904) and Król (2007). Zermelo's aim was to determine the hidden assumptions behind Cantor's conviction that every set can be well-ordered.

Such implicit assertions and non-verbal acts of consciousness can be illustrated using the following example. Consider the star-polyhedron depicted in Figure 1. In Western culture, we usually treat the lines on the picture as representing a three-dimensional polyhedron. But the picture "is flat". In a non-verbal and implicit way we constitute in our minds an implicit assertion to the effect that the figure presents a polyhedron, and we also pose questions concerning, for instance, the number of faces, vertices, etc. In trying to find answers, one makes some hidden non-verbal assumptions: for instance, that the polyhedron is "the same" (is symmetric) on the invisible back side, that it is three-dimensional and continuous, that the faces are flat polygons, etc. We can set out to count in our minds the vertices, edges and faces. These are rational intellectual operations, different from any verbal formulation of their content. A verbal formulation may, however, report the result of the corresponding non-verbal operations. Some people will possibly see something different in the picture: e.g. a surface of a futuristic space-ship, or that the object's interior is empty. Whatever the case, we do not limit ourselves to the physical object that is actually seen and make many implicit assertions ("implicit constructions" or implicit mental operations).

Obviously, such hidden and implicit assertive acts are important in the context of the process of cognition and arriving at knowledge, but it seems unjustified to refer to such acts as beliefs in the sense of, for example, the Justified True Belief definition of knowledge. Such implicit assertions can be true or false, consistent or not, and coherent or not in relation to the explicit act's content, as well as other assertions and beliefs. Moreover, in not all cases can such implicit assertions and beliefs be consciously identified. We usually know much more than we can identify explicitly at a given moment.

Although many assertive acts or beliefs can be well represented by a verbal assertion, it is not possible in all cases to represent (grasp) and transmit (communicate) the content of an assertive act or belief. Thus in principle it is reasonable to treat the verbal counterpart of a given assertive act as an interpretation rather than a "copy" of the original act. On the other hand, we can use intuitive notions and communicate them if other people themselves have a relevant source for the intuition. It is, however, virtually impossible to transmit a genuine intuition of the notion of, for example, a "red colour" to a person who has been born blind. Pre-verbal intellectual acts are nevertheless a *sine qua non* condition of verbal communication and codification of knowledge.

The polyhedron example is an indication of the fact that although it is virtually impossible to formulate a verbal and explicit definition of a polyhedron, the concept of polyhedron can be used, and even a theory of regular polyhedra can be created, and, for example, Euler's hypothesis analysed.[11] This, again, shows that the classical methods of phenomenology have serious limitations, in that they are confined to explicitly given concepts in pure acts of consciousness.

A few comments are in place here concerning the notion of *a priori* knowledge, which is present in classical phenomenological considerations and is understood as resulting from direct, instant cognition, as opposed to *discursive* (indirect and verbal) knowledge. In interpreting the difference, it is useful to introduce a distinction between source-presenting acts and non-source-presenting acts. Source-presenting acts present the object targeted by the intention of the act. The majority of source-presenting acts are implicit assertive acts (recall the star-polyhedron example), and are basic in intuitive reasoning; they are, so to say, self-justifying. Evidence which is important in intuitive reasoning depends on making explicit something that is implicit in source-presenting acts (or depends on a verbalization of this, or of the explicit content of source-presenting acts). Non-source-presenting acts concern objects not presented in the act in question. Assertions and other acts which pertain to the content of some source-presenting acts – i.e. acts justified by evidence stemming from other source-presenting acts – are not, we believe, given due consideration in classical analytical and naturalist approaches. Note that a non-source-presenting act may be the source of a subsequent source-presenting act that reveals the properties and content of a non-source-presenting act. Such source-presenting acts are, we believe, the origin of *a priori* knowledge.

Finally, it should be noted that in the context of the considerations presented in this section, trust *as an attitude* usually amounts to an implicit acceptance of both the implicit and explicit content of a conscious act. The convictions revealed in trust as an attitude are, moreover, fundamental to a subjects' engagement in matters essential to a successful existence, or even its very possibility.

Knowing That Something Exists

Knowing that something exists appears to be a quite complex issue because of the manifold forms of cognition. The complexity of the overall picture which emerges from our considerations is, however, justified if what we are aiming to capture is cognition and knowledge in their full-blooded form. Compared to standard mainstream analyses, our considerations involve quite a few additional notions, and in consequence also many relations between these notions. In particular, we

---

[11] The history of efforts pertaining to the formulation of the definition of the concept of a polyhedron is presented in Lakatos (1976).

have focused on making careful distinctions between acts and non-act-like forms of cognition, explicit and implicit components of cognition, verbal and non-verbal components, assertions and beliefs, beliefs and trust, and also with respect to limitations following from confining cognition to its assertion-based propositional form associated with the assumption that what is known must be true.[12]

The classical phenomenological conception of knowledge, although useful for analysing forms of consciousness, is quite limited in its scope, as it does not refer to many of the important aspects and components explored in the preceding sections. This is probably because the original approach introduced by Husserl was directed towards discovering the essences of things ("back to things themselves"), rather than providing a characterization of cognition and knowledge in their full, context-dependent complexity. This is why we have departed from the classical phenomenological approach at an early stage in the development of our considerations, albeit while retaining some fundamental notions pertaining to forms of consciousness that are phenomenologically inspired.

We believe that our analysis demonstrates that full-blooded forms of knowledge, and especially their non-explicit forms, are, so to say, enablers of our knowing the existence of something. Such knowledge is rooted in implicit beliefs and attitude-based trust. Our sense of conviction – or even *ontological trust* – with respect to the existence of the real world stems from such existential components of knowledge.

The above-mentioned conclusion should, moreover, be considered in the context of the relation between the existence and the essence of beings. It should be noted that Husserl's phenomenology focuses on essences, while the existence of something is bracketed, in the sense of being excluded from the data given directly to a transcendental subject of cognition; for Husserl, existence can be separated from essence. There is, however, no transcendental data which is non-existent; everything that is given transcendentally exists. Of course, to exist in this sense does not mean to exist as a real (spatiotemporal) object: there are many possible kinds of existence, as exemplified by Kant with his distinction between real and imagined thalers. In the context of Husserl's conception, this distinction does not appear, and this is a consequence of his assuming that the essence of imagined

---

[12] By the way, our considerations could prove useful for purposes not necessarily strictly related to philosophy, such as cognitive studies, research pertaining to artificial intelligence, robotics and managerial questions pertaining to the role of tacit and *a priori* knowledge involved in "knowledge creation" (cf. (Wierzbicki & Nakamori, 2007)), to name just a few pragmatically-oriented areas of potential interest.

and real thalers is phenomenologically the same.[13] Our considerations show, we believe, that this assumption is wrong, if considered to constitute a general rule. We should also note in this context that where Heidegger's conception is concerned, this problem is "solved" differently: everything that exists is considered with respect to *Da-sein*; everything that exists, so to speak, resides in *Da-sein* and corresponds to its way of life.

Furthermore, what we have proposed suggests that non-explicit components of an act of consciousness belong to the essence of the act, and in consequence refer to the thing that the act concerns. Moreover, in some cases the affirmation of the existence of an intentional object given in an act belongs to the essence of the object. Mathematical objects are examples of such relations between existence and essence – see Król (2015).

Finally, it should be noted that, as was already mentioned in the Introduction, the general question of the constitutive character of existence itself remains open.

Bibliography

Gettier, E. L. (1963). Is justified true belief knowledge? *Analysis, 23*, 121–123.
Halpern, J. L. (Ed.) (1986). *Theoretical aspects of reasoning about knowledge: Proceedings of the 1986 Conference. IBM Almaden Research Center, Monterey, California*. Los Altos, CA: Morgan Kaufmann.
Hintikka, J. (1962). *Knowledge and belief. An introduction to the logic of the two notions*. Ithaca, N.Y.: Cornell University Press.
Husserl, E. (1970a). *Logical Investigations* (J. N. Findlay, Trans.). London, New York: Routledge & Kegan Paul Ltd. (Original work published 1900/1901)
Husserl, E. (1970b). *The crisis of European sciences and transcendental philosophy* (D. Carr, Trans.). Evanston: Northwestern University Press. (Original work published 1936)
Husserl, E. (1982). *Ideas pertaining to a pure phenomenology and to a phenomenological philosophy – First book: General introduction to a pure phenomenology* (F. Kersten, Trans.). The Hague: Nijhoff. (Original work published 1913)
Husserl, E. (1989). *Ideas pertaining to a pure phenomenology and to a phenomenological philosophy – Second Book: Studies in the phenomenology of constitution* (R. Rojcewicz, & A. Schuwer, Trans.). Dordrecht: Kluwer.
Ingarden, R. (1964). *Der Streit um die Existenz der Welt, Bd. I, II/I, II/2*. Tübingen: Max Niemeyer.
Ingarden, R. (1972). O sądzie kategorycznym i jego roli w poznaniu. In R. Ingarden, *Z teorii języka i filozoficznych podstaw logiki* (D. Gierulanka, Ed.) (pp. 222–259). Warsaw: PWN.
Ingarden, R. (2013). *Controversy over the existence of the world: Vol. I* (A. Szylewicz, Trans.). Bern: Peter Lang. (Original work published 1947)

---

[13] I.e. one grasps the same essence with the use of imaginative variation over what is imagined and what is given in respect of a sensory perceptual *quale*.

Janowicz, K., Schlobach, S., Lambrix, P., Hyvönen, E. (Eds.) (2014). *Knowledge engineering and knowledge management: Proceedings of the 19th International Conference EKAW 2014.* Cham, Heidelberg, New York, Dordrecht, London: Springer.

Król Z. (2007). The emergence of new concepts in science. In Wierzbicki & Nakamori (2007, pp. 417–444).

Król Z. (2015). *Platonism and the development of mathematics. Geometry and infinity.* Warszawa: IFIS PAN.

Lakatos, I. (1976). *Proofs and refutations. The logic of mathematical discovery* (J. Worrall, & E. Zahar, Eds.). Cambridge, London, New York, Melbourne: Cambridge University Press.

Plato (1982). *The Dialogues of Plato: Vol. 4.* London: Oxford University Press.

Poli, R., & Simons, P. (1996). Foreward. In R. Poli, & P. Simons (Eds.), *Formal ontology* (pp. vii–viii). Dordrecht, Boston, London: Kluwer Academic Publishers.

Wierzbicki, A., & Nakamori, Y. (Eds.) (2007). *Creative environments. Issues of creativity support for the knowledge civilization age.* Berlin, Heidelberg: Springer.

Zermelo, E. (1904). Beweis, dass jede Menge wohlgeordnet werden kann. *Mathematische Annalen, 59,* 524–516.

Andrzej Biłat
The World as an Object of Formal Philosophy

Abstract: The purpose of this paper is to reconstruct the classical conception of the subject matter of philosophy (in part I) and to show that this reconstruction is compatible with contemporary logic (in part II) and science (in part III). The reconstruction takes the form of an extensional version of metaphysical realism. In this version the main philosophical concepts of the "world" are clarified and some principles of metaphysics are deduced from scientific knowledge with the use of monadic second-order logic and some interpretative premises.

Keywords: Plato, Aristotle, philosophy, being, class, extensional realism, real world, world of nature, monadic second order logic, interpretative deduction.

> SOCRATES: *It is also the greatest test*
> *of which nature is dialectical and which is not.*
> *For the person who can achieve a unified vision is dialectical,*
> *and the one who cannot isn't.*[1]
> Plato

> *Scientists and philosophers seek a comprehensive system of the world,*
> *and one that is oriented to reference*
> *even more squarely and utterly than ordinary language.*[2]
> Willard Van Orman Quine

Part I

1.1

The classical conception of philosophy, outlined by Plato and developed by Aristotle, can be an important source of inspiration for ontology and formal philosophy

---
[1] Plato (Rep./2004, VII, 537c)
[2] Quine (1981, p. 9)

Andrzej Biłat, International Center for Formal Ontology, Faculty of Administration and Social Sciences, Warsaw University of Technology, Poland.

https://doi.org/10.1515/9783110669411-006

today.³ Its main principle resembles a modern project of axiomatic philosophy built on the ontological foundation:

(CL1) Theoretical philosophy is an axiomatic science concerning the most general properties of beings.

Plato and Aristotle did not explicitly formulate such a project. However, (CL1) is a simple consequence of two fundamental principles of their methodological ideas: firstly, well-structured scientific knowledge has an axiomatic-deductive structure (the main idea of *Posterior Analytics*); and secondly, basic philosophical knowledge is scientific knowledge about all kinds of beings and their most general properties (the major metaphilosophical thesis of *Metaphysics* formulated in the books A and Γ).

Although Aristotle quite clearly pointed to the axiomatic ideal of knowledge, this did not lead to the development of a formal philosophy. Nonetheless, this ideal played a significant role in the development of the ontological paradigm of philosophy (as evidenced even by the structure of the *Summa Theologiae* of Thomas Aquinas or the *more geometrico* style of metaphysical considerations of Baruch Spinoza and Christian Wolff).⁴ It also helped, through his influence on Euclid's work, to develop axiomatic geometry and the foundations of mathematics. This development, particularly in the nineteenth and early-twentieth century, played a key role in shaping contemporary formal philosophy.⁵

---

3 *Formal philosophy* is a philosophy developed with the use of formal methods. Although this expression most often appears in the context of semantics and epistemology, we do not see any obstacles for its wider application, including the analysis of ontological issues, the rational reconstruction of metaphysical positions or the axiomatization of ontological concepts. It is used in this article in this broad sense.

4 The phrase "ontological paradigm of philosophy" (as well as the equivalent phrase "Platonic-Aristotelian model") is used in Schnädelbach (1985). According to H. Schnädelbach, the beginning of a philosophical discourse, characteristic for this paradigm, defines the question "What exists?". Modern and contemporary philosophy also formed at least two other traditions, mentalistic (Cartesian) and linguistic (Wittgensteinian). In the former the right beginning of philosophizing is the question "What can I know?" and in the latter – the question "What can I understand?".

5 Interest in the ontological paradigm of philosophy, especially in the philosophy of Aristotle, has steadily been growing in recent decades. Some authors suggest that this trend may become a "sapiential" component of post-Cartesian European thought (Feser, 2013, p. 1). The main current of these interests is situated within analytical philosophy, and is sometimes referred to as "neo-Aristotelian metaphysics" (Novotny & Novak, 2014, p. 7). At the same time, some of the authors working within this current – e.g. Nino Cocchiarella, Kit Fine, Uwe Meixner, Peter Simons, Barry Smith, Nicholas Rescher, and many others – widely apply formal methods.

Nothing stands in the way of making (CL1) more detailed by distinguishing those beings or their types that we are most interested in. Here is one of the most important examples of such a distinction:
(CL2) Each class of beings is a being.

The (CL2) postulate is consistent with the conceptual realism of Plato and Aristotle; in particular, the intuitive counterparts of the concept of CLASS frequently appeared in the philosophical discourse of both authors (such as the meaning of words γένος, είδος or κατηγορία). (CL2) is also consistent with the thesis of *extensional conceptual realism*, which is a consequence of the philosophical principles underlying modern mathematics and science. According to this thesis, there are various "extensional entities": classes, sets, categories, extensions of terms, domains of theories etc.

(CL1) and (CL2) determine the historically basic, already quite mature, ontological project of philosophy as a science. According to this project, philosophical investigations concern the general properties of beings and their distinguished classes, and their primary method is the axiomatic method.

The presupposition of the classic nature of this conception is quite obvious and in principle does not require justification. However, let us recall two facts. Firstly, it was the first conception of philosophy delineating its historically fundamental, ontological paradigm (which was in force until the beginning of the modern era, see: Schnädelbach, 1985). And secondly, it is still present in some academic textbooks.[6]

(CL2) may provide a convenient basis for further specifications of the form "*X* is a class of beings that satisfy the condition *W*", which are sentences clarifying the description of the domain of theoretical philosophy (from each such specification and principle (CL2) we can derive a sentence of the form '*X* is a being'). Here is an appropriate example:
(CL3) The world as an extensional whole – *summum genus* – is a class (of beings) including all classes.

---

[6] This can be seen, for example, in the first two sentences of the textbook "History of Philosophy", which is very popular in Poland: "Since the beginning of science, efforts have been undertaken to move beyond partial considerations and cover everything that exists in one science; attempts have been made to build a science besides specific sciences, that *yield a view of the world*; this science has been called philosophy. It is the science whose scope is therefore the widest of all the sciences, and whose concepts are *the most general*" (Tatarkiewicz, 1931/2016, p. 7).

(CL3) is an explication of the broadest (not mereological!) meaning of the term "world". (CL2) and (CL3) entail that the world as an extensional whole is a being; and so it is (according to [CL1]) one of the objects of interest of theoretical philosophy.

The possibility of further expanding the list of principles (CL1)–(CL3) in the above (extensional) style seems to be opening up an interesting field of research on the borders of the history of philosophy, ontology and formal philosophy. This, in particular, applies to issues related to the analysis of the philosophical meanings of the term "world".

The aim of this paper is to reconstruct the classical conception of the object of philosophy and to demonstrate that this reconstruction is compatible with contemporary logic and science. This purpose will be achieved through:
- expanding the list of principles (CL1)–(CL3) with principles clarifying the relation between the classical conception of the object of philosophical inquiry and major philosophical concepts of WORLD (in the next two sections),
- expressing these principles in the language of contemporary formal philosophy as a position of extensional realism (in the last section of this part), and
- demonstrating that this position is the philosophical consequence of contemporary logic and of scientific knowledge (in parts II and III).[7]

1.2

Ancient philosophy was usually accompanied by the intention to refer to the whole of reality. Despite the initial limitation of its interest to the world of nature, the term "whole of reality" is used here as widely as possible.[8] Plato's ontology especially had such a universal character comprising both material and "ideal" beings (including numbers and forms).

Sometimes the term "world" is used to denote the widely understood domain of philosophical considerations of the ancient Greeks:

> The question of Being [in the philosophy of Parmenides and Plato] is first of all the question of the nature of reality or the structure of the world, in the very general sense of "the world"

---

[7] Here we use the concept of PHILOSOPHICAL CONSEQUENCE in the sense that was clarified in Woleński (1993, pp. 10–14). In short: proposition α is a philosophical consequence of the set of principles X on the basis of interpretative premises Y if and only if α is deductively derivable from the set $X \cup Y$ and the premises from the set Y have philosophical sense (see also below, section 2.1).

[8] "[Ancient] philosophy claims to know the totality of reality with a rational method and a purely theoretical purpose. [...] At first, the totality of reality, physis, is seen as cosmos" (Reale, 1987, p. 20).

which includes whatever we can know or investigate and whatever we can describe in true or false statements. The question of Being, then, for the Greek philosophers is: How must the world be structured in order for inquiry, knowledge, science, and true discourse (or, for that matter, false discourse) to be possible? (Kahn, 1976/2009b, p. 67)

Since we can easily describe any being as "existing in time" or "not existing in time", the extensions of the Parmenidian-Platonic concepts of BEING and SOMETHING DESCRIBABLE seem to be identical. They are also the extension of the concept of WORLD as an extensional whole.[9]

These concepts are not too old-fashioned. A similar idea appeared at the beginning of the development of contemporary logic and formal philosophy:

> Boole has his universe class, and De Morgan his universe of discourse, denoted by "1". But these have hardly any ontological import. They can be changed at will. The universe of discourse comprehends only what we agree to consider at a certain time, in a certain context. For Frege it cannot be a question of changing universes. One could not even say that he restricts himself to *one* universe. His universe is *the* universe. Not necessarily the physical universe, of course, because for Frege some objects are not physical. Frege's universe consists of all that there is, and it is fixed. (Heijenoort, 1967, p. 325)[10]

We obtain a narrower concept of WORLD, if we limit its extension to the area of *temporal beings* (i.e. beings existing in time, e.g. physical bodies and their systems, events, life processes, people, their experiences, human communities and their histories etc.). As we know, the distinction between temporal and *atemporal beings* (numbers, geometric figures, ideas, genera, species, classes etc.) is fundamental to Plato's philosophy (as expressed in *Timaeus* [Plato, Tim./1888, 37e, 38a-c]).

Aristotle in his *Metaphysics* in turn repeatedly distinguishes natural (or "sensible") substances from general (or "posterior") substances. Here are the relevant quotes (from the beginning of Book H and Book Λ):

> Those [substances] generally recognized are the natural substances, i.e. fire, earth, water, air, etc., the simple bodies; second plants and their parts, and animals and the parts of animals; and finally the physical universe and its parts. [...] But there are the arguments which lead

---

9 We leave open the question of whether the concept of WORLD AS A WHOLE can be identified with the extension of the Platonic ONE OVER MANY. We also omit the question of the relationship of these concepts with further conceptions of the object of ontology (e.g. related to Leibnizian-Wolffian modal philosophy of being, or the theory of objects by K. Twardowski and A. Meinong). In turn, the reference to the conception of G. Frege in the next section has only one goal: to show the similarities of the classical conception of the subject matter of philosophical inquiry with the ideas underlying modern logic and the related formal philosophy.
10 J. van Heijenoort derives this concept from the work of Frege (1882/2014). Given this clarification, the concept of WORLD AS A WHOLE can also be regarded as *Fregean*.

> to the conclusion that there are other substances, the essence and the substratum. Again, in another way the genus seems more substantial than the various species, and the universal than the particulars. And with the universal and the genus the Ideas are connected; it is in virtue of the same argument that they are thought to be substances. (Aristotle, Met./2015, 1042a)

> There are three kinds of substance – one that is sensible (of which one subdivision is eternal and another is perishable; the latter is recognized by all men, and includes e.g. plants and animals), of which we must grasp the elements, whether one or many; and another that is immovable, and this certain thinkers assert to be capable of existing apart, some dividing it into two, others identifying the Forms and the objects of mathematics, and others positing, of these two, only the objects of mathematics. The former two kinds of substance are the subject of physics (for they imply movement); but the third kind belongs to another science, if there is no principle common to it and to the other kinds. (Aristotle, Met./2015, 1069a)[11]

Thus, a more accurate reconstruction of the Platonic-Aristotelian conception of "ontological" philosophy as a science requires a clear distinction between temporal and atemporal beings ("substances"). Let us consider the following explication:
(CL4) The real world is a class of all *and only* temporal beings.

Although Plato and Aristotle probably did not apply the explicit concept of REAL WORLD, its use in the rational reconstruction of their idea of the subject matter of philosophy seems acceptable, taking into account the popularity of this concept in contemporary philosophy. Here is an example of its contemporary application, fully consistent with the (CL4):

> The real world as we grasp it in pre-philosophical, everyday experience appears to be organized in such a peculiar fashion that anything and everything that occurs within its unity is somehow temporal, or is at least bound up with time. (Ingarden, 2013, p. 227)

> All individual entities can be split into two major classes: 1) temporally determined entities; 2) atemporal (in particular, 'ideal') (ibid, p. 229). [12]

According to (CL3) and (CL4), the real world is included in the world as an (extensional) whole. According to Plato and Aristotle, this relationship is a proper inclusion: every temporal being is an element of the world as a whole, but not vice versa. (CL3) and (CL4) therefore define a logical division of the world as a whole

---

[11] Natural substances should undoubtedly be regarded as the Aristotelian "things existing in time"; the phrase appears in turn in the investigations relating to time in Book IV of *Physics*.
[12] The distinction between objects existing in time and not-existing in time also played an important role in the ontological systems of E. Husserl and N. Hartmann.

into two non-empty classes: a class of elements that are temporal beings – the real world – and a class of elements that are not temporal.

1.3

Let us move to the explication of the relationship between the concepts of REAL WORLD and WORLD OF NATURE. Both concepts play an important role in contemporary philosophy; while the former is often used in the description of the object of metaphysics, the latter is used – just as often – in the characteristics of the object of natural philosophy.

In the historical review of the issues in this part of philosophy, it is striking that for the first generations of natural philosophers "to be" meant primarily "to occupy a place":

> I think there can be no disagreement on the close connection between the ideas of existence and location in Greek philosophical thought. We have from Presocratic times the well-established axiom that *whatever is, is somewhere; what is nowhere is nothing*. As Plato puts it (stating not his own view, but that of Greek common sense), "we say that it is necessary for everything which is real to be somewhere in some place and to occupy some space, and that what is neither on earth nor anywhere in heaven is nothing at all" (Plato, Tim./1888, 52b). If existence and location are not identical in Greek thought, they are at least logically equivalent, for they imply one another; that is, they do for the average man, and for the philosophers before Plato. (Kahn, 1966/2009a, p. 32)[13]

Plato made (inspired by Parmenides' thought) a breakthrough in this way of thinking. Firstly, his clarifications show that "to exist" does not always entail "to be somewhere". For example, a number exists, but it does not occupy any place in physical space. And secondly, in *Timaeus* he noted that all the entities occupying certain places are simultaneously temporal (or "perishable") objects (Plato, Tim./1888, 52ab). Therefore, "occupying a certain place" is equivalent to "occupying a certain place at a certain time".[14]

Plato and Aristotle expounded (still in a rather naïve manner) the concept of WORLD OF NATURE as the largest material whole:

---

[13] Cf. also the hypothesis about the linguistically universal character of the distinction between the general and the locative notion of EXISTENCE put forward in (Wierzbicka, 1996, p. 125).

[14] Perhaps the first detailed analysis of the meaning of the term "place" is found in Book IV of *Physics*. Aristotle distinguishes there: (a) a place as that which surrounds the item ("contains that of which it is the place"), and (b) the place immediately occupied by an object within its borders (Aristotle, Phys./1930, 210b, 211a). In further analyses, we shall refer to the clarification (b).

> The maker made neither two universes nor an infinite number; but as it has come into being, this universe one and only-begotten, so it is and shall be for ever. [...] Now the making of the universe took up the whole bulk of each of these four elements. (Plato, Tim./1888, 31b, 32c, 33a)

> But alongside the All or the Whole there is nothing outside the All, and for this reason all things are in the heaven; for the heaven, in a small say, is the All. (Aristotle, Phys./1930, 212b)

> [...] We habitually call the whole or totality "the heaven". [...] Now the whole included within the extreme circumference must be composed of all physical and sensible body, because there neither is, nor can come into being, any body outside the heaven. (Aristotle, Heav./2015, 278b)

In the next part of the paper we will be using the word "concrete" to designate empirically recognizable entities that occupy a certain place at a certain time.

An analysis of the just-quoted statements and earlier comments on "perishable beings" leads to two important explications. Firstly, the world of nature is the largest system of concretes, which itself is concrete.[15] Secondly (the whole) *space* is the area occupied by the (whole) world of nature at a given moment. In other words, the world of nature is a concrete occupying the whole space at any given moment. Taking into account the clarifications (and the quoted statements), the following rule seems to be in line with the spirit of the Platonic-Aristotelian cosmology:

(CL5) The *world of nature* is a concrete occupying the whole space at any given time. Exactly one world of nature exists.

The notion explained above looks like a philosophical generalization of the contemporary concept of UNIVERSE:

> One of the main applications of the general theory of relativity is cosmology, or the science of the global structure of the universe. Universe is understood in this science as the largest of the possible systems to which the laws of physics can be applied. (Heller & Pabjan, 2007, p. 140)

As we know, the laws of the general theory of relativity apply only to the physical systems occupying certain areas of space-time. The universe in turn – the largest such system – occupies the entirety of space-time.

The denotation of the philosophical term "world of nature" may therefore be equated with the denotation of the physical term "universe". We assume that the

---

[15] See the first of Aristotle's sentence quoted in the previous section; it states that the physical world – taken as a whole – is widely regarded as a natural substance (and so also "spatial" and "temporal").

difference in meaning between them applies only to their content: the content of the former is general, and of the latter is particular.

The world of nature is quite a peculiar temporal being (the best illustration of this statement is the fact that modern cosmology ascribes a certain age to the universe [it is probably approx. 13.8 billion years old]). Consequently, it is an element of the real world. This conclusion allows for the specification of the formal relationship between the extensions of the notions of WORLD OF NATURE and REAL WORLD (which will soon be done).[16]

By the *classical conception of the object of philosophy* (abbreviated: CCOP) we will mean the conception of the object of philosophy derived from Plato and Aristotle and reconstructed in the form of postulates (CL2)–(CL5).

The reconstruction seems historically and methodologically correct. It also turns out to be interesting from the perspective of contemporary metaphilosophy, ontology, formal philosophy and philosophy developed in the context of the natural sciences.

1.4

The concepts of CLASS, SET, EXTENSION, LOGICAL VALUE and so on, are typical for the *extensional approach* in formal philosophy, expressed in the recognition of classical logic and the logical theory of classes – or possibly the mathematical set theory – as the primary (though not necessarily the only) tool of philosophical analysis.

The main reasons for choosing this approach at this point are as follows: (1) classical logic (and so extensional logic) is the standard logic; in this sense it is the logic best grounded in modern mathematics and science; similarly (2) the logical theory of classes and mathematical set theory (and therefore the theories of some "extensional entities") are the theories of some kind of universals, which are well grounded in modern mathematics and science; and finally (3) a number of fundamental principles of classical logic and the logical theory of classes correspond well with the CCOP and the ontological paradigm of philosophy. (The clarification of these threads will be developed in the next part of the article).

---

[16] Analytical philosophy often uses a similar concept of ACTUAL WORLD as a possible world that we inhabit. Because the ontology of possible worlds is not directly related to the classical conception of philosophy – although it undoubtedly constitutes an important component of the contemporary analytical version of the ontological paradigm of philosophy – and also because it has an essentially different (modal) character, it will not be considered here. In particular, we leave open the question of whether the actual world can be identified with the real world.

We shall use the symbols "V", "R" and "©" to denote the world as an (extensional) whole, the real world and the world of nature, respectively. We accept the following explications:

(V) V = non-empty class including all classes of beings = *the world as a whole*.
(R) R = non-empty class of all temporal beings = *the real world*.
(©) © = a concrete occupying the entire space at any time = *the world of nature*.

We define the position of *extensional metaphysical realism* as the following set of theses:

MT0 Each class is a being.
MT1 Exactly one world as a whole (V) exists.
MT2 Exactly one real world (R) exists.
MT3 Exactly one world of nature (©) exists.
MT4 $\{©\} \subsetneq R \subsetneq V$.

This position is an explication of CCOP. In particular MT0 is an explicatory counterpart of (CL2), MT1 – of (CL 3), MT2 – of (CL4) and MT3 – the counterpart of (CL5). MT4 states that the world of nature is one of the elements of the real world, which in turn is an extensional part of the world as an (extensional) whole.[17]

The aim of further considerations is to demonstrate, in a manner consistent with the philosophical context of CCOP and the contemporary logic and scientific methodology, that:
- MT0 and MT1 are philosophical consequences of one of the basic systems of the classical logic of quantification (part II),
- MT2 is a philosophical consequence of the fundamental principles underlying modern mathematics and the methodology of empirical sciences (in the first section of part III) and
- MT3 and MT4 are philosophical consequences of the knowledge well grounded in modern physical cosmology (in the final section of part III).[18]

---

[17] We ignore here the question of the specific "qualitative" language of Aristotle's philosophy. We assume that the "mathematical" approach of Plato is the main historical justification for the extensional approach in the analysis of CCOP.

[18] Taking into account the volume limitations and the need to maintain the uniform character of the article, I am not presenting here a formalised theory, the theses of which are MT0–MT4. I presented such a theory in a paper delivered in 2016 at the Polish Contemporary Ontology conference.

Prior to the implementation of this task we shall determine (in the next three sections) the main principles of the adopted research method.[19]

Part II

2.1

The overall assumption of the theses MT0–MT2 is the thesis of the existence of classes. What principles can it be derived from? One of the simplest answers to this question is provided by the following reasoning:
(1) there are many (statistical) populations,
(2) each population is a class,

(3) there are classes.

This reasoning is an example of the use of a formal method, which could be called *interpretative deduction*. Generally, it involves deductive derivation of the analysed proposition from a given set of principles on the basis of a certain additional set of interpretative premises (of a "bridge principles" nature). If X is a set of principles from which the proposition α can be interpretatively derived on the basis of the premise Y, then α is an *interpretive consequence of X* for interpretation Y.[20] In particular X and Y may be singleton sets. For example, the proposition (3) is a consequence of the set $\{(1)\}$ for interpretation $\{(2)\}$.

---

[19] We should notice (to avoid possible misunderstandings) the difference in the nature of the investigations related to the dispute about the existence of the world, resembling the title of the work of Ingarden (2013). Those investigations aim to define the so-called mode of the existence of the world, on the basis of the assumptions that external perception is generally *not* a reliable source of knowledge (and, consequently, that our natural belief in the existence of the real world in philosophy should be "put in brackets"). This assumption is characteristic of the entire Cartesian paradigm ("mentalistic" in the terminology adopted in Schnädelbach, 1985), and it is rejected within the ontological (Platonic-Aristotelian) paradigm. These "programme differences" do not exclude, certainly, the possibility of the use of many interesting intuitions and observations made by the advocates of the "Cartesian" approach.

[20] In the case where Y is a set of premises of philosophical content, we talk about a *philosophical consequence*. The method of "deductive interpretation" is a combination of the so-called regressive analysis (discussed in Beaney, 2014), and Woleński's method of interpretative (philosophical) consequence. Interpretative deduction (just like regressive analysis) corresponds to the Aristotelian procedure of seeking "first principles" from which, on the basis of the logic of syllogisms, the given statements can be deductively derived (see *Posterior Analytics*).

Applying this method can answer the question: "If the proposition α is our true belief, then how – or more precisely, on what basis – can we know this?"; or briefly: "How can we know that α?". If such a question contains specific philosophical content, the correct answer is to indicate to the appropriate principles (philosophical or extra-philosophical) and "bridge" philosophical premises from which this answer can be deductively derived. For example a correct (but not necessarily the only) answer to the question "How can we know that classes exist?" consists in indicating principle (1) and the interpretative premise (2).

This method of interpretative deduction will be repeatedly used in the following sections.

2.2

It seems that any well-defined research approach should indicate the largest logical system in which the relevant theories are built, compared, and – if necessary – formalized.[21] Today *standard logic* (that is the classical first-order calculus with identity) most often plays this role. Given the fact that it is based on the Aristotelian laws of non-contradiction and the excluded middle, and the Leibnizian principle of indiscernibility of identicals, we can conclude that this logic fulfils fairly well the role of a formal tool for the analysis of the basic ontological principles related to CCOP.

As we know from metalogic, a monadic second-order logic (**MSO**) is the weakest extension of the standard logic (regarding the types of quantification) preserving all of these principles and the so-called standard (model theory) interpretation.[22] **MSO** is a formally elegant system with a relatively high – compared to standard logic – expressive power. At first glance it also seems that the extensional conceptual realism that underlies it, and the thesis of the existence of the universal class, make it a useful tool for analysing the propositions MT0 and MT1. The next part will show that this conviction is fully justified. The general reasons for choosing **MSO** as a tool of philosophical analysis are mentioned below.

Typical set theories are powerful tools for analysing the foundations of mathematics. This can be viewed positively from the perspective of the philosophy of

---

[21] A *system* is *the largest* in a given set of deductive systems, when every other deductive system in this set is its subsystem. In this sense, for example, monadic second-order logic is the largest system in the set of all its sub-logics (containing, among others, first-order logic).

[22] A simple (so-called "pure") version of **MSO** is equivalent to the logical theory of classes, which is accessibly presented in the handbook: (Tarski, 1941/1994, pp. 63–80). For basic informations about metalogical properties of **MSO** see: (Shapiro, 1991, p. 221–226).

mathematics and philosophical semantics, but it also raises important metaontological questions. The application of such strong theories in general ontology raises the question of the undecidability of some important problems that can be expressed in the languages of these theories (e.g. the problem of the continuum hypothesis), as well as the question of the costs of introducing ontologically questionable entities with features of mathematical artefacts: sets with a power higher than the continuum, sets which we cannot construct, etc.

This argumentation is quite convincing. Therefore, the main subject of our interest is – much weaker in its expressive power – monadic (second-order) logic.

This logic generates the "existentially light" theory of classes; its set of specific existential theses boils down to the thesis that there are at least two classes (an empty and a universal one). In particular, **MSO** does not assume that there are multi-element classes, power classes, series, or ordered pairs.[23]

The most common argument against treating monadic logic as a proper tool for philosophical analysis relates to the semantic incompleteness of **MSO**, which means that not all of its tautologies can be proven in it. This weakness does not apply to standard logic, which is a complete system. However, a similar argument can be formulated against the latter: standard logic is, as opposed to **MSO**, a non-categorical system (i.e. there does not exist a first-order theory with an infinite model, all models of which are isomorphic).[24] At the same time we cannot see any methodological criterion for determining the hierarchy of the properties of deductive systems, which would justify the conviction that completeness is a "more important" property than categoricity (and – vice versa).[25]

We exclude from this "metalogical ranking" stronger systems of higher order because of their undecidability, heterogeneity of logical types and their ontological status, that are not well-enough defined (raising questions like: if relations are classes, then why do they constitute separate non-monadic logical types? And if they are not, then why do they meet the extensionality condition, which is characteristic for classes?). Of course, such an exclusion of higher order systems does not limit the expressive power of a language for ontology built on the basis of **MSO**. It only implies the need to build the relevant theory of relations or the

---

[23] G. Boolos rightly therefore pointed out (in Boolos, 1975/1998), that it is a theory too weak to be considered – as Quine said in his well-known criticism (in Quine, 1986) – a "set theory in sheep's clothing".
[24] This is a consequence of the Löwenheim-Skolem-Tarski theorem.
[25] See (Boolos, 1975/1998). From the ontological point of view, categoricity could be considered even "more important" then completeness, taking into account some of the consequences of the so-called Skolem paradox (which does not apply to categorical second-order theories). Some authors (e.g. H. Putnam) believe that these consequences lead to a certain version of antirealism.

theory of higher order classes as an extralogical theory (e.g. within the monadic set theory).[26]

2.3

We do not thus notice major, *methodological* difficulties related to the acceptance of monadic logic as a tool of philosophical analysis and formalization of theories that – in one way or another – are already involved in the ontology of classes. There are, however, a number of *philosophical* arguments for its recognition as a useful tool for the analysis of the rules relating to CCOP and its ontological paradigm. We can identify at least three such arguments.

Firstly, **MSO** corresponds well with Aristotle's idea of *ontological dualism*, according to which two types of categoremata exist, one of them being a primary one; each of them involves similar ("analogous") concepts of EXISTENCE.[27] Their equivalents are the meanings of two types of existential quantification expressible in the language of monadic logic "over objects" ($\exists x(\ldots x \ldots)$) and "over classes" ($\exists X(\ldots X \ldots)$).

Secondly, **MSO** includes complete, realistic (in the sense of conceptual realism) versions of the Aristotelian laws of non-contradiction and the excluded middle, and an extensional version of the Leibnizian principle of the identity of indiscernibles:

$$\sim\exists xY(x \in Y \wedge \sim x \in Y),$$
$$\forall xY(x \in Y \vee \sim x \in Y),\ [28]$$
$$\forall Z(x \in Z \Leftrightarrow y \in Z) \Rightarrow x = y.\ [29]$$

These versions are strict explications of the fundamental principles underlying rational cognition, characteristic of the ontological paradigm of philosophy.

---

[26] MSO has also an important application in automata theory (see e.g.: Grädel et al., 2002, pp. 207–302).

[27] "Of things there are some universal and some individual or singular, according. I mean, as their nature is such that they can or they cannot be predicates of numerous subjects, as 'man' for example, and 'Callias'" (Aristotle, Interpr./1934, 17ab). "So species and genera only are rightly designated as substance, first substances only excepted" (Aristotle, Cat./1938, 2b). "For such propositions have always individuals or species for subjects" (Aristotle, Cat./1938, 3a).

[28] The "completeness" of these versions means that they are closed sentences, not schemes of sentences as in standard logic. In turn, their realistic nature (in terms of conceptual realism) is due to the fact that classes are quantified in them, and so constitute a kind of universals.

[29] This principle allows the formulation of the definition of identity (Leibniz-Russell): $x = y \Leftrightarrow \forall Z(x \in Z \Leftrightarrow y \in Z)$.

And thirdly, a set of complete (non-schematic) claims of **MSO** constitutes a contemporary extensional version of the traditional theory of universals. This version is in fact a very weak theory of classes with the *extensionality axiom for classes*:

$$\forall x(x \in X \Leftrightarrow x \in Y) \Rightarrow X = Y$$

(classes are identical, if they consist of exactly the same elements) and with an unlimited (schematic) *comprehension axiom*:

$$\exists X \forall y(y \in X \Leftrightarrow \alpha), \text{ if } X \text{ is not free in } \alpha$$

(for each condition there exists a class of objects that meet it). This scheme generates the thesis about the existence of an empty class (through the substitution of formula "$y \neq y$" for α) and a universal class (through the substitution of formula "$y = y$" for α):[30]

$$(\exists X \sim \exists y)\ y \in X,$$
$$(\exists X \forall y)\ y \in X.^{31}$$

The application of the extensionality axiom to the above also leads to the conclusion that exactly one empty class exists, and – to the conclusion that:
MT1*   exactly one universal class exists.

MT1* will soon (at the end of the next section) become the basis for the interpretive deduction of the proposition MT1.

These arguments lead us – given the existence of classes – to consider **MSO** as a proper tool for analysing the principles associated with CCOP and the ontological paradigm.

2.4

Like other objectually interpreted systems of quantifier logic, **MSO** provides a strict explication of *the general notion of existence*, intuitively expressed in the French

---

30 It is worth adding that principles – identity, extensionality and comprehension – have no equivalent in standard logic.
31 An even more general explication of the notion of the "universal class" is possible outside the context of monadic logic: in the Boolean algebra of classes it is just a complement of the empty class. In turn, the theory of classes and sets (derived from J. von Neumann) provides a more detailed explication; V in it is a class of all the elements of certain classes (and V is not an element of any class).

Il y a, German *Es gibt*, English *There is*, or in the wording "There exists something about which it is true what follows". The key component of this explication is the existential quantifier; binding the general notion of EXISTENCE with this quantifier determines its (intuitive) *existential interpretation*.

It is worth noting that Aristotle sometimes used the general notion of EXISTENCE in his philosophical inquiry (in the form of one of the idiomatic uses of ἐστι).[32] One of the most interesting examples of these applications is the "Kantian" observation of the Stagirite that existence is neither the kind nor the essence of anything (*Posterior Analytics*, 92b). Including such examples, as well as the fact that the notion under consideration is broadly applied in natural language,[33] leads to the conclusion that there are no major obstacles for applying a contemporary quantifier explication of the concept of EXISTENCE in the formal analysis of the ontological principles related to CCOP.[34]

The immediate application of the existential interpretation of the "small" quantifier is the justification for the accuracy of the following rules of the *ontological interpretation of monadic logic* (i.e. the rules of intuitive interpretation of the formulae of monadic ontological theory):[35]

(OI1) "$\exists x(\ldots x \ldots)$" means

"At least one object x exists such that $\ldots x \ldots$",

(OI2) "$\exists X(\ldots X \ldots)$" means

"At least one class X exists such that $\ldots X \ldots$".

This leads us to consider how in this context the meta-predicate "X is a being" should be understood (important for the strict expression of MTO). Now, with regard to the traditional meaning of the term "being" ("something that exists"), it can be read: "X is something that exists". A simple clarification of this phrase is

---

[32] Owen (1965), Kahn (1966/2009a, p. 40). An example of such use is found in the first sentence of the second quotation from *Metaphysics*, given in section 1.2.

[33] The evidence of the widespread use of the general notion of EXISTENCE in natural language is provided by the fact that the following hypothesis was formulated in contemporary ethnolinguistics: a primitive notion expressed in *There is* is distinguishable from the locative notion of EXISTENCE in every language of the world (Wierzbicka, 1996, p. 125).

[34] We should distinguish between the general concept of EXISTENCE outlined above and Quine's criterion of the ontological commitments of a theory (to which we shall not refer here directly).

[35] The phrase "monadic theory" denotes here any theory axiomatized on the ground of **MSO**. In the rules OI1–OI4 below the word "means" is treated as an abbreviation for the phrase "means in the language of any formalized, monadic ontological theory." The issue of the complete explication of the notion of MONADIC ONTOLOGICAL THEORY – with the use of this type of rules – we leave open here.

the paraphrase "Something that is identical with X exists", formally "$(\exists Y)X = Y$".
According to this method of formalization, we set the rule:
(OI3) "$(\exists Y)X = Y$" means "X is a being".

As a result, the following **MSO** formula states that each class is a being:
(1) $(\forall X \exists Y)X = Y$.

This statement is equivalent to MT0. At the same time (1) is a thesis of **MSO**.

On the basis of the ontological interpretation of the **MSO** system – as we accept it – all the ontological theses generated by this interpretation we regard as true propositions, in the intuitive sense of the word "true". In particular, we regard the propositions "There exists exactly one universal class" and "Each class is a being" as ontological paraphrases of the relevant theses of **MSO**. Therefore, recognizing **MSO**, we are, at the same time, obliged to recognize: (a) that there exists exactly one universal class and (b) that each class is a being.[36]

To derive MT1, it is enough to accept the rule of the ontological interpretation:
(OI4) "$\forall Y \forall x(x \in Y \Rightarrow x \in X)$" means "$X$ includes all classes of beings".

The universal class satisfies (on the ground of **MSO**) the formula on the left side of (IO4), thereby it includes all classes of beings. Moreover, the universal class is non-empty (as a result of the assumption of the non-emptiness of the domain); therefore, it is identical with the world as a whole (as defined in [V] formulated in section 1.4). This conclusion, together with the thesis about the uniqueness of the universal class, is equivalent to MT1: exactly one world as a whole exists.

It turns out that MT0 and MT1 are philosophical consequences of the relevant laws of monadic logic. In particular, MT0 is such a consequence of law (1) for interpretation (OI3), and MT1 is a philosophical consequence of MT1* for interpretation (OI4) and definition (V).

---

[36] The same reasoning leads to the conclusion that the formula "$(\exists y)x = y$" means "x is a being", and that every object is a being. We note in passing that the metalanguage predicate "is a being" is here, as in the Aristotelian system, systematically ambiguous: each of the two meanings is associated with a different type of quantification (in the sense or [OI1] or [OI2]).

Part III

3.1

If you have well-defined concepts of TIME and EXISTENCE IN TIME, then the definition of the real world should probably be like this (the value of the variable "X" here is any class):

(R') X is a *real world* $=_{df}$
    $\forall x(x \in X \Leftrightarrow \exists y(y \text{ is a moment } \wedge x \text{ exists at } y)) \wedge X \neq \emptyset$.

In addition to the general arguments considered earlier (in section 1.2), there are at least three other reasons why it is worth introducing the definition (R') to general ontology.

Firstly, there are axiomatic theories providing good explications of the aforementioned temporal concepts (TIME and EXISTENCE IN TIME).[37]

Secondly, it does not seem reasonable to treat the real world as a mereological whole, whose parts are objects as diverse as supernova explosions, petrol stations, and thought processes. The recognition that the world of temporal beings is an extensional "whole" is much more convincing; this "whole" is either a class (domain) of these beings, or a relational system (model), whose domain is this class.

And thirdly, if we decided that the real world is a relational system and not a "simple" class, then a difficulty would immediately emerge related to the issue of a non-arbitrary choice of such a unique system. We can easily get around this difficulty by introducing the definition (R'). This decision leads to distinguishing the (unique) real world from one or another of its structures (spatio-temporal, causal etc.). Note that it does not follow from the fact that in the definition of class R no relations between temporal beings have been distinguished that such relations do not actually exist. This definition is so general that it allows us to skip this issue (e.g. as it is an issue of specific ontology).

The thesis of the existence of temporal beings can be justified – on the basis of two basic principles of the methodology of empirical sciences – by means of the following deduction:

(1) Each observable object is a being existing at a certain time.
(2) Observable objects exist.
(3) A temporal being = a being existing at a certain time.

(4) Temporal beings exist.

---

37 See note 18.

In line with the previous findings, including the definition of the real world and the comprehension schema, the thesis about the existence of the real world can be derived from the following premises:

(4) Temporal beings exist.
(5) If temporal beings exist, then exactly one non-empty class of them exists.
(6) The real world = the non-empty class of temporal beings.

(7) Exactly one real world (R) exists.

This conclusion authorizes us to "lawfully" use the symbol "R" to denote the real world:

$$R = \{x : \exists y(y \text{ is a moment} \land x \text{ exists at } y)\}.$$

The proposition MT2 is equivalent to thesis (7) which is a philosophical consequence of observation (2), "bridge" premise (1), definitions (3), (6), and principle (5) directly using the axioms of **MSO** (comprehension and extensionality axioms).[38] These axioms have their counterparts in each version of mathematical set theory. MT2 is thus a philosophical consequence of well-established principles in the foundations of mathematics and the empirical sciences.

3.2

According to the recognized findings (in section 1.3), the denotations of the terms "world of nature" and "universe" are identical. Consequently, if the universe exists, then the world of nature exists. According to knowledge well-established in modern physical cosmology, the universe exists; then, there exists the world of nature.

This reasoning can be presented in the following "standardized" form:
The class of designata of the term "universe" = the class of designata of the term "the world of nature".
At least one universe exists.

At least one world of nature exists.

This raises the question of whether we can derive the thesis of the uniqueness of the world of nature in a general way, on the basis of philosophical assumptions

---

[38] The philosophical "bridge" premises are: the definition of "temporal object", definition (5) and thesis (6). The last of these premises results from the assumption that no class is a temporal being.

alone. Such proofs usually employ a suitable identity criterion for a given category of objects. We presently need such a criterion for the category of concretes.

Let us adopt the *space-temporal criterion*, frequently used in these types of considerations: concretes are identical when they occupy exactly the same space at the same time moment. Some physicists claim that fields and bosons can occupy the same areas of space at the same time. If they are right, these physical objects are obviously not concretes.

On the basis of this criterion the proof of the uniqueness of the world of nature becomes quite simple:

(1) If x, y are concretes occupying exactly the same space at the same time, then $x = y$.
(2) x is a concrete occupying the whole space at any time.
(3) y is a concrete occupying the whole space at any time.

(4) $x = y$.

In light of the above analysis it is clear that the proposition MT3 is a philosophical consequence of the cosmological thesis about the existence of the universe. An additional interpretative premise, allowing for the derivation of this proposition is a space-temporal identity criterion and definition of WORLD OF NATURE as an ontological generalization of the physical concept of UNIVERSE.

Denoting the class of concretes by "C", the class of moments (time) by "T" and the class of points (space) by "S", we can – taking into account the thesis about the existence and uniqueness of the world of nature – give the following definition of the constant "©":

$$© = x =_{df} x \in C \wedge \forall y\, (y \in T \Rightarrow x \text{ occupies S at } y).$$

Hence we directly obtain the thesis:
©1 © ∈ C,
©2 $\forall y (y \in T \Rightarrow$ © occupies S at y).

We obtain the thesis MT4 from ©1, the inclusion "C ⊂ R" (which is true under the adopted terminological findings) and two additional premises, quite obvious on the basis of common sense knowledge, scientific knowledge and the ontological paradigm of philosophy:
– concretes non-identical with the world of nature exist,
– atemporal objects (e.g. mathematical objects) exist.

Taking into account the final conclusion of part II, we can assume that the propositions MT0–MT4 are the philosophical consequences of the set of principles of monadic logic and principles well grounded in the foundations of empirical sciences. Pointing to these principles yields the following correct answers to the questions: "How can we know that there exist classes?", "How can we know that the world as a whole exists?", "How can we know that the real world exists?", "How can we know that the world of nature exists?", and "What are the basic relationships between the concepts WORLD AS WHOLE, REAL WORLD and WORLD OF NATURE?". Consequently, we can treat both the world as an extensional whole, and the other two designates of the (ambiguous) word "world" as objects of the formal ontological philosophy.

At the same time the main assumptions of our considerations are consistent with many principles typical for CCOP and its philosophical context. This allows us to believe that the position of the extensional metaphysical realism is consistent with the classical conception of the object of theoretical philosophy.

Acknowledgment: The article is a revised and expanded version of the first three parts of the four-part paper "How do we know that the world as a whole exists?", delivered in May 2016 at the *Contemporary Polish Ontology* conference in Warsaw. I would like to thank all those who took part in the discussion on the paper for their inspiring remarks. My special thanks go to Zbigniew Król, Marek Kuś, Józef Lubacz, Francesco Orilia, Alexander Pruss, Bartłomiej Skowron, and the anonymous reviewer. The project was funded by the National Science Centre, granted under Decision No DEC-2011/03/B/HS1/04586.

Bibliography

Aristotle (1934). *Categories* (H. P. Cooke, Trans.). In H. P. Cooke, & H. Tredennick (Trans.), *Aristotle, Categories. On interpretation. Prior analytics* (pp. 12–111). William Heinemann & Harvard University Press.

Aristotle (2015). *On the heavens* (J. L. Stocks, Trans.). Retrieved from https://ebooks.adelaide.edu.au/a/aristotle/heavens

Aristotle (1934). *On interpretation* (H. P. Cook, Trans.). In H. P. Cooke, & H. Tredennick (Trans.), *Aristotle, The categories. On interpretation. Prior analytics* (pp. 114–181). William Heinemann & Harvard University Press.

Aristotle (2015). *Metaphysics* (W. D. Ross, Trans.). Retrieved from https://ebooks.adelaide.edu.au/a/aristotle/metaphysics

Aristotle (1930). *Physics* (R. P. Hardie, & R. K. Gaye, Trans.). In W. D. Ross (Ed.), *The works of Aristotle* (Vol. 2). Oxford: Clarendon Press.

Beaney, M. (2014). Analysis. In E. N. Zalta (Ed.), *The Stanford Encyclopedia of Philosophy (Summer 2016 edition)*. Retrieved from http://plato.stanford.edu/archives/sum2016/entries/analysis

Boolos, G. (1998). On second order logic. In G. Boolos, *Logic, logic, and logic* (pp. 37–53). Cambridge, Mass.–London: Harvard University Press. (Reprinted from *Journal of Philosophy*, 72(16), pp. 509–527, 1975)

Feser, E. (2013). Introduction: an Aristotelian revival? In E. Feser (Ed.), *Aristotle on method and metaphysics*. Pallgrave Macmillan.

Frege, G. (2014). Über den Zweck der Begriffsschrift. In G. Frege, *Begriffsschrift und andere Aufsätze*. Georg Olms. (Original work published 1882)

Grädel, E., Thomas, W., & Wilke, T. (Eds.). (2002). *Automata, logics, and infinite games. A guide to current research*. Berlin: Springer.

Heijenoort, van J. (1967). Logic as calculus and logic as language. *Synthese, 17*, 324–330.

Heller, M., & Pabjan, T. (2007). *Elementy filozofii przyrody* [Elements of the philosophy of nature]. Tarnów: Biblos.

Ingarden, R. (2013). *Controversy over the existence of the world*: Vol. I (A. Szylewicz, Trans.). Bern: Peter Lang. (Original work published 1947)

Kahn, Ch. H. (2009a). The Greek verb 'to be' and the concept of being. In Ch. H. Kahn, *Essays of being*, pp. 16–40. Oxford University Press. (Reprinted from *Foundations of Language*, 2(3), pp. 245–265, 1966)

Kahn, Ch. H. (2009b). Why existence does not emerge as a distinct concept in Greek philosophy. In Ch. H. Kahn, *Essays of being*, pp. 62–74. Oxford University Press. (Reprinted from *Archiv für Geschichte der Philosophie* 58(4), pp. 323–334, 1976)

Owen, G. E. L. (1965). Aristotle on the snares of ontology. In R. Bambrough (Ed.), *New essays on Plato and Aristotle*. London: Routledge.

Plato (2004). *Republic* (C. D. C. Revee, Trans.). Indianapolis: Hackett Publishing.

Plato (1888). *Timaeus* (R. D. Archer-Hind, Ed.). London–New York: MacMillan.

Novotny, D. D., & Novak, L. (2014). *Neo-Aristotelian perspectives in metaphysics*. New York–London: Routledge.

Quine, W. V. (1986). *Philosophy of logic* (2nd ed.). Cambridge, Mass.–London: Harvard University Press. (Original work published 1970)

Quine, W. V. (1981). *Theories and things*. Cambridge, Mass.–London: Harvard University Press.

Reale, G. (1987). *A history of ancient philosophy I. From the origins to Socrates* (J. R. Catan, Trans.). New York: SUNY Press.

Schnädelbach, H. (1985). Philosophie. In E. Martens, H. Schnädelbach (Eds.), *Philosophie. Ein Grundkurs*. Rewohlt Taschenbuch Verlag.

Shapiro, S. (1991). *Foundations without foundationalism. A case for second-order logic*. New York–Oxford: Oxford University Press.

Tarski, A. (1994). *Introduction to logic and to the methodology of the deductive sciences*. New York–Oxford: Oxford University Press. (Original work published 1941)

Tatarkiewicz, W. (2016). *Historia filozofii: Tom 1. Filozofia starożytna i średniowieczna* [History of philosophy: Vol. 1. Ancient and medieval philosophy] (22nd ed.). Warszawa: PWN. (Original work published 1931)

Wierzbicka, A. (1996). *Semantics. Primes and universals*. Oxford–New York: Oxford University Press.

Woleński, J. (1993). *Metamatematyka a epistemologia* [Metamathematics and epistemology]. Warszawa: Wydawnictwo Naukowe PWN.

Urszula Wybraniec-Skardowska
Logic and the Ontology of Language

Abstract: The main goal of this paper is to outline a general formal-logical theory of language construed as a particular ontological being. The theory itself will be referred to as an ontology of language, because it is motivated by the fact that language plays a special role: it reflects ontology, and ontology reflects the world. Linguistic expressions will be regarded as having a dual ontological status: they are to be understood as either *concreta* – i.e. *tokens*, in the sense of material, physical objects – or *types*, in the sense of *classes of tokens* – i.e. abstract objects. Such a duality will then be taken into account in the logical theory of syntax, semantics and pragmatics presented here. We point to the possibility of constructing the latter on two different levels, one stemming from *concreta*, construed as linguistic *tokens* of expressions, the other from their classes – namely *types*, conceived as abstract, ideal beings. The aim of this work is not only to outline such a theory with respect to the dual ontological nature of the expressions of language in terms that take into account a functional approach to language itself, but also to show that the logic based on it is ontologically neutral in the sense that it is abstracted from the level at which certain existential assumptions relating to the ontological nature of these linguistic expressions and their extra-linguistic ontological counterparts (objects) would have to be embraced.

Keywords: formal logic, ontology, ontology of language, syntax, expression-*token*, expression-*type*, semantics, meaning, denotation, ontic category.

Introduction

This section has a preliminary character. It discusses the principal aspects and concepts pertaining to the descriptive, representational and referential functions of language, and the dual ontological nature of its expressions, given certain assumptions and logical foundations. The theory of language thus construed is outlined in the main part of the paper (Section 2), and some summary results and conclusions are included in Section 3.

Note: This article is a modified version of the work entitled *Logiczna koncepcja języka wobec założeń egzystencjalnych* [The logical conception of language towards existential assumptions], dedicated to Professor Jacek J. Jadacki on his 70th birthday, Wybraniec-Skardowska 2017.

Urszula Wybraniec-Skardowska, Department of Philosophy, Cardinal Stefan Wyszyński University, Warsaw, Poland.

https://doi.org/10.1515/9783110669411-007

Knowledge-Language-Reality

This section introduces the issue of linguistic adequacy as this relates to the function of language as an ontological being used on the one hand to describe the world (which is what ontology deals with as the theory of being), and on the other to represent our knowledge of this world. I seek here to furnish a justification for thinking that our theory of language should be an ontological one. I also argue in favour of the logical conception of language.

For the most general definition of ontology, we shall refer to the definition proposed by Perzanowski:

> Ontology is the general theory of possibility, i.e. the realm of all possibilities – the ontological space. Metaphysics, on the other hand, is an ontology of the world, i.e. the reality of all existing items, called facts. [...] Real philosophy, however, is about being. (Perzanowski, 2012, p. 45)

By *being*, we understand here everything that exists, that can exist, that is not contradictory in itself. The task of ontology – as we understand it here – is to describe the structure of being or reality. Language, while at the same time serving as a tool for constructing the theory of being itself, is put at the service of such a description.

Studies of language can certainly prove helpful when it comes to producing a descriptive account of this kind. For language to be able to perform this basic function of providing a faithful description of reality and its structure, a specific sort of compatibility obtaining between the elements of the following triad is called for,

Language – Knowledge – Reality,

which I shall refer to for short as *linguistic adequacy*, where this in turn is to be described within a theory of language.

Language serves to represent human knowledge acquired in the process of cognizing reality. It is simultaneously a means of describing that cognized reality. Operating with language in ways that involve logic and thinking enables us to transform and enrich our knowledge in order to better get to know and discover the world. It is thus also a tool for expanding our cognition of reality on the basis of the knowledge we already possess – which, by the way, need not be confined to the domain of ontological matters.

In order for language to fulfil its descriptive function, it should reflect the structure of being, reality, with its own structure. The structure of language is undoubtedly connected with that of the cognized world. It is conditioned by the formation of knowledge obtained in the process of cognizing the reality, framing the structure of this reality. This structure is described, as we know, by ontology. Knowledge of the structure of reality allows us to speak in a uniformly consistent way about the world, making inter-human communication more effective.

The relevant components of knowledge correspond to the elements that compose the reality. In language, we speak about both the former and the latter by means of its expressions. They have their counterparts in language: in its components, its expressions. The components of reality belong to appropriate *ontological categories*, and those of knowledge to appropriate *categories of components of knowledge*, whereas components of language belong to appropriate *syntactic categories* or *semantic categories* – i.e. to certain defined categories of linguistic expression. The language of such expressions serves to faithfully describe the world and the given domain of knowledge.

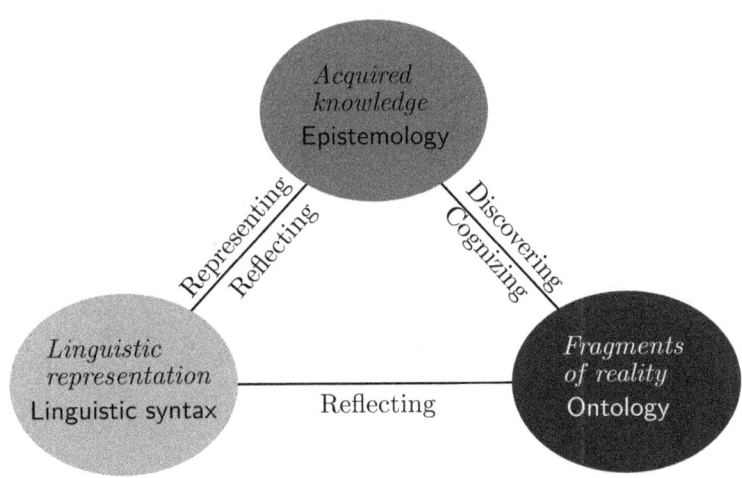

Diagram 1

Linguistic adequacy is achieved when the syntax of language faithfully reflects its bi-aspectual semantics: i.e. on the one hand the existing fragments of cognized or discovered reality (extensional semantics), and on the other the acquired knowledge resulting from the cognition or discovery of these (intensional semantics) (see Diagram 1). Language should thus reflect both some defined portion of reality and our knowledge of it – knowledge, that is, that has been acquired, but which is also in the process of being expanded.

As can be seen, language and its syntax are connected with both the ontology of the world – i.e. with everything that exists – and with epistemology, which deals with cognition of the world, whose result is the acquisition of knowledge.

Since language exerts such a considerable impact on ontology, it becomes vital to work out a general theory of language: language, that is, construed as a particular ontological being. This theory will be called the *ontology of language*. Using metalanguage, it will set out to describe the structure of language and its properties.

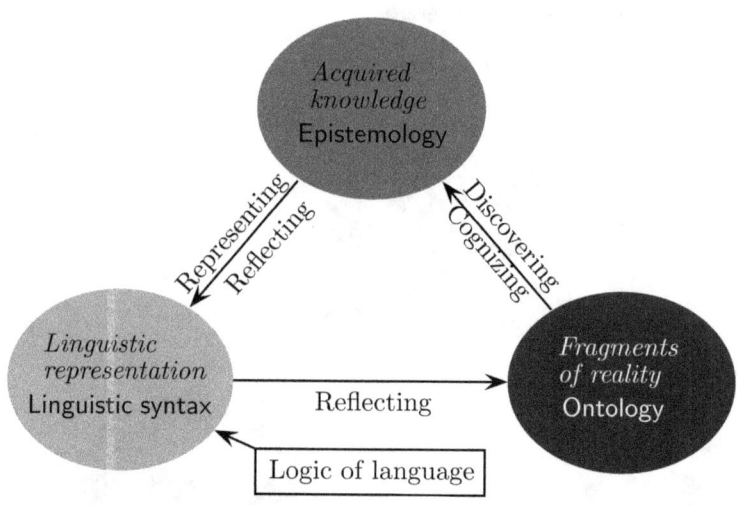

Diagram 2

In the same way as there exist a great number of conceptions of being, so there exist plenty of conceptions of language, and many theories of language. Here, the conception of language in the framework proposed by Ajdukiewicz (1974, p. 12, 13) will be of interest to us, and the theory to be constructed according to this conception will be the logical theory of language (logical semiotics) formalized on the basis of classical formal logic and set theory (see Diagram 2). Its assumptions are presented below.

The Logical Conception of Language

In the framework proposed by Ajdukiewicz (Ajdukiewicz, 1974, p. 13, note 6), the logical conception of language assumes that "in order to describe a language we have (i) to list its expressions, and (ii) *univocally* to assign specified meanings to these expressions." Ajdukiewicz (1974, p. 13) also wrote that

> By drawing attention to the difference between the logical concept of language and those concepts of language which are being used by linguists we wish to emphasize that the logical concept of language is much simpler than the linguistic one, and that its analysis prepares that set of concepts which is indispensable to give clarity to the research done by linguists.

Ajdukiewicz's logical conception of language will figure in our own proposal for shedding light on how language should be understood on the basis of logic. In that context, language is to be conceived as a system of conventional signs. In compliance with such a conception, the following will be the basic elements which make up language (as a system of signs):

1. vocabulary,
2. rules of syntax:
   (a) qualifying – establishing which objects qualify as simple expressions (words) of the vocabulary;
   (b) constructive – determining how to form other signs from simpler signs – complex expressions of language;
3. semantic rules: settling down what the signs a) mean, b) designate and denote;
4. pragmatic rules: determining the relations between linguistic signs and their users when communicating and cognizing reality.

Language, on such a conception, is an ideal creation. All real languages known to us are "logically defective". By contrast, in this case it is an idealizing reconstruction of real languages. It is this that constitutes the subject of formal-logical description in the present work: a description that, as a matter of fact, does not apply to languages as they are normally used. The formal-logic theory of language that will be sketched in due course (in Section 2) sets out to frame problems pertaining to the foundations of the theory of language in the most general terms possible, narrowing the problem area while providing a simple set of concepts and solutions relevant to issues connected with the ontological nature of linguistic expressions, their meaning (*intension*), and their denotation (*extension*). In proceeding thus, the theory reflects various assumptions, including certain existential ones, which are not satisfied in full by authentic languages, since the actual conditions in which the latter function most often depend on extra-linguistic factors and are not entirely neutral.

Language characterized according to the logical conception will consist of verbal signs, these being expressions in compliance with the rules of syntax: i.e. so-called well-formed expressions, which have a single meaning and denotation assigned to them, and which at the same time perform the function of representing knowledge acquired about a cognized reality, all the while playing the role of an intermediary in the process of transferring and exchanging information. Because language, as an ontological being, consists of expressions, and their ontological nature in turn can be of two kinds, our theoretical considerations pertaining to language must include certain initial assumptions regarding the ontological status of the expressions of that language. This problem will be discussed in the following subsection of the present work.

The Dual Ontological Status of Linguistic Expressions

What is the ontological status of expressions – of the objects making up a language? The question of the ontological status of such objects comes down to the following two bipolar (yes-no) questions:
1) Are linguistic objects, including words or expressions, concrete, real objects of a defined shape, extended in time and space?
2) Are linguistic objects, including words or expressions, abstract, objects or ideal beings of some sort?

The ontological status of the linguistic objects figuring in the above-mentioned questions is different: they belong to two different ontological types. At the same time, in logico-semiotic practice they are treated as possessing equal status.

Most often, in compliance with the differentiation made by Peirce (1931–1935, sec. 4.537), inscriptions, words or expressions are understood as either *concreta*, meaning *tokens* (events) that are material objects perceivable through the senses, or *types*, meaning *classes of* (uniform and, in a broad sense, identifiable) *tokens*, whose relevance is as abstract objects. Such a duality in respect of our understanding of linguistic inscriptions first showed up in the famous monograph by Tarski (1933, pp. 5, 6, note 5, p. 24, note 19; 1956, p. 156, note 1; p. 173, note 1), before gaining popularity particularly thanks to the work of Carnap in the 1940s (see Carnap 1942, sec. 3: Sign-Events and Sign-Designs).

Expressions such as perform on the one hand the function of representing knowledge acquired about a cognized reality, and on the other hand the role of intermediaries in the process of transferring and exchanging information, are sign-*tokens*, meaning specimens of sign-*types*, which in turn are classes of sign-*tokens* that are in some respect *identifiable* – i.e. *equiform*.[1] Any *meaning* or *denotation* is, on the other hand, assigned only to expression-*types*, which in contrast to their tokens (these being their physical representations) are thus object-*concreta* (e.g. inscriptions or sounds), these being object-*abstracta*.[2] Here, it must be said immediately that although, in logical semantics, explanations of the notions of 'meaning' and 'denotation' require the use of expression-*types*, in the very defining of these notions themselves it is expression-*tokens* that are used.

We also encounter the dual ontological character of linguistic expressions when taking into account the so-called functional approach to language, as in the framework proposed by Jerzy Pelc (1979). This obliges us to take into consideration two manners of usage of expressions. As regards the first of these, the manner of usage (*use*) appears exclusively in defined conditions, in determinate language-situational contexts, and concerns expression-*tokens*, whereas in the case of the other the manner of usage (*use*) characterizes the meaning of an expression as an expression-*type* seen in isolation from any situational-context: e.g., as in an entry in a dictionary edited in traditional book form.

The logical theory of language should thus assume the existence of both expression-*tokens* (language-based *concreta*) and expression-*types* (language-based *abstracta*). The dual ontological character of *linguistic objects*, and their being employed in a dual manner, point to the necessity of giving a bi-aspectual characterization of language in the theoretical context of the logical conception of language: as the language of expression-*tokens* and of expression-*types*.

At the same time, no definite elaboration of a theory of language can remain uninfluenced by the two main currents of linguistic ontology that have emerged in the light of the fundamental opposition associated with the controversy over universals: i.e. nominalism and realism.

Taking the nominalistic and concretistic position, it will be assumed that the basic plane of language consists of expression-*tokens*, and thus *concreta*. Abstract expressions, that is types of expressions, are then constructs emerging from a secondary level of analysis. On the other hand, if we assume that the basis for linguistic studies consists of ideal objects, in the sense of *abstracta* understood as types of expression, with expression-*tokens* available thanks to cognition through the senses being seen as constructs emerging from a secondary level of analysis, then we are opting for a platonizing standpoint.

---

[1] Equiformity is treated here as *cum grano salis* (Jadacki, 1997). Carnap (1934) refers to the relation as one of syntactical equality. (See the translation in 2001, p. 15).
[2] The differentiation between *sign-token* and *sign-type* was introduced into semiotics by Peirce (1906, pp. 506, 512; 1931–35, CP 4.537; 1998, pp. 125, 480, 488).

Nevertheless, when it comes to constructing a formal-logical theory of linguistic syntax (Section 2.1), we are obliged to determine whether the primary linguistic beings are sign-*concreta* and the secondary ones sign-*abstracta*, or the other way round. In this way, theoretical questions pertaining to the logic of language intertwine with problems of a philosophical nature, especially ontological ones. This concerns not only the logical syntax of language, but also its logical semantics and pragmatics, as well as the problem of the linguistic adequacy of knowledge itself as it relates to reality.

Any notions introduced and rendered precise within parts of a logical theory of language must have their existence secured by means of the relevant axioms and definitions. However, should logic settle anything as regards the existence of the extra-linguistic entities that linguistic expressions relate to? We shall attempt to answer this question in due course (Section 2.2), before discussing what the categories of linguistic expressions and of their extra-linguistic counterparts are.

Categories of Linguistic Expression and Ontic Categories of Object

The general idea of a category as a predicative subset of a given set, as articulated by Jadacki (Jadacki, 2003, pp. 109ff), permits one to speak of both linguistic categories (picked out from within the set of all expressions) and ontic ones (picked out from within the set of all objects).

Linguistic expressions performing a determinate syntactic function and constructed according to the rules of linguistic syntax (i.e. well-formed expressions) will occupy a place in the appropriate syntactic categories. Broadly speaking, expressions playing the same role in the construction of complex expressions will likewise belong to the same categories. At the same time, when we abandon the purely syntactic point of view on expressions and take into account their semantic counterparts (components of knowledge) or what count in turn as counterparts of the latter (i.e. beings which the expressions relate to or denote), then these expressions will also be included in the appropriate semantic categories (which will be intensional or extensional, respectively). The compatibility of the appropriate syntactic and semantic categories of linguistic expression is an indispensable condition of *linguistic adequacy of knowledge relative to cognized reality* (Wybraniec-Skardowska, 2015, sec. 4). This compatibility entails the compliance of linguistic categories with the appropriate ontological categories, where these latter include the extra-linguistic counterparts of linguistic expressions. This will, of course, be relative to whatever ontology is embraced. We thus embark on our investigation on the basis of a structure of prior ontological commitment, taking into account only the substantive counterparts of expressions of linguistic categories, meaning extra-linguistic objects – i.e. beings.³ Such objects will be placed in appropriate ontic categories, the typology for the latter having been established by the ontology adopted. Depending on the ontological conception involved, one or several ontic categories will be distinguished. Hence, the following may serve as ontic categories: individuals, sets of individuals, properties, relations (in particular, single-argument or multi-argument operation-functions), periods, areas and states of affairs. Distinguishing these or other ontic categories is obviously connected with the issue of which beings we attribute existence to, as being real or not (e.g., as intentional or ideal).⁴

Beyond this, when it comes to presenting certain semantic foundations for the formal-logical theory of language (Section 2.2.), we shall accept the postulate of the "democratic nature of beings": all beings are equally empowered, being treated in the same way when it comes to existence and deciding something about them.

Outline of a Formal-Logical Theory of Language

The formal-logical considerations we are seeking to address pertain to both syntax and a bi-aspectual, intensional and extensional semantics of language characterized categorially in the spirit of the theory of syntactic categories proposed by Leśniewski and Ajdukiewicz (see Leśniewski (1929, 1930) and Ajdukiewicz (1935, 1960)) while remaining in compliance with the ontological canons of Frege (1892, pp. 31–36, 36–38; 1997, pp. 155–159, 159–171), the motto of Bocheński (1949) to the effect that *syntax mirrors ontology*, and some ideas of Suszko (1958, 1960, 1964, 1968) asserting that language is supposed to function as a linguistic schema for ontological reality and, at the same time, as a tool for its cognition. The theory of language outlined here is intended to offer a framework for understanding the development (and some of the explication) of the ideas put forward in the works of

---

3 Following Jadacki (1992), we can accept that everything that has at least one property vested in it counts as an object (an entity). (The relation of vesting here will be a binary relation, whose domain will be a set of properties and counter-domain just a set of objects; the relation of vesting is therefore probably a primitive notion of ontology.) Following Łukasiewicz, we can also assume that everything that can both have and not have a certain property, where this is non-contradictory, counts as an object.
4 See Ajdukiewicz (1949–1950), Bocheński (1974), Augustynek & Jadacki (1993).

the above-mentioned authors.[5] In Section 2.1 we outline our version of the logical theory of linguistic syntax, and in Section 2.2 its extension to the theory of linguistic semantics and pragmatics.

On the Logical Theory of Linguistic Syntax

Each and every language can be more or less adequately captured by a determinate grammar. In the Polish tradition it is categorial grammar that serves this purpose. The latter originated from the work of Kazimierz Ajdukiewicz (1935, 1960), and grew under the influence of Husserl's idea of pure grammar (Husserl, 1900, 1901), as well as Leśniewski's theory of semantic/syntactic categories (Leśniewski, 1929, 1930).

The logical syntax of any language L will be characterized formally on two dual levels, one of them concerning the language of expression-*tokens*, the other that of expression-*types*. If one were to accept the view that expression-*concreta*, i.e. expression-*tokens* (physical objects), make up the fundamental layer of language, whereas the secondary layer of L is made up of expression-*abstracta*, i.e. expression-*types* (ideal objects), then one would be adhering to a concretistic philosophical view about the nature of linguistic entities (as was endorsed by, among others, Leśniewski). On the other hand, in supporting the view that expression-*types* are the basic layer, while expression-*concreta* are secondary, we ourselves shall be adopting the opposite standpoint, which is a platonizing one.

2.1.1. In the first case, on the *level of tokens*, language L is generated in the most general way by grammar:

$$G = \langle U_L, \sim, c, V, \varrho, E; S \rangle, \text{ where}$$

U_L is a non-empty universum containing all sign-*tokens* of L,
$\sim$ is a two-argument relation of identifiability of signs of universum U_L,
V is a vocabulary of *word-tokens* of language L,
c is a three-argument relation of concatenation defined in U_L,
ϱ is an n-argument relation of forming complex expression-*tokens* ($n > 1$),
E is the smallest set of all expression-*tokens* containing V and closed under the relation ϱ,

---

[5] The theory will be constructed on the basis of classical logic and set theory. Its outline is based on ideas presented in my previous contributions (Wybraniec-Skardowska, 1991, 2006, 2007, 2009, 2015). *Tokens* of linguistic expressions will be represented by the variables $e, e', e_1, e_2, \ldots$, *types* of such expressions by the variables $t, t', t_1, t_2, \ldots$.

S   the set of all well-formed expression-*tokens* of language L.

The notions U_L, $\sim$, V, c, ϱ, are primitive notions of the theory, characterized axiomatically. When G is a classical categorial grammar, each expression-*token* e of set S has a categorial index $i(e)$ of some non-empty set I assigned in an unambiguous way, and each complex expression of set S is constructed on a functor-argument basis, so that it is possible to distinguish within it a constituent, the so-called *functor*, which, together with the remaining constituents of that expression, called *arguments of the functor*, forms this expression. The notion of a constituent of a complex expression is defined inductively. Categorial indexes serve *inter alia* to establish the syntactical role of expressions and examine their *syntactic connectivity*. Set S is formally defined as the smallest set of expressions, containing vocabulary V and closed with respect to relations linked to Ajdukiewicz's *principle of syntactic connection*. All the sets and relations of system G are non-empty sets – hence the resulting primary existence of expression-*tokens* in particular.

On the second level, the *level of types* of expression, language L is characterized through a system of notions which is the dual of system G:

$$\underline{G} = \langle \underline{U_L}, =, \underline{c}, \underline{V}, \underline{\varrho}, \underline{E}; \underline{S} \rangle, \text{ where}$$

$\underline{U_L}$ is the set of all linguistic sign-*types* in language L,
$=$ is a relation of common identity of signs of universum $\underline{U_L}$,
$\underline{V}$ is a vocabulary of word-*types* of language L,
$\underline{c}$ is a relation of concatenation defined on types of sign of $\underline{U_L}$,
$\underline{\varrho}$ is a relation of forming complex expression-*types*,
$\underline{E}$ is the set of all expression-*types* of language L,
$\underline{S}$ is the set of all well-formed expression-*types* of language L.

All the notions of grammar $\underline{G}$ are derivative constructs, defined with reference to the dual notions of grammar G. Any set $\underline{Z}$ of types of system $\underline{G}$ is a quotient set of set Z of *tokens* of the first level, due to the relation of identifiability $\sim$, i.e.,

$$\underline{Z} = Z/\sim.$$

Thus, any set $\underline{Z}$ of *types* of expressions is composed of equivalence classes of *tokens* of set Z, i.e.,

$$t \in \underline{Z} \Leftrightarrow \exists e \in Z(t = [e]_\sim = \{e' \in Z \mid e' \sim e\}).$$

The relation of concatenation $\underline{c}$ on *types* of sign is defined by means of the relation of concatenation c on *tokens* of signs of language L:

$$\underline{c}(t_1, t_2; t) \Leftrightarrow \exists e_1, e_2, e \in U_J(t_1 = [e_1]_\sim, t_2 = [e_2]_\sim, t = [c(e_1, e_2; e)]_\sim).$$

The concatenation relation $\underline{c}$ is a two-argument function on types of sign in language L.

It is proved that each dual counterpart of the thesis of the theory of syntax initially constructed on the *level of concreta* is a thesis of this theory developed on the *level of types* – on the second level of formalization of the syntax of language L.

The concretistic approach to the formal-logical theory of the syntax of language L has been set out in previous works by the present author (Wybraniec-Skardowska, 1991, 2006).

2.1.2. The opposite standpoint – the *platonizing* one – is founded on the assumption that types of signs in language L are ideal signs, conceived as independent and objective beings, and are primary in relation to the linguistic *tokens* that are their representatives. The primitive notions of the syntactic theory are thus the following notions of system $\underline{G}$: $\underline{U}_L$, $\underline{c}$, $\underline{V}$. The other notions of this system are defined subsequently. Obviously, the axiom stating the existence of sign-*types*, assuming that any type t is a non-empty set, is then accepted.

On the other level of formalization, the *level of tokens*, we find *tokens* of signs of language L that are introduced through axioms and definitions:
(1) $e_1 \in t_1 \wedge e_1 \in t_2 \Rightarrow t_1 = t_2$,
(2) $e \in U_L \Leftrightarrow \exists t \in \underline{U}_L (e \in t)$.

The above definition (2) can be considered under the general schema of the definition of subsets Z of set U_L:
(DZ) $e \in Z \Leftrightarrow \exists t \in \underline{Z}(e \in t)$.

The relation of *identifiability* is defined as follows:
(D$\sim$) $e \sim e' \Leftrightarrow \exists t \in \underline{U}_L(e, e' \in t)$.

The relation of concatenation on *tokens* of signs is determined by the following definition:
(Dc) $c(e_1, e_2; e) \Leftrightarrow \exists t_1, t_2, t \in \underline{U}_L(e_1 \in t_1, e_2 \in t_2, e \in t \wedge \underline{c}(t_1, t_2; t))$.

We determine relation ϱ in a similar fashion.

It is proved that each dual counterpart of the thesis of the syntactic theory initially constructed on the *level of types* is a thesis of this theory on the second level of its formalization – i.e. on the *level of tokens*.

2.1.3. The two dual approaches to the two-level syntactic theory of language given in Subsections 2.1.1 and 2.1.2 are logically equivalent (see Wybraniec-Skardowska, 1988). Within the scope of the linguistic syntax, the two conceptions deriving from different existential assumptions are equivalent. This statement is of philosophical significance, as it proves that *in the context of theoretical syntactic considerations pertaining to language, the assumption of the existence of abstract linguistic beings can be passed over.*[6]

The Foundations of the Formal-Logical Theory of the Semantics and Pragmatics of Language

2.2.1. The logical theory of syntax allows us to determine sets S and $\underline{S}$ of all well-formed expressions of language L. Being in line with the logical conception, its characterization requires an unambiguous assignment of meanings to its expressions.

Only an efficient, precise and clear language can become a tool for describing the world, enabling us to properly transmit information and communicate about reality. The expressiveness of language consists specifically in the unambiguous character of its expressions, both as regards their structure and their meaning (*intension*) and denotation (*extension*). The absence of syntactic and semantic ambiguity where linguistic expressions are concerned is a condition of its logical meaningfulness. It entails the categorial compatibility of language, which is different from that mentioned earlier: i.e. the compatibility of syntactic categories of linguistic expression with semantic ones, be they semantic (*intensional*) or denotational (*extensional*). This compatibility, in turn, entails the syntactic and semantic structural compatibility of language, described in the form of three principles of compositionality for complex language expressions, mutually corresponding to one another. Of these, one is syntactic and two semantic, with the latter pair consisting of the compositionality of meaning and the compositionality of denotation.

However, since absence of ambiguity is such an important aspect of linguistic adequacy, we first need to establish what the *meaning* of the composed expressions of language L is, and what this unambiguous character of its expressions consists in.

There exist quite a number of conceptions relating to the nature of *meaning*, as well as different theories concerning this notion, in the literature dealing with the philosophy of language. So far, however, none of these has gained widespread acceptance. Moreover, none of them can be said to constitute a general theoretical conception. I myself have offered a sketch of such a conception in previous work

---

6 The proof of this theorem (see Wybraniec-Skardowska, 1988, 1989) is, however, based on standard Platonic set theory. The applied formalism is not thus in fact ontologically neutral. This remark was formulated by Jerzy Perzanowski.

(Wybraniec-Skardowska, 2007) and will be making reference to that in this part of the present article.

2.2.2. Since the time of Frege, the notion of 'meaning' has been differentiated from that of 'denotation'. Frege (1892, p. 31; 1997, p. 156) distinguished, respectively, *Sinn* (English: *intension*) and *Bedeutung* (English: *extension*), while we owe the *intension-extension* distinction to Carnap (1947, Ch.I, sec. 5, sec. 6, pp. 26, 27; sec. 9, pp. 40-41). On the other hand, the literature devoted to linguistics and semiotics does not always differentiate between these two notions.

The notions of 'meaning' and 'denotation' are used with reference to expression-*types* of language L. They are "assignments" of meanings and denotations to these expressions, respectively. As such, they are operations (functions) on expressions of set $\underline{S}$, yet not on all expressions of the set – rather just their non-empty sub-types, meaning elements of the set. Thus:

$$\underline{S}^* = \{t' \subseteq t \mid t' \neq \emptyset \land t \in \underline{S}\},$$

with the "assignments" construed as functions on any non-empty sets of identifiable expression-*tokens* of set S.

We define these operations, making use of some ideas connected with the understanding of the notion of 'meaning' put forward by Ajdukiewicz (1931, 1934) and Wittgenstein (1953, third edition 1967, paragraphs: 20, 349, 421-2, 508, p. 184, 190), as the *manner of usage* of an expression. In order to be able to determine what being the use of an expression consists in, we must invoke certain semantico-pragmatic notions.

2.2.3. We shall therefore enrich the theory of syntax of language L with some new primitive notions: the set *User* of all users of language L, the set *Ont* of all extra-linguistic objects which expressions of language L relate to, and the two-argument operation *use* of using expression-*tokens* of language L.

The sets *User* and *Ont* will be conceived in a very broad way. *User* may consist not only of current users of language L, but also past and future ones. Meanwhile, nothing will be assumed as regards the ontological nature or existence of objects of *Ont*: they may be *concreta*, *abstracta*, ideal, intentional (quasi-objects), fictional objects, etc. It will be merely axiomatically assumed about these objects that they are non-empty sets. Nothing will be assumed about the ontic categorization of *Ont*. The ontic categories can – but need not – consist of the following: the category of individuals satisfying certain properties, categories of various relations and functions, category of states of affairs, and the like.

The relation *use* of usage of expression-*tokens* will also be conceived in the broadest terms: e.g., as an operation of invoking, exposing and forming expression-*tokens* to indicate appropriate objects of the set *Ont*. The operation *use* will also be said to be a *function of the objective references of expression-tokens* made by users of language *L*. This function can also be conceived as the set of all physical activities of users of *L* that were, are or will continue to be activities used in determinate situations with the aim of referring concrete *tokens* of language *L* to objects of *Ont*. It will be axiomatically assumed that the function *use* is a set-theoretical function, partially mapping the Cartesian product *User* × *S* onto the set *Ont*, whose primary domain is the whole set *User*, while its secondary one will be a proper subset of the set of expression-*tokens S*.

We shall read the expression $use(u, e) = p$, where $u \in User$, $e \in S$ and $p \in Ont$, as follows: user *u* uses expression-*token e* with reference to object *p*. When $use(u, e) = p$ takes place, then object *p* is to be called the *object reference of token e* indicated by user *u* of language *L*. We say about expression *e* that it *has an object reference* when used by a user with reference to some object. Two expression-*tokens* have – at the same time – the same manner of usage *use*, when they have the same object reference.

The relation *use* of using expression-*types* is determined by means of the operation *use* of using *tokens of expressions*. It is axiomatically assumed about it that it is a non-empty relation defined in terms of the Cartesian product $User \times S^*$ and by the following formula:

D0. $u \underline{use} \, t \Leftrightarrow \exists e \in t \, \exists p \in Ont \, (use(u, e) = p)$.

It follows from the already accepted assumptions or definitions that each user of language *L* uses at least one expression-*token* with reference to any object, and hence *uses* at least one expression-*type* of set $\underline{S}^*$.

Defining the *meaning* of an expression-*type* as a common way of *using* types of expression requires that we introduce the notion of a *relation* ≈, *the same manner of usage of these expressions*. In the definition of this relation, however, it is necessary to employ the notion *use* of using expression-*tokens*.

2.2.4. The formal definition of relation ≈ is introduced in the following way:

D1. $t \approx t' \Leftrightarrow \forall u \in User[(u \underline{use} \, t \Leftrightarrow u \underline{use} \, t') \land \forall p \in Ont(\exists e \in t(use(u, e) = p) \Leftrightarrow \exists e' \in t'(use(u, e') = p))]$.

In accordance with definition D1, two expression-*types* will have the same manner of usage *use* if and only if each user of language *L* uses, in the sense of *use*, one of them if and only if he or she uses the other of them, and uses, in the sense of *use*, a *token* of one of them with reference to any object if and only if he or she uses a *token* of the other of them with reference to the same object.

For instance, the word "rain" and the expression "an atmospheric fall in the form of drops of water falling down from a cloud" have the same manner of usage _use_. Similarly, the expression "a public concert" and the expression "a public performance of pieces of music" have the same manner of usage _use_.

It can easily be determined that if two expression-*types* have the same manner of usage ≈, then there exist *tokens* of one and of the other of them, respectively, which have the same manner of usage in the sense of *use*.

2.2.5. The relation ≈ of having the same manner of usage of *types* of expression is an equivalence relation in set $\underline{S}^*$ of expression-*types*. Operation *m of assigning a meaning* to these expressions can thus be defined as the function:

D2. $m: \underline{S}^* \to 2^{\underline{S}^*}$, where $m(t) = [t]_\approx$ for any $t \in \underline{S}^*$.

Thus, the *meaning m(t)* of expression-*type t* is the equivalence class of relation ≈ of possessing the same manner of usage of *types* determined by *type t*. Intuitively, this may be conceived as the *common property of all expression-types having the same manner of usage as t*. It is this property which we shall call the *manner of usage of expression-type t*.

Meaning *m(t)* of expression-*type t* is thus an abstract being (a non-empty set), whose existence is guaranteed by set theory.

2.2.6. In Ajdukiewicz's logical concept of language, each of its expressions is to have an unambiguously assigned meaning. Type *t* may, however, include subtypes, the meaning of which differs from the global meaning *m(t)* as established by definition D2. For instance, the subtype "key[1]" of the expression-*type* "key", composed only of the identifiable tokens of the expression-*type* "key" whose object references are musical clefs, has a meaning that differs from the global meaning of the word "key", which does not have an unambiguously assigned meaning.

D3. Expression-*type t has a meaning assigned unambiguously* ⇔ no proper subtype of expression *t* has a meaning that differs from the meaning of expression *t*; i.e., symbolically:

$$\neg \exists t' \subseteq t(t' \neq t \wedge m(t') \neq m(t)), \text{ i.e. } \forall t' \subseteq t(m(t') = m(t)).$$

2.2.7. An expression-*type* possessing an unambiguously assigned meaning in language *L* should be an *unambiguous expression* of this language. A formal definition of an unambiguous expression can be introduced by means of the notion of *denotation*, which makes reference to the idea of the designation of objects of set *Ont* by *types* of expression of language *L*.

D4. *t designates p* ⇔ $\exists u \in User \exists e \in t(use(u, e) = p)$, where $p \in Ont$.

Thus, expression-*type* $t \in \underline{S}^*$ designates an object *p* iff at least one user of language *L* uses some *token* of expression *t* with reference to the object *p*.

By way of example, the word "laptop" designates each and every laptop, and the expression "intention" each and every intention.

Objects designated by an expression-*type* are called *denotata* of this expression. When the denotata of some such expression are *object-concreta* (things, persons, etc.),[7] we shall call them *designata* of this expression.

We call *denotation* $d(t)$ of expression-*type* t the set of all its denotata. Formally, $d(t)$ is a value of *denotation* function d defined in the following way:

D5. $d: \underline{S}^* \to 2^{Ont}$ and $d(t) = \{p \in Ont \mid t \text{ designates } p\}$, for any $t \in \underline{S}^*$.

It follows from the already accepted assumptions or definitions that each expression-*type* which is used by someone in the sense <u>use</u> has (denotes) a non-empty denotation (a set of denotata), and therefore designates an object of set *Ont*. If, then, we speak about so-called *empty names* as having an empty denotation, we mean just that the set of designata (*concreta*) is an empty set. Such names are thus not used by users in the sense <u>use</u>, as their *tokens* do not make reference to any object (material, physical); the set *Ont* will consist for them exclusively of real *concreta*.

It should also be noted that not every well-formed expression-*type* has a non-empty denotation. For instance, the expression "the ceiling writes hot ice" is a syntactically correct one, but as a piece of semantic nonsense is not used and has no denotatum. Moreover, let us also take note of the fact that subtypes of a given expression-*type* can have a different denotation: one which is 'smaller' than the expression itself does.

Let us state here two theorems resulting from definitions D5 and D4, as well as from the theorems of algebra of sets:

T1. If t' is a subtype of expression-*type* t (i.e., $t' \subseteq t$), then $d(t') \subseteq d(t)$.
T2. If t_1, t_2 are subtypes of expression-*type* t and $t = t_1 \cup t_2$, then $d(t) = d(t_1) \cup d(t_2)$.[8]

It can also be proved that the denotation of the sum of a finite number of subtypes forming a given type will be the sum of the denotations of these subtypes.

2.2.8. The relations between meaning, absence of ambiguity, and denotation are given in the theorems below.

The basic relation between meaning and denotation is described by the following theorem (cf. Wybraniec-Skardowska, 2007, pp. 127–128):

---

[7] Here we should emphasize that we are thinking of actually existing *concreta*, in that one can also speak about non-existent *concreta* (e.g., *thought-based* ones; see Jadacki, 1992).
[8] The proof of this theorem is given in the *Appendix* at the end of this article.

T3. $m(t) = m(t') \Rightarrow d(t) = d(t')$, for any $t, t' \in \underline{S}^*$.

According to T3, two expression-*types* have the same denotation when they have the same meaning; therefore, if the denotations of these expressions are different, their meanings will also be so.

The theorem that is the converse of T3 does not hold true as, e.g., the expressions "an equilateral triangle" and "an equiangular triangle" have the same denotation, yet different meanings.

The notion of an expression being unambiguous (i.e. having just one meaning) is introduced by means of the following definition:

D6a. t is *unambiguous* $\Leftrightarrow \neg \exists t' \subseteq t(d(t \setminus t') \neq \emptyset \wedge d(t') \cap d(t \setminus t') = \emptyset)$,
i.e. $\forall t' \subseteq t(d(t \setminus t') = \emptyset \vee d(t') \cap d(t \setminus t') \neq \emptyset))$.

D6b. t is *ambiguous* $\Leftrightarrow t$ is not unambiguous.

Thus, an expression-*type* t is *unambiguous* iff there does not exist any such subtype of t as would have some *denotatum* in common with a non-empty denotation of the difference between the expression t and this subtype;[9] when such a subtype does exist, expression t is *ambiguous*.

By way of example, the expression "a key" is ambiguous, as its subtype "a key^2", designating only keys to open doors, does not have *denotata* in common with the denotation of the expression "a key" \ "a key^2", designating all other keys (e.g., clefs in music, keys for decoding encrypted texts, or controls on mechanical devices).

We shall now state several theorems that serve to characterize unambiguous (and therefore also ambiguous) expressions with the help of the notions and theorems introduced earlier.

T4. t is *unambiguous* $\Leftrightarrow \neg \exists t' \subseteq t[(d(t \setminus t') \neq \emptyset \wedge d(t \setminus t') = d(t) \setminus d(t')]$.[10]

A direct conclusion following from Theorem T4 is

T5. $\forall t' \subseteq t(d(t') = d(t)) \Rightarrow t$ is *unambiguous*.

The implication that is the converse of T5 is not obviously true. For example, if t' is a *singleton* and has the following inscription as its only token,

laptop

---

[9] Let us remind ourselves that expression-*types* are sets of *tokens*; hence the difference between two expressions here will be that between two sets.

[10] The proof of this theorem is given in the *Appendix*.

whose object reference is my own laptop, while the expression *t* is a set of all *inscription-tokens* identifiable with this inscription (and acknowledged to be an unambiguous expression, according to D6a), then the denotation of subtype *t'* of expression *t* is not equal to the denotation of expression *t*.

T6. *t* has an unambiguously assigned meaning ⇒ *t* is unambiguous.

Proof T6 follows directly from D3, T3 and T5.

Thus, the possession of unambiguously assigned meanings by expression-*types* of language *L* is a sufficient condition of their unambiguity.

Obviously, it also follows from T6 that ambiguous expressions do not have unambiguously assigned meanings, and that language as it figures in the logical conception should be free of ambiguous expressions.

The condition for the non-ambiguity of expression *t* is not, however, a sufficient one for *t* to have an unambiguously assigned meaning, as when, for example, *t* = "a book" is an unambiguous expression, in compliance with D6a, then there will exist some expression-*type* *t'* = "a book^1" which is a set of tokens identifiable with words, whose object reference will be books by Jacek Jadacki, such that $t' \subset t$ and $d(t') \neq d(t)$ (because $d(t') \subset d(t)$), whence on the basis of theorem T3 we have $m(t') \neq m(t)$, and *t* does not have an unambiguously assigned meaning, as it does not satisfy definition D3.

On the Ontological Neutrality of Logic

In this part I offer a summary, recapitulating the basic assumptions and results presented in Section 2 with the aim of showing the extent to which the principal objectives pursued have been realized.

3.1. Subsections 2.2.1 and 2.2.2 of this work have sketched and discussed formal-logical theories of language, constructed in accordance with the logical conception of language. These theories are based on classical logic, together with set theory. The theoretical considerations addressed have been rather general, but also quite far-reaching. They do not depend on any particular symbolism or notations of expressions, or concrete grammatical rules, of the language being described.

3.2. In discussing in Subsection 2.2.1 the theory of linguistic syntax, we pointed to the possibility of building it on two different levels, one of which stems from *concreta*, i.e. linguistic *tokens* of signs, the other from their classes, i.e. *types* of linguistic sign, conceived as abstract beings.

- The outcome of our theoretical considerations has been a statement of complete analogousness obtaining between the syntactic notions of the two levels.
- Thus, logic does not settle here which view pertaining to the nature of linguistic objects – the concretistic one or the idealistic, platonizing one – is correct.
- Since, however, the two dual-aspected theoretical approaches to linguistic syntax are equivalent, in formalizing language initially on the level of concreta we are not impoverishing the resources offered in the form of theorems for the linguistic syntax being described, and we do without postulating the existence of ideal beings of the sort that types of language expression are.
- Hence, a philosophical thesis is entailed, concerning the possibility of eliminating assumptions regarding the existence of ideal beings in the context of considerations pertaining to syntax, as long as these beings are treated as classes of identifiable sign-*tokens* (linguistic *concreta*).[11]

3.3. By sketching, in Subsection 2.2.2, the semantic-pragmatic theory of language, we showed that:
- a meaning can be assigned to its well-formed expression-*types* (through function *m*),
- these expressions have a meaning (D2),
- a meaning (D3) can be unambiguously assigned to them,
- while being used, they designate some objects (D4),
- they denote (have a denotation), since
- a denotation can be assigned to them (through function *d*),
- designated objects belong to the set *Ont*.

As regards the set *Ont* of *extra-linguistic objects* (beings) designated by expression-*types*, we have simply assumed that it is a non-empty set, inseparably bound up with the structure of its beings and their ontological categorization. A full characterization of language, considered from an ontological point of view and in terms that comply with Ajdukiewicz's logical conception, can be furnished using this formal-logical theory.

For the purposes of our description we have not made use of any other existential assumptions, apart from those imposed by set algebra: neither when it came to the existence of linguistic expressions, nor for their extra-linguistic counterparts, was this the case. In this regard we may assert that the logic applied here (using set theory) has been ontologically neutral.

---

[11] The formalism leading to this statement, however, is based on Platonist set theory, and so is not really ontologically neutral (see note 7).

Appendix

We give proofs here of theorems T2 and T4, using the (assumptive) method of natural deduction put forward in the work of J. Słupecki and L. Borkowski (1967).

T2. $t = t_1 \cup t_2 \wedge t_1 \subseteq t \wedge t_2 \subseteq t \Rightarrow d(t) = d(t_1) \cup d(t_2)$.

Proof.
1. $t = t_1 \cup t_2$ {assum.}
2. $t_1 \subseteq t \wedge t_2 \subseteq t$ {assum.}
3. $d(t_1) \subseteq d(t) \wedge d(t_2) \subseteq d(t)$ {2, T1}
4. $d(t_1) \cup d(t_2) \subseteq d(t)$ {3}
1.1. $p \in Ont \wedge p \in d(t)$ (additional assumption)
1.2. $p \in d(t_1 \cup t_2)$ {1, 1.1}
1.3. $\exists u \in User \, \exists e \in t_1 \cup t_2 (use(u, e) = p)$ {1.2, D5, D4}
1.4. $u_1 \in User \wedge (e_1 \in t_1 \vee e_1 \in t_2) \wedge use(u1, e1) = p$ {1.3}
1.5. $e_1 \in t_1 \Rightarrow \exists e \in t_1 \exists u \in User(use(u, e) = p) \Rightarrow$
$p \in d(t_1) \Rightarrow p \in d(t_1) \cup d(t_2)$ {1.4, D5, D4}
1.6 $e_1 \in t_2 \Rightarrow \exists e \in t_2 \exists u \in User(use(u, e) = p) \Rightarrow$
$p \in d(t_2) \Rightarrow p \in d(t_1) \cup d(t_2)$ {1.4, D5, D4}
1.7. $e_1 \in t_1 \vee e_1 \in t_2 \Rightarrow p \in d(t_1) \cup d(t_2)$ {1.5, 1.6}
1.8. $p \in d(t_1) \cup d(t_2)$ {1.4, 1.7}
5. $p \in Ont \wedge p \in d(t) \Rightarrow p \in d(t_1) \cup d(t_2)$ {1.1 → 1.8}
6. $\forall p \in Ont(p \in d(t) \Rightarrow p \in d(t_1) \cup d(t_2))$ {5}
7. $d(t) \subseteq d(t_1) \cup d(t_2)$ {6}
$d(t) = d(t_1) \cup d(t_2)$ {4, 7}
□

T4. t is *unambiguous* $\Leftrightarrow \neg \exists t' \subseteq t[(d(t \setminus t') \neq \emptyset \wedge d(t \setminus t') = d(t) \setminus d(t')]$.

Proof. Proof by contradiction (⇒).
1. t is *unambiguous* {assum.}
2. $t_1 \subseteq t \wedge d(t \setminus t_1) \neq \emptyset \wedge d(t \setminus t_1) = d(t) \setminus d(t_1)$ {indirect assump.}
3. $d(t_1) \cap (d(t) \setminus d(t_1)) = \emptyset$ {set algebra}
4. $t_1 \subseteq t \wedge d(t \setminus t_1) \neq \emptyset \wedge d(t_1) \cap d(t \setminus t_1) = \emptyset$ {2, 3}
5. $\exists t' \subseteq t[d(t \setminus t') \neq \emptyset \wedge d(t') \cap d(t \setminus t') = \emptyset]$ {4}
6. t is *not unambiguous* {D6, 5}
contradiction {1, 6}

Proof by contradiction (⇐).
In the proof, we use the following theorem of set algebra:
T(*). If $A = A' \cup B \wedge A' \cap B = \emptyset \wedge A = A' \cup C \wedge A' \cap C = \emptyset$, then $B = C$.

1. $\neg \exists t' \subseteq t[(d(t \setminus t') \neq \emptyset \land d(t \setminus t') = d(t) \setminus d(t')]$ {assum.}
2. t is *not unambiguous* {indirect assump.}
3. $t_1 \subseteq t \land d(t \setminus t_1) \neq \emptyset \land d(t_1) \cap d(t \setminus t_1) = \emptyset$ {D6a, 2}
4. $d(t_1) \subseteq d(t) \land t = t_1 \cup (t \setminus t_1)$ {3, T1}
5. $d(t) = d(t_1) \cup d(t \setminus t_1) \land d(t_1) \cap d(t \setminus t_1) = \emptyset$ {4, T2, 3}
6. $d(t) = d(t_1) \cup (d(t) \setminus d(t_1)) \land d(t_1) \cap (d(t) \setminus d(t_1)) = \emptyset$ {4}
7. $d(t \setminus t_1) = d(t) \setminus d(t_1)$ {5, 6, T(*)}
8. $\exists t' \subseteq t[(d(t \setminus t') \neq \emptyset \land d(t \setminus t') = d(t) \setminus d(t')]$ {3, 7}
   contradiction {1, 8}

□

Acknowledgment: The author expresses her sincere gratitude to her colleagues Gabriela Besler, Alex Citkin and Zbigniew Bonikowski for an access to source publications and data needed to successfully complete this work.

The author would also like to thank Bartłomiej Skowron – the editor of this volume – and the Referees who offered a number of suggestions which greatly enhanced this paper.

Bibliography

Ajdukiewicz, K. (1931). O znaczeniu wyrażeń [On the meaning of expressions]. In *The commemorative book of Polish Philosophical Society in Lvov* (pp. 31–77).
Ajdukiewicz, K. (1934). Sprache und Sinn. *Erkenntnis, 4*, 100–138.
Ajdukiewicz, K. (1935). Die syntaktische Konnexität. *Studia Philosophica, 1,* 1–27. English translation: *Syntactic connection* in S. McCall (Ed.), *Polish logic 1920–1939*, pp. 202–231, 1967. Oxford: Clarendon Press.
Ajdukiewicz, K. (1949–1950). On the notion of existence. *Studia Philosophica*, vol. IV, 7–22.
Ajdukiewicz, K. (1960). Związki składniowe między członami zdań oznajmujących [Syntactical relations between constituents of declarative sentences]. *Studia Filozoficzne, 6*(21), 73–86.
Ajdukiewicz, K. (1975). *Pragmatic Logic*. Synthese Library, vol 62, Dordrecht-Boston-Warsaw: Reidel-PWN.
Augustynek, Z. & Jadacki, J. J. (1993). *Possible ontologies*. Amsterdam-Atlanta: Rodopi.
Bocheński, J. M. (1949). On the syntactical categories. *New Scholasticism, 23,* 257–280.
Bocheński, J. M. (1974). Logic and ontology. *Philosophy East and West, 24*(3), 275–292.
Carnap, R. (1934). *Logische Syntax der Sprache*. (Schr.z. wiss. Weltauff), Wien: Springer. English translation: *The logical syntax of language*, 1937, London and New York: Horcout and Kegan; Reprinted in 2000 and 2001, London: Routledge.
Carnap, R. (1942). *Introduction to semantics*. Cambridge, MA: Harvard University Press.
Carnap, R. (1947). *Meaning and necessity*. Chicago: University of Chicago Press.
Frege, G.(1892). Über Sinn und Bedeutung. *Zeitschrift für Philosophie und pilosophishe Kritik, 100,* 25–50. English translation: *On Sinn and Bedeutung*, in H. Feigel, & W. Sellars (Eds.), *Readings in philosophical analysis*, 1949, New York: Appleton-Century-Crofts, and also in M. Beaney (Ed.), *The Frege reader*, pp. 151–171, 1997, Oxford: Blackwell.
Husserl, E. (1900). *Logische Untersuchungen*, vol. I. Halle.

Husserl, E. (1901). *Logische Untersuchungen*, vol. II. Halle.
Jadacki, J. J. (1992). Change, action, and causality. *Dialogue and Humanism*, vol. II No. 3, pp. 87-99.
Jadacki, J. J. (1997). Troubles with categorial interpretation of natural language. In R. Murawski & J. Pogonowski (Eds.). *Euphony and logos. Essays in honour of Maria Steffen-Batóg and Tadeusz Batóg*. Amsterdam-Atlanta: Rodopi.
Jadacki, J. J. (2003). *From the viewpoint of the Lvov-Warsaw School*. Amsterdam-Atlanta: Rodopi.
Jadacki, J. J. (2011). What semantics is and what purpose it serves. In A. Brożek, J. J. Jadacki & B. Żarnić (Eds.), *Theory of imperatives from different points of view*. Warszawa: Wydawnictwo Naukowe Semper.
Leśniewski, S. (1929). Grundzüge eines neuen Systems der Grundlagen der Mathematik. *Fundamenta Mathematicae, 14*, 1–81.
Leśniewski, S. (1930). Über die Grundlagen der Ontologie. *Comptes rendus des séances de la Société des Sciences et des Lettres de Varsovie, Classe II, 23*, 111–132.
Peirce, Ch. S. (1906). Prolegomena to an Apology for Pragmaticism. *Monist*, 16, 492-546.
Peirce, Ch. S. (1931–1935). Hartshorne C., & Weiss P. (Eds.). *Collected papers of Charles Sanders Peirce: Vols. 1–5*. Cambridge, MA: Harvard University Press.
Peirce, Ch. S. (1998). *The Essential Peirce. Selected Philosophical Writings*, vol. 2 (1893-1913), Bloomington and Indianapolis: Indiana University Press. Peirce Edition Project (N. Houser, general Editor).
Pelc, J. (1979). A functional approach to the logical semiotics of natural languages. In *Semiotics in Poland 1894–1969* (selected and edited with an introduction by Jerzy Pelc.). Synthese Library, *Studies in Epistemology, Logic and Methodology of Science*, vol. 119 (pp. 342–375). Dordrecht – Boston: PWN – Reidel.
Perzanowski, J. (2012). Towards Combination Metaphysics. In J. Sytnik-Czetwertyński (Ed.), *Art of Philosophy. A selection of Jerzy Perzanowski's Works*, (pp. 45-67). Frankfurt, Ontos Verlag. (Originally published in 2004.)
Słupecki, J., Borkowski, L. (1967). *Elements of mathematical logic and set theory*. In *International Series of Monography in Pure and Applied Mathematics*. Oxford–New York–Toronto – Warsaw: Pergamon Press – PWN.
Suszko, R. (1958). Syntactic structure and semantical reference, Part I. *Studia Logica, 8*, 213–144.
Suszko, R. (1960). Syntactic structure and semantical reference, Part II. *Studia Logica, 9*, 63–93.
Suszko, R. (1964). O kategoriach syntaktycznych i denotacjach wyrażeń w językach sformalizowanych [On syntactic categories and denotation of expressions in formalized languages]. In *Rozprawy logiczne* [Logical dissertations in memory of Kazimierz Ajdukiewicz] (pp. 193–204). Warsaw: PWN.
Suszko, R. (1968). Ontology in the *Tractatus* of L. Wittgenstein. *Notre Dame Journal of Formal Logic, 9*, 7–33.
Tarski, A. (1933). *Pojęcie prawdy w językach nauk dedukcyjnych*. Warszawa: Nakładem Towarzystwa Naukowego Warszawskiego. English translation: The concept of truth in formalized languages. In J. Corcoran (Ed.), *Logic, semantics, metamathematics: Papers from 1923 to 1938*, 1956. Oxford: Oxford University Press; second edition 1983. Indianapolis, Indiana: Hackett Publishing Company.
Wittgenstein, L. (1953). *Philosophical investigations*. Oxford: Blackwell. Third edition in 1967, reprinted 1968-1986.

Wybraniec-Skardowska, U. (1988). Logiczne podstawy ontologii składni języka [Logical foundations of ontology of language syntax]. *Studia Filozoficzne, No 6–7 (271–272)*, 263–284.

Wybraniec-Skardowska, U. (1989). On the eliminability of ideal linguistic entities. *Studia Logica, 48*(4), 587–615.

Wybraniec-Skardowska, U. (1991). *Theory of language syntax. Categorial approach.* Dordrecht–Boston–London: Kluwer Academic Publisher.

Wybraniec-Skardowska, U. (2006). On the formalization of classical categorial grammar. In J. J. Jadacki, & J. Paśniczek (Eds.), *The Lvov-Warsaw School — the new generation.* In *Poznań Studies in the Philosophy of Sciences and Humanities, vol. 89.* (pp. 269–288). Amsterdam–New York: Rodopi.

Wybraniec-Skardowska, U. (2007). Meaning and interpretation, Part I. *Studia Logica, 85*, 105–132.

Wybraniec-Skardowska, U. (2009). On metaknowledge and truth. In D. Makinson, J. Malinowski, & H. Wanshing (Eds.), *Trends in logic: Towards mathematical philosophy* (pp. 319-343). Berlin–Heidelberg: Springer.

Wybraniec-Skardowska, U. (2015). On language adequacy. *Studies in Logic, Grammar and Rhetoric, 40*(53), 257–292.

Wybraniec-Skardowska, U. (2017). Logiczna koncepcja języka wobec założeń egzystencjalnych [Logical conception of language towards existential assumptions]. In *Myśli o języku, nauce i wartościach. Series two. To Professor Jacek J. Jadacki on his seventieth birthday* [Thoughts on language, science and values] (pp. 299–313). Warsaw: Semper.

Krzysztof Śleziński
Benedict Bornstein's Ontological Elements of Reality

Abstract: Bornstein arrived at an original mathematical system of relationships and categorical-ontological structures: i.e. a general ontology, or a metaphysics in the broader meaning of that term. Metaphysics, as a theoretical and mathematical science, is concerned with all of being and/or being generally. It is, therefore, a universal and pure mathematical science – *mathesis universalis* – of the kind sought after by Plato, Descartes, Leibniz and Hoene-Wroński. In this article, Bornstein's algebraico-geometrical logic, known as "topologic", will be treated as a spatial representation of algebraic logic. The representation is effected through an application of Descartes's co-ordinates to logic, and by making use of the correspondences between duality in logic and in (projective) geometry. The spatialization of logic enables us to give it a clear structural and architectonic character – one which brings out the "order" internal to this domain. The foundations of geometrical logic as such are dealt with, and the architectonics responsible for governing its elements is highlighted. The second half of the 20th century saw work being undertaken on spatial logic that is still ongoing today, and whose precursor is undoubtedly Bornstein, making it all the more worthwhile that we pay attention to the results of his own ontological research.

Keywords: Benedict Bornstein, ontological categories, ontological structures of reality, categorical geometric logic, topologic, scientific metaphysics.

Introduction

To gain a clear view of the ontological conception of reality developed by Benedict Bornstein (1880-1948), I will just present an outline of his qualitative and structural research project. In the present paper, I follow his original and symbolic notation of algebraic and logical formulas, although I am aware of their anachronistic character. I do so in order to ensure a proper understanding of Bornstein's analyses of the relevant ontological and formal issues. This symbolic notation should be updated in the future if Bornstein's concept of research is to be further pursued. In my opinion, the project justifies undertaking these activities.

Krzysztof Śleziński, Institute of Educational Science in Cieszyn, University of Silesia in Katowice, Cieszyn, Poland.

https://doi.org/10.1515/9783110669411-008

With reference to the ontological assumptions of the formal sciences, Bornstein developed a scientific metaphysics which turned out to have the character of a general theory of being. However, before developing this theory, he focused on detailed issues – including, primarily, epistemological ones – that refer to the relationship between the logical and intuitive elements in Immanuel Kant's philosophy,[1] as well as to ontological elements in the philosophy of mathematics.

Theoretical research into the relationship of rational and sensory, logical and intuitive elements in the context of Kant's philosophy and mathematical philosophy confirmed Bornstein in his conviction that there is a basic harmony existing between the world of non-spatial thought and the world of spatial beings. In this article, I will omit any detailed description of the construction of Bornstein's philosophical research tools. However, I would like to underline the most important questions that come with the elaboration of this new formal approach, as well as the results of its implementation in respect of deepening our knowledge of the structural unity of the world of thought and of spatial objects. The paper initially focuses on his ontological solutions to the many philosophical problems in the field of the philosophy of mathematics. It then goes on to furnish ontological and structural solutions relevant to a general theory of being.

The Ontological Foundations of Mathematics and Their Role in the Overcoming of Contradictions

Bornstein was searching for a precise and objective grounding for mathematics: one which would guarantee the systematic consistency of its notions and intuitions. He pointed to the possibility of developing mathematics on the basis of ontological foundations.

The Ontological Presuppositions of the Philosophy of Geometry

Thanks to the fact that the spatiality of geometrical objects directs our attention to the spatiality of reality, Bornstein was able to bring to our notice the possibility of developing the edifice of geometry on ontological foundations. Knowing that in experience, no geometrical object or spatiality counts as a given for us, he

---

[1] Bornstein (1910), and also, Bornstein (1907, pp. 261–303).

invented the notion of geometrical spatiality. He also identified the modes and types of existence of the elements in this spatiality, such as point, line and plane. Simultaneously, he arrived at the objective assumptions underpinning geometrical axioms.

The results achieved by Bornstein remain important for our comprehension of the essence of mathematics, and allow us to shed quite new light on the effectiveness of using mathematics to describe phenomena in the real world.

As the starting point of his research, Bornstein accepted Kant's transcendental aesthetics as the *a priori* science of all of the principal rules of sensory experience. He also implemented the theory of ideas and notions worked out by Kazimierz Twardowski (1898). The author of the work entitled *Prolegomena filozoficzne do geometryi* [*Philosophical Prolegomena to Geometry*] distinguishes between the image of physical space and the notion of geometrical space. He defines geometric space as pure, unlimited space, or as pure, unlimited extension.[2] In an attempt to uphold the bonds between geometrical space and physical space, he assumed that the basic image of geometrical space must be an image whose object exists and is perceived in reality.[3] The features common to a notion such as that of geometrical space and the object of this image are ones that are founded on experiential reality. They do not refer only to the world of subjective images.[4] In Bornstein's opinion, the sought-after image of spatiality could therefore be neither imitative nor productive, but instead had to be actually perceived.

Having learnt about the inner structure of geometrical space, Bornstein identified certain contradictions whose source was the statement that length and spatiality are made up of discontinuous points. He pointed out that geometrical spatiality is something different from algebraic continuity. There is no real length in the field

---

2 It is possible to develop a notion of geometrical space on the basis of a founding image of colourful and limited physical space by removing those elements of vividness and limitations stemming from imaginary construals prompted by one's experience, in order to leave just spatiality and its features, such as three-dimensionality and continuity (Bornstein, 1912, pp. 2–14).

3 In his analysis, Bornstein implemented the theory of contents and presented objects worked out by Kazimierz Twardowski in 1894 in his work entitled *Zur Lehre vom Inhalt und Gegenstand der Vorstellungen*.

4 Bornstein's image of spatiality is neither imitative nor productive, but perceptual. If it were productive, one could imagine a fourth dimension and base the notion of geometrical space on it. Still, this notion would be subjective, not founded in the objective world. By contrast, if it were imitative, in the sense of being based on acceptance of a founding image of space in the form of the heavenly vault as a plane, we would arrive at a notion of two-dimensional geometrical space. However, this notion would not possess subjective importance. Perceived images of the heavenly vault are therefore not observed or experienced in the strict sense, as they do not reach objects in reality, and remain in the subjective field (Bornstein, 1912, pp. 6–7).

of numbers. Both numbers and their creations are non-spatial. The creations of numerical continuity have no spatial nature. Bornstein proposed his own ontological interpretation of mathematical objects, whose purpose was to furnish an undisputable foundation for geometry.

The Polish philosopher worked on the ontological interpretation of objects in geometrical space, such as point, straight line and plane. Among these, a distinguished place is occupied by the straight line, as it is the simplest object that is actually spatial. The solution to the problem of the existence of the geometrical line is to understand it as a one-dimensional, coherent, continuous, precise object, lacking any sensory qualities and corresponding to an analytical function. He presented his proposal in a paper entitled *Problemat istnienia linji gieometrycznych* [*The Problem of the Existence of Geometrical Lines*] (Bornstein, 1913).

Bornstein proved that of three *a priori* possible answers, only one is true. It should not be assumed either that a geometrical line corresponds to no continuous functions, or that such a line corresponds to every such function. The only true answer remains that geometrical lines correspond to some continuous functions, but not others. It should be assumed that geometrical curves without tangents do not exist, and as all geometrical curves have tangents, only functions with derivatives relate to them. This conclusion is in accordance with the possibility of describing the movement of bodies in space.

Hence, movement on lines without defined tangents is not possible, from which it follows in turn that not all arithmetical functions can be geometrical in nature.

As was shown by Bornstein, all uncritical attempts to transfer theorems from the field of pure notions to the spatial field without taking into consideration the differences between them must be ruled out. Failure to do so may lead to errors or even contradictions in analyses.

The Existence of Actual Infinity and the Real Problem of Movement

In Bornstein's opinion, infinite systems can only be given to us with the help of something possessing the form of a whole that encompasses within itself an infinite multitude of elements. Actual infinity has never been given to us as an infinity of specific elements. An actually infinite system can have an uncountable but always limited number of updated elements. Stating that "a straight geometrical segment is a multitude of infinite points" is not contradictory in itself, since the straight geometrical segment exists, but at the same time it is not a multitude of actually existing points.

Bornstein rejected the notion of actual infinity in mathematics, and accepted the existence of potential infinity and actual finiteness. Concurrently, he defined the geometrical conditions of movement according to which the ancient paradox of the Eleatics can be solved. Analysing the phenomenon of movement, he defined it as follows:

> The movement of an object from location A to location B following the path AB on which actual indirect locations exist, is only possible if the multitude of these locations is finite. (Bornstein, 1913, p. 10)

In his analysis of movement, the philosopher assumes the actual existence of indirect locations on the path of movement between the two locations of the object. These locations constitute a multitude that includes the first and last elements, and so also a finite multitude *in statu essendi* (Bornstein, 1916, p. 9). If there were no last element, the addition of new elements would be unlimited, and the process of passing through the indirect locations would be infinite, so it would be impossible to complete a particular path. We would be faced with a process of a multitude forming *in statu fiendi*. The assumption that a given path is completed continues to contradict speculation about the non-existence of the last indirect element – i.e. the infinite nature of this multitude.

As we can see, Bornstein, in solving the issue of movement, did not pursue his analysis at a logical level, but rather drew attention to an ontological solution applicable to this phenomenon. In his assessment, the statement that a finite multitude of indirect locations must be present for movement to be possible is an absolutely true one.

For Bornstein, a straight line is a spatial object which does not exhibit granular structure. The points on this line do not exist until they are designated. The points do not pre-exist on the line, and they do not exist on it in the strict sense of the word. Initially, there must be a line on which other points can later exist, not the other way round. The points are not independent components of the line, but elements dependent on it. Movement of an independent and separately existing point on a straight line is always movement along an already independently moving object.[5] The *aporic* solution offered by Bornstein is, on the one hand, consistent with the accepted principle of continuity, while on the other, it is also consistent with the principle of non-spatiality – i.e. with locations accepted discretely in the description of actually occurring movement.

---

[5] There will be a three-dimensional object in three-dimensional space, and a one-dimensional object in one-directional space, but no zero-dimensional one (Bornstein, 1916, pp. 23–25).

The Ontological and Structural Investigation of the Logico-Geometrical System

Mathematics includes fields subject to both quantitative and qualitative research. With the aim of making mathematics a tool for philosophical research, Bornstein implemented the logical algebra developed by George Boole, together with the existing projective geometry. He imparted a categorial form to these fields and simultaneously brought them closer to philosophical research (Bornstein, 1934, pp. 62–76).

Having uncovered the structure of categorial algebraic logic and the geometry of locations, he noted their structural similarity and developed a system for categorial geometrical logic (logo-topics) or, in other words, categorial logical geometry (topo-logic).[6] The two dimensions of this system constitute one field (λόγος – τόπος), examined under two aspects. It has turned out that this system carries profound philosophical significance. With its help, ontology and the study of the categories of being become a science comparable to mathematics. This logico-geometrical system revealed the qualitative aspect of reality, and enabled the development of a general theory of being. For the author of the work entitled *Architektonika świata* [*The Architectonics of the World*], this system would become a *novum organum philosophiae* (see Śleziński, 2009, pp. 101–102).

Categorial Algebraic Logic and Qualitative Geometry

In his development of algebraic logic (the algebra of logic) (Bornstein, 1935, pp. 10–29), Bornstein referred to the following system of axioms listed by Edward Huntington in 1904:

1a) for any element a there will be an element 0 such that $a + 0 = a$
1b) for any element a there will be an element 1 such that $a1 = a$
2a) $a + b = b + a$
2b) $ab = ba$
3a) $a + bc = (a + b)(a + c)$
3b) $a(b + c) = ab + ac$
4a) for some element a there will be an element a' such that $a + a' = 1$
4b) for some element a there will be an element a' such that $aa' = 0$.

---

[6] An interesting research task would be to compare Bornstein's metaphysical results with the results of modern mereotopology (cf. Skowron, 2017).

This system assumes the occurrence of the two different elements a and b, the three operations of logical negation ($'$), sum (+) and multiplication, commutation and distribution of the sum and product, plus the two elements 0 and 1.

A logical sum is based on the consolidation of two or more elements in their total. In this way, consolidating the notion of "man" marked as (a) and the notion of "good" as (b), we obtain the notion "good man" as ($a + b$). While Bornstein's sum is an operation reminiscent of that of quantitative algebra, logical multiplication possesses a different nature from the quantitative multiplication of mathematics. This is because logical multiplication is based on defining the largest common element of two or more notions. For example, the maximum common element for the notion of "animal" as (a) and the notion of "plant" as (b) is the notion of "organism" (ab).

In algebraic logic, the element 0 possesses the poorest logical content, as added to any element it does not change its content, while constituting a module of summation. In ontology, the element 0 articulates the content of the notion "something" or "object in general". In turn, the element 1 presents the richest notion in terms of content as "whole", "everything", and is simultaneously a module of multiplicity. The logical 1 constitutes the upper limit of the field of notions, while the logical 0 is its lower limit.

The relation of inclusion does not occur as the foundation of any algebraic statements of logic. However, it can be introduced as the symbol "<", taking into consideration logical operations based on the equivalence quotient (the symbol "="), and linking it at the same time to the axioms mentioned earlier. As a result, we get the following:

(Ia) $a < b = (b = a + b)$
(Ib) $a < b = (a = ab)$
(Ic) $a < b = (ab' = 0)$
(Id) $a < b = (a' + b = 1)$.

In other words, formula (Ia) states that if content a is included in content b, then in adding content b, we do not by adding the content a already included in content b change content b. If $a < b$, then $b = a + b$ – and vice versa. Each notion poorer in content is included in the notion richer in content. The essence of all cases of equivalence based on the mutual inclusion of elements of that equivalence can be presented in the following definition:

$(a = b) = (a < b) + (b < a)$.

The relations of comprising that obtain between elements of content logic should not be represented synthetically in the form $a < b$; instead, they should be represented analytically in the form $ab < a$ or $a < a + b$. The relation of comprising obtains only between the product (ab), the sum ($a + b$), and the elements a and b. It does not obtain among elements themselves, in that these do not stand in relations of either inferiority or primacy towards each other. The law of duality also applies to this type of relation. According to this law, one should exchange elements of the resulting relation for dually respective expressions. For example, the formula $a < a + b$ corresponds to the formula $ab < a$, which means that the common part of elements a and b is included in element a. With his introduction of the comprising relation, Bornstein expanded the field of mathematical logic.[7]

In his work, Bornstein also made use of projective geometry, introducing categorial form into this area, too. As a result of the transformation of projective geometry into categorial geometry, all categories of location and direction came to be derived on a plane. Dividing the plane into four quarters, he achieved all possible basic categories and types of appropriate straight line and point, as well as inappropriate ones, such as are found in only one, two or three quarters of the plane or in all quarters or outside of these quarters (Bornstein, 2011).

In the wake of his categorisation of qualitative geometry, Bornstein noted that the relations of this geometry correspond to those of algebraic geometry. Still, in algebraic logic various ontological interpretations of categorical elements are allowed as classes, ranges, types and contents, while for categorical algebraic geometry only categorical points and straight lines are considered.

---

[7] Bornstein also demonstrated the possibility of transferrals from the notation of the logic of classes and ranges to that of content logic. There are at least two modes of transferral from the logic of class to content logic. One of them will be discussed here. In this case, the element 0 in content logic corresponds to everything, any object, and thus is the broadest class (1) in range logic. The content defined as 1 in turn constitutes the upper limit of all content, the notion including all possible content. It is the most defined and determined notion. The range of the upper content limit is the smallest limit class (0). This assumes that the formulas for multiplicity and logical summation in content logic do not possess the same meaning as in the logic of classes. The content described as ab is not the common content of a and b, as it is richer than their content, due to it being the accumulation of these contents. Thus, the essence of the multiplicity of content becomes identical here to the essence of the summation of classes. Similarly, the summation of contents corresponds to the formula for the multiplicity of classes. On this basis, the range of the class:

$$0 < ab < a < a + b < 1$$

will correspond to the range of contents

$$1 > a + b > a > ab > 0.$$

However, the formulas in the logic of classes described as $a + a' = 1$ and $aa' = 0$ will become $aa' = 0$ and $a + a' = 1$ in the logic of content. All formulas for the logic of classes can be changed into formulas for the logic of content in this manner.

If categories *a* and *b* are presented in the form of different straight lines, then the crossing point of the straight lines *a* and *b* will not comprise a common entity of these categories. It will form new content derived from the accumulation of the categorial contents corresponding to the content of categories *a* and *b* (see Diagram 1). The creation described as *a* + *b* is not included in *a* and *b* as is observed in the field of nominal multiplicity. On the contrary, notions *a* and *b* are included in their creation defined as *a* + *b* – thus, *a* < *a* + *b*, *b* < *a* + *b*. The contents of the notions *a* and *b*, taken separately, constitute only the possibility of forming *a* + *b*, as they are less defined or determined than their creation. Points *a* and *b* in this respect reveal a richer notion with regard to content: a notion corresponding to something qualitative that is more differentiated and specific than straight lines *a* and *b*, which reveal a poorer notion in this respect, corresponding to a kind and to a difference in genre. This means that the creation described as *a* + *b* determines *a* and *b*, which in turn constitute it and as such are included in it. Further, the content *ab* reveals what points *a* and *b* share in common, which is the closest qualitative element specified by them both. The less determined content described as *ab* is included in both content *a* and content *b*.

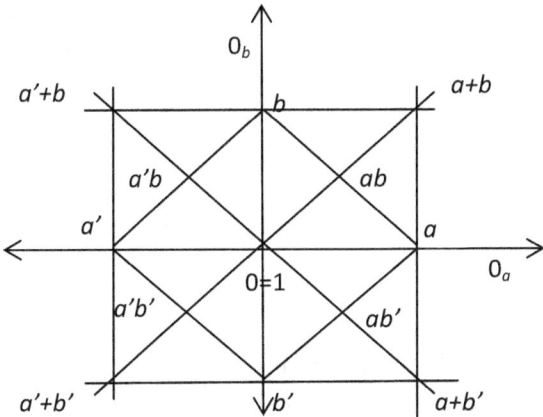

Diagram 1: Structure of logical and geometrical quality categories.

The basic duality of logical formulas is clearly outlined in the above diagram (Diagram 1). It was defined by Charles Sanders Peirce in 1897 as "the principle of duality". Simultaneously, it revealed a connection with the basic duality of geometrical activities, cutting and projection. That is to say, two straight lines described as a and b determine point $a + b$, and at the same time the two points a and b determine the straight line described here as ab. Bornstein was the first to note the appropriateness of duality for logical and geometrical activities. Undoubtedly, this was an important scientific achievement, and it led him to geometrising, to the shifting of logic into spatiality, and from there to scientific metaphysics.

Bornstein's solutions hold great importance in terms of ontology and semantics, as the boundary case of the logical sum $a + b$ in the form of the formula $a = a + 0$ points to the importance of the general notion a. Therefore, this notion will be one of the relevant quality. It should be noted that the content of notion 0 means "any object" or "object in general". On account of this, there is a point on the species of straight line a described as a. This point is a logical boundary point, and is revealed as a quality of undetermined differentiation as regards its status as a possible genre or *in potentia* genre. This means that there is a point on the species of straight line defined as a, with the same degree of determination as the straight line a. The point defined as a constitutes a species boundary point, being only the possibility of a kind, and thus a species.

Subsequent to this, the boundary case of the logical multiplicity defined as ab, where b is in infinity, will therefore have a greater value than each finite b, and can be presented in the following formula: $ab = a1$. In terms of geometry, we will experience a straight line crossing a parallel to 0_b. A straight line crossing point a on axis 0_a, parallel to 0_b, constitutes a straight line defined as $a = a1$. Therefore, both in this boundary case of logical multiplicity and in the above case, in bringing the logical sum to a boundary form we overcome the difficulty associated with the process of giving spatial meaning to content logic dual formulas. It should be noted as well that the richer in content and more determined a categorial element is, the smaller its "stretching" and "range". For example, observing a sequence such as $a + b > a > ab$, it can be perceived that the process of stretching of the particular elements increases from "uni-quarterness" for $a + b$ through "double-quarterness" for a to "tri-quarterness" for ab. Meanwhile, point O will be "quadri-quarterness", as the logical minimum and least determined element.

It can be said that Bornstein uncovered an extremely simple and pure geometrical schematics for two-element logic formulas. This simultaneously ensures and safeguards the spatial interpretation of the remaining statements of logic. In this way, the constitutive structure of *logica situs* – or topologic as it was called by Bornstein for the first time in 1926 in his work entitled *Geometria logiki kategorialnej i jej znaczenie dla filozofii* [*The Geometry of Categorial Logic and its Importance for Philosophy*] – was successfully elaborated, and Leibniz's unfulfilled dream about each step in abstract comprehension having its own spatial analogue was allowed to come true.

Harmonic Structures

The geometrising of logic not only allows one to read off logical statements from Bornstein's diagram, but also to identify properties in the world of content logic of a kind inaccessible to strictly logical research. One example of such properties is undoubtedly the existence of harmonic structures.

In the second volume of his work entitled *The Architectonics of the World* (Bornstein, 1935, pp. 99–131), Bornstein devoted a great deal of his attention to deriving harmonic relations directly from projective geometry and introducing them into the world of geometrical logic. Every element in the categorical topologic is a foundation or a vertex of the harmonic four. When we speak of a vertex, we are thinking of harmonic bunches. For example, the straight lines b, $a'b$, ab and $0_{bb'}$ (see Diagram 2) will be the harmonic bunch for vertex (point) b.

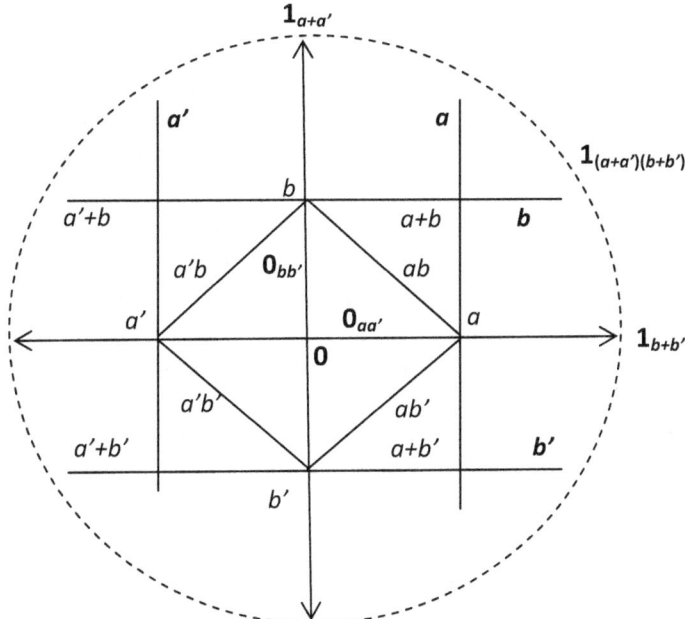

Diagram 2: Categorial plane with the infinite elements of zeros and unities.

Thanks to the existence of harmonic relations, it is possible to reach a more profound comprehension of the rules and statements applicable in the world of geometrical logic. Taking into consideration a harmonic four such as $1_{a+a'}$, b, 0, b', with the founder straight line marked as $0_{bb'}$, we can give a deeper sense to the process of defining the negative element with the help of the positive one plus the elements 0 and 1 in the defining form $bb' = 0$ (for limit element 1) and $b + b' = 1$ (for limit element 0).

The four elements display a harmonic system such that the negative element marked as b' can be defined as an element linked with the positive element b in this system, where the second pair of linked elements are 0 and 1. In this way, due to b and b', the development of 0 and 1 involves the defining of pairs of harmonically linked elements in a harmonic group, of which the second pair are the developing elements 0 and 1. It is also worth paying attention to the dichotomous formula for element a, which instead of taking the form $a = ab + ab'$, can be presented as $a + 0 = ab + ab'$. This becomes clear on account of the harmonic four that form the group 0, a, ab, ab'. This formula depicts not only the logical sum of two logical elements, but also the logical sum of two logical intervals, on account of the fact that in geometrical logic this sum is composed of bunch elements with the vertex marked as a.

The formula of dichotomy defines types of dichotomy as dividing and connecting. In the dividing dichotomy $a = ab + ab'$ the segments of this division are less determined than the element a ($ab < a$, $ab' < a$). The respective dual defining dichotomy form $a = (a + b)(a + b')$ includes elements that are richer in terms of content and more determined compared to the element a ($a + b > a$, $a + b' > a$). In this case, we experience determination, not division, of content. The difference between these two forms of dichotomy comes down to the fact that content a is referred to boundary content 0 in the dividing dichotomy, but to boundary content 1 in the defining dichotomy. In Bornstein's opinion, this means that in the case of the defining dichotomy element a does not reach 1 but restricts 1 to the content $a + b$ and $a + b'$. Transferring this to the realm of species content and qualitative content, we may observe a differentiation of the qualitative content into two species contents.

The Ontology of Geometrical Metaphysics

In *The Architectonics of the World*, Bornstein distinguished certain absolute elements amongst general categorial ones. When his attention began to focus on metaphysical problems, he returned to these elements, submitting them to detailed analysis. At the same time, he changed his research methods. The place of geometrical logic was taken by logical geometry, which led to geometrical metaphysics (metaphysical geometry). As a result of this, a model of the wholeness of being was developed, presented most thoroughly in *Teoria absolutu* [*The Theory of the Absolute*] (Bornstein, 1948, pp. 64–106). Thanks to algebraic logic becoming geometrical, different types of zero and unity principles were revealed, as well as the absolute structures linking them. By implementing the archeoscopic method, which boils down to a process of placing an algebraic, a logical, and an ontological layer on the geometrical category plane, Bornstein developed the concept of a scientific metaphysics. Within its framework, the philosopher discusses the categories of concretum and abstractum, in place of speaking about regional categories of space such as points, or straight lines and regional categories of logic. In this way, he realised Leibniz's ideal of being able to present elements of the real world *more geometrico* (Bornstein, 2014).

It should be noted that topologic combines different fields which are poles apart and have different substrata. During the transition from one field to another, the appropriate form and structure is preserved. The only explanation for the preservation of identical structures across these fields is the profound similarity between them, as pointed to by Bornstein. Such similarity is ultimately a matter of their qualitative character, so that what is revealed is the universal nature of those structures.[8]

From the architectonic perspective of being construed as a whole, the categorial world is contained between two borders, which will be absolute elements. It is in these elements that the categorial world reaches its ultimate form of interconnectedness: that of "one as a whole" (1) and "one as a commonality" (0), where these in turn constitute the double form of the Absolute.[9] What is more, the elements of the Absolute occupy a prominent place within the spatial picture of the categorial world. Three distinct unities and three zeros can be identified on the two-dimensional categorial diagram, which also depicts the structure of pan-logic as the most complete categorial development of the world of content logic. All the unanimous elements lie on the infinity of the plane, while the differentiated zero elements constitute two axes and the centre point of the system of coordinates.[10] The whole world of finite elements lies between these elements (Bornstein, 1948, p. 64). In contrast to finite categories, Bornstein perceived uncompleted elements as rules with a dialectical nature. In his view, these were universal boundary rules in an absolute sense (Bornstein, 1948, p. 65). Moreover, Bornstein accepted the dynamic nature of being. In his analysis, he demonstrated how finite elements

---

8 Cf. Bornstein, 1948, pp. 41–42.
9 Cf. Bornstein, 1934, p. 201.
10 Since the unities are dual according to the dual and zero system of coordinates, the points in infinity can be regarded as two poles, out of which the straight lines a, a', b and b' extend coordinately.

emerge from the absolute rules. The categorial metaphysical plane constitutes a model for the whole of being, which was interpreted ontologically by Bornstein in accordance with Aristotle's concept of being. In his opinion, horizontal straight lines are images of the substratum, matter, while vertical lines are images of forms of being. The absolute element in infinity $1_{b+b'}$ constitutes a whole, the whole of abstract matter. The straight line b is the image of physical matter, while the straight line b' is the image of the psychical substratum, and the absolute straight line $0_{a a'}$ being in the centre of the image of matter and the psyche constitutes the image of the "life" substratum. In turn, the absolute straight line in infinity denoted by $1_{(a+a')(b+b')}$ is the spiritual substratum. The particular antithetical types of the finite being and the boundary types of being consolidate together in absolute unity as $1_{b+b'}$, the wholeness of matter. This unity is called the "substratum source"[11] by Bornstein.

In the categorial metaphysical plane, vertical straight lines present two finite forms shaping objects a and a' as well as two absolute border forms, $0_{bb'}$ and $1_{(a+a')(b+b')}$. The form giving the minimum of definition to the matters is the image of the absolute axis. A straight line in infinity becomes the image of the highest form and constitutes the whole that such forms make up (Bornstein, 1948). The vertical straight lines that cross horizontal straight lines determine points that are images of particular beings. In this way, the points a and a' are pure forms. Similarly, points b and b' are pure matter. However, the centre for the system of coordinates as the least defined object will be the lowest matter-form conjunction. The straight line in infinity will be the highest matter-form conjunction and thus the most defined image of the object.

The correlation and specification of zeros and units form specific elements of the world (Bornstein, 1948, pp. 86–103). The organizing power $1_{a+a'}$ develops dichotomously, giving abstract polar elements in the form of straight lines a and a', specifying themselves on the horizontal axis in the form of specific elements, the points a and a'. The dual 0_{aa}, in relation to $1_{a+a'}$, develops dichotomously into specific elements – the points a and a', whereas the subtractive content of being develops dichotomously into abstract elements in the form of straight lines b and b', which specify themselves in specific elements – the points b and b'. In turn, $0_{bb'}$ gives us the specific elements b and b'. After further analysis, one can conclude that every absolute straight line is, on the one hand, directed towards the outside in the direction of the real world, while on the other hand, it is directed inside, according

---

[11] Bornstein (1948, p. 83). The harmonic four thus achieved has two pairs of straight lines linked so harmoniously that the space between the life substratum and the spirit substratum is harmoniously divided by harmoniously linked finite straight lines, depicting the physical substratum and the psychological one.

to Bornstein, heading towards the Absolute.[12] It should be noted that in his work on scientific metaphysics, Bornstein pointed to the mutual interconnectedness of absolute rules and finite categories.

★ ★ ★

Bornstein's area of research has been presented here with reference to the thought that such investigations could fruitfully be continued. Algebraic, logical and geometrical research that includes categorial and ontological analysis constitutes an original proposal for the description and qualitative explanation of the structure of reality. It includes many innovative ideas, which continue to prove inspiring for the contemporary pursuit of analysis in the field of formal ontology.

Bibliography

Bornstein, B. (1907). Preformowana harmonja transcendentalna jako podstawa teorji poznania Kanta [Pre-formed Transcendental Harmony as a Foundation for Kant's Theory of Cognition]. Przegląd Filozoficzny, 10(3).
Bornstein, B. (1910). Zasadniczy problemat teoryi poznania Kanta [The Fundamental Problem of Kant's Theory of Cognition]. Warszawa: Skład Główny w Księgarni G. Centnerszwera i S-ki.
Bornstein, B. (1912). Prolegomena filozoficzne do geometryi [Philosophical Prolegomena to Geometry]. Warszawa: Wyd. E. Wende i S-ka.
Bornstein, B. (1913). Problemat istnienia linji gieometrycznych [The Problem of Existing Geometrical Lines]. Przegląd Filozoficzny, 16(1), 64–73.
Bornstein, B. (1916). Elementy filozofii jako nauki ścisłej [Elements of Philosophy as a Strict Science]. Warszawa: Skład Główny w Księgarni E. Wendego i S-ka.
Bornstein, B. (1934). Architektonika świata: Tom 1. Prolegomena do architektoniki świata [The Architectonics of the World: Vol. 1. Prolegomenon to An Architectonics of the World]. Warszawa: Skład Główny Gebethner i Wolff.
Bornstein, B. (1935). Architektonika świata: Tom 2. Logika geometryczno-architektoniczna [The Architectonics of the World: Vol. 2. Geometrico-Architectonic Logic]. Warszawa: Skład Główny Gebethner i Wolff.
Bornstein, B. (1948). Teoria absolutu. Metafizyka jako nauka ścisła [The Theory of the Absolute. Metaphysics as a Strict Science]. Łódź: Łódzkie Towarzystwo Naukowe.
Bornstein, B. (2011). Co to jest kategorialna geometria algebraiczno-logiczna? [What is Categorial Algebraico-Logical Geometry?]. In K. Śleziński, Filozofia Benedykta Bornsteina oraz wybór i opracowanie niepublikowanych pism [The Philosophy of Benedict Bornstein and a Selection and Elaboration of His Unpublished Writings] (pp. 63–77). Katowice – Kraków:

---

12 When discussing the absolute, Bornstein does not consider God. The term "absolute" was introduced into projective geometry to define a straight line in infinity as the absolute of a plane by the mathematician Arthur Cayley (1821-1895), the originator of the algebraic theory of invariables.

Uniwersytet Śląski – Wydawnictwo Scriptum. [Bornstein's typescripts with handwritten corrections and supplements are in the Jagiellonian Library under the number 9026 III].

Bornstein, B. (2014). *Wstęp do metafizyki jako nauki ścisłej* [*Introduction to Metaphysics as a Strict Science*]. In K. Śleziński, *Benedykta Bornsteina niepublikowane pisma z teorii poznania, logiki i metafizyki. Wybór i opracowanie oraz wprowadzenie i komentarze* [*Benedykt Bornstein's Unpublished Writings from the Theory of Knowledge, Logic and Metaphysics. A selection of Papers with an Introduction and Commentary*] (pp. 239–250). Katowice: Uniwersytet Śląski – Wydawnictwo Scriptum.

Skowron, B. (2017). *Mereotopology*. In Seibt J., Gerogiorgakis S., Imaguire G., Burkhardt H.: *Handbook of Mereology*. Philosophia Verlag GmbH. (pp. 354–361).

Śleziński, K. (2009). *Benedykta Bornsteina koncepcja naukowej metafizyki i jej znaczenie dla badań współczesnych* [*Benedykt Bornstein's Concept of Scientific Metaphysics and its Importance for Contemporary Research*]. Kraków: Uniwersytet Śląski – Wydawnictwo Scriptum.

Twardowski, K. (1898). *Wyobrażenia i pojęcia*. Lwów: Księgarnia N. Altenberga. English translation: *Imageries*. Axiomathes, vol. 6 (I), 79-104 (1995).

Twardowski, K. (1894). *Zur Lehre vom Inhalt und Gegenstand der Vorstellungen. Eine Psychologische Untersuchung*. Wien, Holder. English translation: *On the Content and Object of Presentations. A Psychological Investigation*. Mortinus Nijhof, The Hague (1977).

Janusz Kaczmarek
On the Topological Modelling of Ontological Objects: Substance in the *Monadology*

Abstract: In this paper I explore a methodological problem: how can we use topology and topological concepts as a basis for making sense of ontological concepts and problems? I try to show that it is possible to describe some fundamental concepts of Leibniz's *Monadology* using topology. Therefore, I shall treat monads as topologies, and substance as a set of topologies, with a certain topology distinguished as being the so-called dominant monad. This, as we shall see, furnishes some interesting theorems, comparable with those of systems theory and Leibniz's own theory of substance.

Keywords: topology, topological ontology, substance, monad, dominant substance, system.

What is Topological Ontology?

We usually describe ontology in accordance with Greek philosophy (Aristotle) as a branch of philosophy concerned with objects *qua* objects, or being *qua* being. What this means is that we try to define what is or what exists: i.e. being and some fundamental properties of being. Philosophers, particularly ontologists, have pointed to different collections of categories as constituting the fundamental characteristics of being. For example, in Aristotle's papers, we find a system of categories in which primary substance is distinguished as a category from nine others: qualities, quantities, relations, place, time, actions and affections. Having these nine categories at our disposal, we can predicate thus: Socrates (primary substance) is righteous (quality), gives a lecture for his students (relation), in the Athenian agora (place), at noon (time), etc.

We can also propose, for example, Husserl's categories as articulated in his *formal ontology*, i.e. his ontology of objects *qua* objects (i.e. objectuality in general). Husserl suggested the following categories (i.e. concepts or terms): object, state of affairs, property, relation, number, unit, plurality, etc. Apart from these, Husserl introduced his so-called *material ontology* with its three general regions: nature (including physical objects and events, but also the world of what is alive), culture (artefacts, social entities and values) and consciousness (cf. Husserl, 1913/1982).

Janusz Kaczmarek, Department of Logic, University of Lodz, Poland.

https://doi.org/10.1515/9783110669411-009

In an earlier study (Kaczmarek, 2008a), I proposed an array of terms and notions that are important for ontological investigations, pointing to the following levels and relevant terms:
a) the level of individuals: individual, property, essential and attributive property, positive and negative property, complete object, the extension of an idea, but also state of affairs, fact and relation;
b) the level of ideas: general object, species, genera, the hierarchy of general objects, species difference, the property of an idea and the property given in the content of an idea;
c) the level of concepts: concept, the structure of concepts, the content of a concept, positive and negative content, the extension of a concept.

I have presented the definitions of the terms and notions in question, together with some theorems, in set-theoretical language (cf. Kaczmarek, 2008a,b).

But now we face the following problem: what is topological ontology? Here we are interested not in formal ontology in general, but in an ontology which tries to use topological concepts, topological theorems and structures and – perhaps – also a topological point of view.[1] In a private conversation, my colleague Bartłomiej Skowron suggested that I speak about the modelling of ontological concepts, objects and theorems using topological ones. And indeed, we can encounter such a perspective in Salamucha's papers. This Polish analytic philosopher sought to represent the so-called "geometric point of view" because, in his opinion, after the discovery of Euclidean geometry this point of view would be dominant in science and philosophy (cf. Salamucha, 1946). Let us also emphasize that this approach was already present in Aristotle's metaphysics.

On Some Topological Concepts

Topological Space

Let X be a set (not necessarily non-empty), and T_X a family of subsets of X. A pair (X, T_X) is a topology or a topological space on X, iff the following conditions are fulfilled:
a) $\emptyset \in T_X$ and $X \in T_X$,

---

[1] The terms "topological ontology" and "topological philosophy" (in Polish: "ontologia topologiczna" and "topologiczna filozofia") have been proposed by Mirosław Szatkowski and Bartłomiej Skowron.

b) if $A_1, A_2, \ldots \in T_X$, then $A_1 \cup A_2 \cup \ldots \in T_X$,
c) if $A_1, A_2 \in T_X$, then $A_1 \cap A_2 \in T_X$.

Examples of topologies

τ1. If $X = \emptyset$, then $(\emptyset, \{\emptyset\})$ is a topological space.
τ2. If $X = \{1, 2\}$, then $(X, \{\emptyset, \{1\}, X\})$ is a topological space. It is known as a *Sierpiński space*.
τ3. If $X = \mathbb{R}$, $\mathbb{R}$ is the set of real numbers, and any set of $T_\mathbb{R}$ is a union of sets of the form $(r_1; r_2)$, for $r_i \in \mathbb{R}$, $(\mathbb{R}, T_\mathbb{R})$ is a topological space to be referred to as a *natural topology on $\mathbb{R}$*.
τ4. If $X = \mathbb{R}$ and $\emptyset \neq A \subset X$, then $(X, \{\emptyset, A, X \setminus A, X\})$ is a topological space.
τ5. For any set X, the *discrete topology* on X is the topology T_d, such that $T_d = \{U : U \subseteq X\}$, so the collection of open sets of T_d equals the power set of X, i.e. $T_d = P(X)$. Next, the *indiscrete topology* (or *trivial topology*) on X is the topology T_{triv}, such that $T_i = \{\emptyset, X\}$.
τ6. For $X = \{1, 2, 3\}$, we can define 29 topologies on X.[2] Here are some of them:
  τ6.1. $T_X = \{\emptyset, \{1, 2\}, X\}$,
  τ6.2. $\{\emptyset, \{1\}, \{1, 2\}, X\}$,
  τ6.3. $\{\emptyset, \{1, 2\}, \{3\}, X\}$,
  τ6.4. Naturally, we can define on X the topology T_d and T_i.

In the present paper, I try to explain why these topologies are important for ontological investigations. Initially, let us note that all topological spaces on a given set X can be ordered by the "weaker than" or "stronger than" relation. We may define matters thus: for two topologies T and T' on X, we say that T is *weaker* (or *coarser*) than T' (equivalently: that T' is *stronger* or *finer* than T) if $T \subset T'$ (we write: $(X, T) \leq (X, T')$). This means that each open set from T will also be an open set in T'. Of course, for any set X, and any topology T on X, we will have:

$$T_i \subset T \subset T_d \quad (\text{or: } (X, T_i) \leq (X, T) \leq (X, T_d)).$$

Separation Axioms

In general topology, so-called *separation axioms* are introduced. They define which kinds of topological object can be separate: for example, points or closed sets. So let us recall the axioms T_0, T_1, T_2, T_4.
Let (X, T_X) be a topological space. Then:

---

[2] The reader can calculate this for himself or herself, or find it in Warren (1982).

SA. 0. (X, T_X) will be a T_0 space (or *fulfils condition T_0*), iff for any $x, y \in X$, $x \neq y$, there exists either an open set U such that $x \in U$ and $y \notin U$, or an open set U such that $x \notin U$ and $y \in U$.

SA. 1. (X, T_X) will be a T_1 *space*, iff for any $x, y \in X$, $x \neq y$, there exists an open set U such that $x \in U$ and $y \notin U$ and there exists an open set V such that $x \notin V$ and $y \in V$.

SA. 2. (X, T_X) will be a T_2 *space* (or *Hausdorff space*), iff for any $x, y \in X$, $x \neq y$, there exist open sets U and V with $U \cap V = \emptyset$ such that $x \in U$ and $y \in V$.

SA. 4. (X, T_X) will be a T_4 *space* (or *normal space*), iff (X, T_X) is a T_1 space and for any closed sets $E, F \subseteq X$, $E \cap F = \emptyset$, there exist open sets U and V such that $E \subset U$, $F \subset V$ and $U \cap V = \emptyset$.[3]

The following simple theorems are true.

Fact 1. *If (X, T_X) is a T_i space, then (X, T_X) will also be a T_{i-1} space, for* $i \in \{1, \ldots, 6\}$.

Fact 2. *If (X, T_X) is a T_i space and $(X, T_X) \leq (Y, T_Y)$, then (Y, T_Y) will be a T_i space, for $i = 0, 1, 2$.*[4]

Examples. The topology given in τ6.1, i.e. $T_X = \{\emptyset, \{1, 2\}, X\}$, is not even T_0, because for 1 and 2 there is not any open set U such that $1 \in U$ and $2 \notin U$ or $1 \notin U$ and $2 \in U$. Sierpiński spaces are T_0 but not T_1. The topology ($\{1, 2\}, \{\emptyset, \{1\}, \{2\}, \{1, 2\}\}$), which is stronger than the Sierpiński space ($\{1, 2\}, \{\emptyset, \{1\}, \{1, 2\}\}$), is T_1. In turn, the natural topology on $\mathbb{R}$ is normal, and hence a Hausdorff space.

Topological Subspaces

Let (X, T_X) be a topological space and $A \subseteq X$. Then (A, T_A) will be called a *subspace* of X iff $T_A = \{A \cap B : B \in T_X\}$. T_A is usually referred to as the *subspace topology* on A.

Let us now note:

Fact 3. *If (X, T_X) is a topological space fulfilling the axiom T_i, $i = 0, 1, \ldots, 6$, and (A, T_A) is a subspace of (X, T_X), then (A, T_A) fulfils T_i.*

---

[3] Other separation axioms (like T_3, $T_{3\frac{1}{2}}$, T_5 and T_6) have been left out here.

[4] Proofs of these theorems can be found in Kuratowski (1972). The same goes for other simple facts that I propose in this paper.

Examples. Consider the natural topology $(\mathbb{R}, T_{\mathbb{R}})$ and the set $\langle 0, 1\rangle$. Then the family
$$T_{\langle 0,1\rangle} = \{A : A = U \cap \langle 0, 1\rangle \text{ and } U \in T_{\mathbb{R}}\}$$
is the topology on $\langle 0, 1\rangle$. Let us note, for example, that the set $\langle 0, \frac{1}{2})$ is open in the given subspace, but neither open nor closed in the natural topology on $\mathbb{R}$. Next, the set $\langle 0, 1\rangle$ is closed in the natural topology and both open and closed in the subspace on $\langle 0, 1\rangle$. If, now, we consider the topology $T_X = \{\emptyset, \{1, 2\}, X\}$, where $X = \{1, 2, 3\}$, then the subspace on the set $\{1, 2\}$ is trivial and the subspace on $\{1, 3\}$ is isomorphic to a Sierpiński space with the topology $\{\emptyset, \{1\}, \{1, 3\}\}$.

Leibniz on Monads and Substance: Substance as a Set of Topologies

Leibniz's *Monadology*

In his famous work the *Monadology*, Leibniz set out a theory of monads and substances. I propose to present just some of the more important theses of Leibniz's theory.

A monad is a simple substance (*Monadology*, Point 1). This means that it has no parts (Point 1). It can, however, itself be a part of something composite (1). As something that has no parts, it cannot be extended, have a shape, or be split up (3). So monads are true atoms of Nature – the elements out of which everything is made (3) (cf. Leibniz, 1714).

Next (a fragment of the *Monadology*):

> Monads, although they have no *parts*, they must have some *qualities*. There are two reasons why this must be so. (1) If they didn't have qualities they wouldn't be real things at all. (2) If they didn't differ from one another in their qualities, there would be no detectable changes in the world of composite things [...]. (Leibniz, 1714, Point 8)

> [...] every created thing can change, and thus that created monads can change. I hold in fact that every monad changes continually. [...] *natural changes* in a monad [...] come from an internal force, since no external cause could ever influence its interior. (Leibniz, 1714, Points 10 and 11)

Leibniz also defines the concepts of perception and appetition. In Points 13 to 15 and 17 he explains:

> [...] So although there are no parts in a simple substance, there must be a plurality of states and of relationships. [...] The passing state that incorporates and represents a multitude within a unity – i.e. within the simple substance – is nothing but what we call perception. The action of the internal force that brings about change – brings the monad from one perception to another – can be called appetition [...].
> And that is all that can be found in a simple substance – perceptions and changes in perceptions: and those changes are all that the internal actions of simple substances can consist in.

Let us complete the theory with the definition of composite things or, more precisely, of composite substances, such as, for example, a living body or an animal. This theory is partly given in the *Monadology* and partly in other works (including letters). For Leibniz, a composite thing is a collection of monads with a so-called dominant monad (central monad) or dominant entelechy. This dominant monad can be understood philosophically as a vegetative soul in the case of a plant, or a sensitive soul in the case of an animal. In Points 63 and 70, he writes:

> 63. What we call a "living thing" is a body that has a monad as its entelechy or its soul, together with that entelechy or soul. And we call a living thing "an animal" if its entelechy or central monad is a soul. Now this body of a living thing or animal is always highly organized [...].
> 70. We can see from this that every living body has one dominant entelechy, which in an animal is its soul; but the parts of that living body are full of other living things, plants, animals, each of which also has its entelechy or dominant soul.[5]

The Topological Definition of Substance

We shall now put forward some sort of formal definition of the concept of a composite, having recourse to topological concepts.

Let $D = (X, T_X)$ be a topology on X. I propose to define a *substance S in Leibniz's sense* as a family $\mathcal{F}_S$ such that:
1. $D \in \mathcal{F}_S$,
2. if $M \in \mathcal{F}_S$, then M is a subspace of D or $M \leq D$.[6]

In this definition, we call topology D a dominant substance. In a well-known work of historical commentary on the relevant area of philosophy (Copleston, 1958, Chapter XVII, paragraph 4), we meet with an example of bodily substance. For instance, some particular sheep construed as a substance, i.e. as a bodily and an

---

[5] Of course, it is evident that we ourselves cannot interpret Leibniz's propositions as the German philosopher would have, insofar as he can be thought of as having held that these "other living things, plants or animals" had to be something like animalcules or homunculi.
[6] Let us mention that $M \leq D$ means that M is topologically weaker than D.

organic substance, is understood as a collection of monads with one dominant monad. This dominant monad will be a counterpart of a sensitive soul. But let us now give a topological example of a substance.

Let $D = (X, T_x)$ be a dominant monad, and T_x^1 and T_x^2 two topologies weaker than D and such that $\sim(T_x^1 \le T_x^2)$ and $\sim(T_x^2 \le T_x^1)$. Moreover, let (A, T_a), (B, T_b), (C, T_c) be subspaces of $D = (X, T_x)$, and $A \subset X$, $B \subset X$, $C \subset A \cap B$. At this point our substance will take the following form:

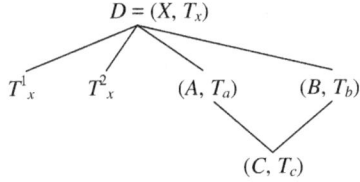

Here one can readily see that the collection or aggregate of the above monads (topologies) amounts to a partially ordered set.[7] A substance given in that form looks like an object in the theory of objects as systems (Bocheński, Bunge, Ingarden). In systems theory, moreover, one encounters the following theorems:

(TS1) A system consists of real and/or ideal elements.
(TS2) Each system has at least two elements.
(TS3) Each system has some unifying ingredient (or principle of synthesis).
(TS4) A system can be a part of another (larger) system.[8]

For examples of systems, consider Euclidean geometry (consisting of ideal elements only, such as definitions, axioms and theorems) or a university (consisting of both real and ideal elements – such as buildings or scientific equipment, and a university statute or scientific ideas, respectively).

---

[7] An order $\le_S$ on $\mathcal{F}_S$ is defined by the condition: $M_1 \le_S M_2$ iff M_1 is a subspace of M_2 or $M_1 \le M_2$.
[8] See, for example: Bocheński (1988).

Substance Theorems and Interpretation

Let us first note that on the basis of the topological definition of substance and Facts 1 to 3, we can prove the following theorems for substances, these being theorems that pertain to matters of ontological significance and value.[9]

Fact 4. *Each monad (topology) is a substance in Leibniz's sense.*

Fact 5. *A monad is a one-element substance. In this case it is dominant for itself.*

Fact 6. *In a substance $\mathcal{F}_S$ having two or more elements, one can distinguish other substances.*

Example. Taking into account the above picture, the families:
(1) D, (A, T_a), (C, T_c) and
(2) (A, T_a), (C, T_c)
are substances.

Fact 7. *Two substances S_1 and S_2 are identical iff $\mathcal{F}_{S_1} = \mathcal{F}_{S_2}$.*

Fact 8. *Each substance $\mathcal{F}_S$ has only one dominant monad.*

Let us now deal with the interpretation and modelling of ontological objects by means of topological concepts. I propose to define (and interpret) matters in the following way:
1) the set X of a given topological space (X, T_X) is a set of *atomic perceptions* (or, perhaps, a unified body of information or an atomic state of affairs);
2) I call any set $A \in T_X$ a *perception* of a given monad (i.e. our perceptions are compound, complex, and also in the form of $\{x\}$);
3) let *int* and *cl* denote interior and closure operators; *intA* (interior of A) will be defined as the largest open set included in A; *clA* (closure of A) will be defined as the smallest closed set including A; then, because $A \in T_X$ is an open set (i.e. *intA* = A) and *intA* $\subseteq A \subseteq$ *clA*, for any subset $A \subseteq X$, I shall call *intA*, and hence any $A \in T_X$, an *essential perception* of a given monad; we should add that what we are treating as atomic perceptions are not perceptions *sensu stricte*, but if $x \in X$, then $\{x\}$ will be a perception;
4) if $A = clA$, then I shall call the set A a *complete* or *closed perception*;

---

[9] These facts are easy to understand, so I shall pass over the relevant proofs.

5) I consider all operations on sets of a given topology, such as union, intersection, set difference, complement, interior, closure and others, to be *elements* or *components* of the interior force of the relevant monad (appetition); so the action of the monad will be, for example, to transition from A and B to $A \cup B$, $A \cap B$, $X \setminus A$, clA, etc.; the result of this action will also constitute a perception;
6) if some topology (i.e. some monad) (A, T_a) is weaker than (A, T'_a), then I propose to establish that (A, T'_a) has *highly seasoned* or *more distinct* or *more individualized* perceptions (cf. Leibniz, 1714, points 25 and 19). In fact, for example in a topology T_2 on X, for any $x, y \in X$, one can find two perceptions A and B, such that $x \in A$ and $y \in B$, whereas in a T_1 topology this is impossible;
7) in the case of subspaces we observe that when (A, T_a) is a subspace of (X, T_X), then each perception $B \in T_a$ can be represented by a perception $C \in T_X$, and set $A \subset X$ as $C \cap A$. Hence, in this case, also, we can say that what is observable in (A, T_a) is observable as well in (X, T_X) (the problem being, what kind of perception is $C \cap A$ in (X, T_X), if C is neither open nor closed in (X, T_X)?); from **Fact 3** we get that in (A, T_a) we have no more and no less distinct perceptions than in (X, T_X), but in the former we have perceptions reduced to A.

To depict the action of a given monad-topology in the sense being presented here, let us consider the following:

Example. Let $X = \{1, \ldots, 10\}$ and T_X be a topology on X defined as

$$T_X = \{\emptyset, X, A, B, C, A \cup B, C \cup A, C \cup B\},$$

where $A = \{1, 2, 3\}, B = \{4, 5\}, C = \{6, 7, 8, 9, 10\}$.

In this case, we say, *inter alia*, that X can be presented as a union of A, B and C. This means that in X we can distinguish some perceptions A, B, C, so that we can understand X as the sum of A, B and C. (In this way it becomes our knowledge about X: X is something that consists of A, B and C). We can also say that A is not B. Why? Because $A \cap B = \emptyset$. This situation is similar to that of recognising, say, a tree, and constructing knowledge about it. We distinguish the roots (A), the trunk (B) and the crown (C). Then we see that the trunk is not the crown (because $B \cap C = \emptyset$). This amounts to a possible interpretation of some monad's acting: an interpretation of why that monad maps a fragment of the world. Here I have given an example of an impoverished topology that is not even T_0, but the way this monad acts is transferable to other monads and their ways of acting.

To complete this part of the present paper, let us note that one can prove the following:

Fact 9. *A dominant monad has the most distinct perceptions (in the sense of separation axioms); this means precisely that if (X, T_x) is a dominant monad of $\mathcal{F}_S$ and $(Y, T_y) \in \mathcal{F}_S$, then if (Y, T_y) is a T_i space, for $i = 0, 1, 2$, then (X, T_x) will be at least a T_i space.*

Moreover, one can also argue (because it is not a formal theorem) that:

Fact 10. *The structure of substance is univocally given by properties of monad-topologies, the ordering relation "being weaker than", and the relation "being a subspace of".*

I hope that the results given above are what one would expect. Even **Fact 6** – although odd-looking – can be interpreted as a counterpart of the main proposition included in Point 70 of the *Monadology*.

Conclusions

The brief analysis presented in this paper shows us how we can use topology to investigate and model the concept of substance with certain correlated concepts (compound substance, monad, perception, appetition, etc.). It is possible, of course, that some of those concepts, definitions, and theorems will be corrected or even changed. I think that for topological ontologists the following issues show up as important:
1. the exploitation of continuous functions and homeomorphisms in ontological analysis;
2. the use and ontological interpretation of different kinds of set (defined in topology) – such as dense sets, nowhere dense sets, the boundary of a set, or compact and connected sets;
3. the comparing of different complex substances – i.e. comparing families $\mathcal{F}_S$ of topologies.

Acknowledgment: This paper was supported financially by the National Science Centre, Poland, Grant No. 2017/27/B/HS1/02830.

Bibliography

Aristotle (1933). *Metaphysics, Books I–IX* (H. Tredennick, trans.). In *Loeb Classical Library* (No. 271). Cambridge: Harvard University Press.
Aristotle (1935). *Metaphysics, Books X–XIV* (H. Tredennick, trans.). In H. Tredennick & C. Armstrong (trans.) *Metaphysics, Books X–XIV. Oeconomica. Magna moralia*. In *Loeb Classical Library* (No. 271). Cambridge: Harvard University Press.
Aristotle (1938). *Categories* (H. P. Cooke, trans.). In H. P. Cooke & H. Tredennick (trans.), *Aristotle, Categories. On interpretation. Prior analytics*. In *Loeb Classical Library* (No. 325). Cambridge: Harvard University Press.
Bocheński, J. M. (1988). *Analyse der industriellen Unternehmung*. In J. M. Bocheński. *Autorität, Freiheit, Glaube. Sozialphilosophische Studien*, (pp. 119–150). München/Wien: Philosophia Verlag.
Copleston F. (1958). *History of philosophy: Vol.4. Descartes to Leibniz*. Newman Press.
Husserl, E. (1982). *Ideas pertaining to a pure phenomenology and to a phenomenological philosophy – First book: General introduction to a pure phenomenology* (F. Kersten, trans.). The Hague: Nijhoff. (Original work published 1913)
Kaczmarek, J. (2008a). *Indywidua. Idee. Pojęcia. Badania z zakresu ontologii sformalizowanej* [Individuals. Ideas. Concepts. Investigating Formalised Ontology.]. Łódź: Wyd. Uniwersytetu Łódzkiego.
Kaczmarek, J. (2008b). What is a formalized ontology today? An example of IIC. *Bulletin of the Section of Logic, 37*(3–4), 233–244.
Kaczmarek, J. (2016). Atom ontologiczny: atom substancji [Ontological atom – atom of substance]. *Przegląd Filozoficzny, 4*(100), 109–124.
Kuratowski, K. (1977). *Introduction to set theory and topology. Containing a supplement on elements of algebraic topology by R. Engelking*. Warsaw: PWN. (See also: Kuratowski K., *Topology*, vol. I, 1966, vol. II, 1968).
Leibniz, G. W. F. (1714). *Monadology*. See: J. Bennett (trans.), *The principles of philosophy known as monadology*. Retrieved from http://www.earlymoderntexts.com/authors/leibniz
Salamucha, J. (1946). Z historii jednego wyrazu ("istota") [From the history of one word ("essence")]. *Tygodnik Powszechny, 7*(48), 3–4.
Warren, R. H. (1982). The number of topologies. *Houston Journal of Mathematics, 8*(2), 297–301. See also: http://oeis.org/A001930/a001930.pdf

Krzysztof Wójtowicz
Does Mathematical Possibility Imply Existence?

Abstract: In the article I discuss the use of modal notions in the philosophy of mathematics, in particular the (seemingly) uncontroversial "maximum principle", according to which the mathematical universe is very rich as it contains "implementations" of all coherent mathematical notions. In particular, I discuss the more general problem of ontological reductions and argue that completeness theorems are not the appropriate formal paraphrase of this intuition. I also discuss Gödel's and Quine's versions of mathematical realism in this context. Finally, I argue that this issue is important for both the realism-antirealism debate in the philosophy of mathematics and for the problem of mathematical explanation.

Keywords: mathematical realism, mathematical explanation, set theory, multiverse, ontological reduction, mathematical possibility.

Several philosophical theses concerning mathematics are expressed with the use of modal notions. According to the widespread view, mathematical truths are necessary and the standard explication of this notion is "true in all possible worlds". This sounds plausible as the properties of abstract mathematical objects should not depend on some contingent state of affairs (e.g. on the value of Planck's constant, the age of the universe or the average temperature on Mount Everest on 1.01.2000).

Another thesis which seems to be rather uncontroversial is a kind of "maximum principle" concerning the mathematical universe. The general idea is that the consistency of the definition of a mathematical object is a sufficient condition for its existence. So, the mathematical universe is extraordinarily rich and has no overt restrictions (apart from logical ones). This point of view is coherent with the feeling of conceptual freedom in mathematics: we are allowed to invent new mathematical notions, introduce new definitions, formulate new theories etc. (we have only to obey the underlying logic).

Both these statements can be accepted provided we accept some kind of mathematical realism.[1] So, in this article, mathematical realism is presupposed as the

---

[1] A nominalist will interpret these modal claims differently (or definitely reject them). Hellman, for example, uses modal notions in order to defend a version of antirealism, i.e. modal structuralism,

Krzysztof Wójtowicz, Institute of Philosophy, Warsaw University, Poland.

https://doi.org/10.1515/9783110669411-010

working hypothesis: I do not analyse the arguments for and against realism but concentrate on some internal problems that arise from its use of modal notions.

I will focus on the second thesis, which concerns the existence of possible mathematical objects. The underlying ontological intuition might be expressed as the claim that in the mathematical world there are implementations of all possibilities, i.e. that every mathematical notion has an ontological counterpart. In short:

(PIE) All possible mathematical objects exist.

Throughout this article I use the acronym PIE (Possibility Implies Existence) to denote this thesis. Possibility Implies Existence is a much stronger statement than "the existence of mathematical structures is possible". Hellman claims, for example, that the existence of mathematical structures is possible, but he never claims that possible structures exist (Hellman, 1989). Of course, some consistent mathematical conceptions are regarded by mathematicians as not worth investigating, but this is another problem of a methodological rather than an ontological character. PIE seems to be very close to the attitude of a typical mathematician: it would be quite strange if a mathematician invented a consistent mathematical conception but abandoned his investigations because of an ontological (i.e. philosophical) claim that the corresponding mathematical structures do not really exist. In this sense, negative existential claims have no support within mathematics.

So, PIE looks *prima facie* extremely attractive – it has a genuine explanatory value and is (at least) worth investigating. In the article I discuss some difficulties with this view, its possible interpretations, and the problem of the presuppositions that are necessary in order to consider it plausible.

The plan of the article is as follows:

In Section 10, I discuss the problem of conceptual reductions within mathematics (the main example is set-theoretic reduction) and discuss some problems concerning PIE in this context. In particular, I claim that both the reductionist and non-reductionist standpoints face serious problems.

In Section 10, the problem of the ontological presuppositions of statements like "a consistent property has an exemplification" (or "a consistently defined object exists") is discussed.

---

in which mathematical claims are interpreted as claims concerning the possibility of the existence of mathematical structures (Hellman, 1989). Modal notions can be attractive for antirealists, especially when they are treated as primitive; this frees them from any ontological commitments (I consider this move too philosophically doubtful for other reasons).

In Section 10, I investigate the hypothetical formal paraphrase of PIE. It seems quite natural to claim that the appropriate paraphrase has the form of a completeness theorem. I argue that this interpretation is not compatible with the general ontological intuition that underlies PIE.

In Section 10, PIE is discussed in the context of set theory.

Finally, in Section 10 I show that PIE is compatible with neither Gödel's nor Quine's versions of mathematical realism.

A brief summary follows. The general conclusion is that PIE expresses a natural ontological intuition, but it has to be reformulated and made much more precise in order to become an interesting proposal for the realist. I hope that this paper can serve as a step toward clarifying this problem and as a contribution to the discussion of the problem of explanation in mathematics.[2]

Reductions in Mathematics

In this paragraph PIE is analysed in the context of the problem of conceptual reduction(s) within mathematics. PIE appeals to the notion of a mathematical object, so some preliminary clarifications are necessary. What is the class of entities called "mathematical objects"? *Prima facie*, there are many kinds (natural numbers, complex functions, geometrical figures, function spaces, topological manifolds, C^*- algebras, uncountable subsets of [0, 1], etc.) but it is also quite natural to suppose that they really are (or can be reconstructed as) entities of one type, so that both the conceptual structure and the ontology of mathematical theories become uniform.

According to a widespread view, the best candidate for such a fundamental theory is set theory. In principle we can translate every mathematical statement φ into a set-theoretic statement φ_{set}, and this translation preserves its logical structure: if T is a mathematical theory and T_{set} is its set-theoretic counterpart (translation), then the following implication is true:

φ follows from T $\rightarrow$ φ_{set} is a logical consequence of T_{set}.[3]

---

[2] Philosophical claims are often explained by finding appropriate formal counterparts (paraphrases), and some formal results are given philosophical explanations. The analyses given here can serve as an explication of the notion of mathematical possibility.

[3] There need not be equivalence, as T_{set} may use some stronger set-theoretical means, therefore proving sentences φ_{set}, such that φ is not provable in T. This translation need not always be conservative, but this seems not to be an important problem in this context.

The notion "φ follows from T" has been used in the non-formal way, which conforms to mathematical practice. We understand perfectly well that the Hahn-Banach theorem follows from the principles of functional analysis, even if we do not know the formalized version of the proof. If we accept such a translation (φ into φ_{set} and T into T_{set}), it means that we have reconstructed mathematical notions as set-theoretical notions and translated informal mathematics into set theory. Plane geometry, for example, translates into (a fragment of) the theory of real numbers, which are really Dedekind cuts on rational numbers, which are equivalence classes... which are sets.

Another important example of a reduction is provided by reverse mathematics, where mathematical concepts are reconstructed within Z_2, i.e. second-order arithmetic.[4] The aim of reverse mathematics is to discover the strength of set-theoretic assumptions necessary to prove theorems from ordinary mathematics (real and complex analysis, differential equations etc.).[5] We can also consider other foundational reductive programs, e.g. category-theoretic. In general, we are faced with the following situation:

1. There is a mathematical conception K and a mathematical domain S_K, such that S_K is the ontological counterpart of K.[6]
2. There is a mathematical (foundational) theory T_K (e.g. set theory), which is precise, has clear semantics and a clear notion of proof, etc.
3. There is a translation (paraphrase/reconstruction/...) of K into T_K: if C is a claim within K, then there is the corresponding translation α_C within T_K. Some of the important aspects to consider are:
   (a) The "entailment-structure": whenever a claim C follows from K, then T_K proves α_C.
   (b) Semantic counterparts: "T_K describes S_K" (in other words: S_K is a realization of T_K) has as its counterpart "M_K is a model for T_K" (where M_K is some formally defined structure intended to be the counterpart of S_K).

---

[4] The reconstructions of mathematical notions (like continuous function, derivative, manifold etc.) within Z_2 are then more complicated than within set theory ZFC.

[5] Simpson gives the following interpretation of the term "usual mathematical practice": "We identify as *ordinary* or *non-set-theoretic* that body of mathematics which is prior to or independent of the introduction of abstract set-theoretic concepts. We have in mind such branches as geometry, number theory, calculus, differential equations, real and complex analysis, countable algebra, the topology of complete separable metric spaces, mathematical logic, and computability theory" (Simpson, 1999, p. 1).

[6] E.g. K = Naïve Number Theory, and S_K = The Genuine Natural Numbers; K = Our Conception of the Continuum, and S_K = the Real Numbers. All these notions are informal, of course, but they have a meaning clear enough for mathematicians.

(c) Ontology: the claim that K has an ontological correlate S_K is understood as the claim that T_K has a model M_K. There are some subtleties here: T_K can have many models, M_K being just one of them. But the general thesis is clear: the notion of an ontological correlate has as its counterpart the notion of a model.

An important case is set theory. Many (perhaps all) mathematical concepts can be reconstructed within set theory, and mathematical objects can be treated as sets. From this point of view, the discovery of set theory was a discovery concerning the nature of mathematical objects (just as the discovery of atoms and elementary particles was a discovery concerning the nature of material objects). According to this view – even if we did not identify mathematical objects as sets at first – the objects of study **really are** the universe of sets. This is a very strong claim, of course. A general question arises: imagine we have found a satisfactory paraphrase of a mathematical conception (or even of a large part of mathematics) within a foundational framework F. So, we might ask whether we have discovered the true nature of the reduced (interpreted) notions, or rather just made a (more or less arbitrary) stipulation that we shall give an interpretation of these notions within F (e.g. set theory) and treat mathematical objects as the objects of F?

In my opinion, this ontological reductive thesis is coherent neither with the basic intuition underlying PIE, nor with mathematical practice.[7]

"The PIE Intuitions" – Objection

One of the important intuitions underlying PIE is the vision of a rich mathematical world in which many inhabitants of various kinds live. We cannot expect this complex world to be describable within a single conceptual scheme.[8] In particular, the claim that mathematical objects exist *per se* regardless of their set-theoretic representation is reasonable. From this point of view, mathematical sentences apply directly to these entities (e.g. arithmetical sentences apply to numbers), without involving their set-theoretic representations. Of course, sets also exist – they are the objects of set theory – but they are just one of many species of mathematical objects living within the mathematical realm. There are some drawbacks to this point

---

[7] There are also some internal subtleties of the "set-theoretic ontologism" concerning the universe/multiverse issue; this will be discussed in Section 10.
[8] An example of a system of concepts and ideas deliberately intended **not** to be of set-theoretical character is category theory. Objects and arrows are meant to be objects and arrows – **not** set-theoretic constructs.

of view: we cannot use set-theoretic notions (in particular the notion of the model, or logical consequence) to investigate logical relations between mathematical sentences or theories. Not having set theory at our disposal, we would have to rely on our pre-theoretic understanding and "grasp" the logical relationships as providing uniform semantics (and ontology) for mathematics would be problematic.

We encounter a general problem here: what does the existence of reductions (interpretations) say about the ontological status and the nature of objects? Consider two theories:

T_1 = PA (i.e. Peano Arithmetic)

T_2 = ZF with ¬Inf (the negation of the Axiom of Infinity) instead of Inf (the Axiom of Infinity).

The intended model of T_1 is natural numbers, and the intended model of T_2 is hereditarily finite sets (i.e. V_ω).[9] These two theories are mutually interpretable. The interpretation of T_1 in T_2 is straightforward: natural numbers are simply interpreted as ordinal numbers. The encoding of hereditarily finite sets as natural numbers is slightly more tedious: we have to encode finite sets as natural numbers in such a way that the membership relation between sets is "mimicked" as a certain number-theoretic relation between natural numbers; however, the possibility of encoding hereditarily finite sets as natural numbers will not convince us that they **really** are natural numbers. We have the fundamental intuition that the membership relation (and the other definable set-theoretic relations: inclusion, being the power set of a given set etc.) are quite different from number-theoretic concepts (multiplication, primeness, Diophantine equation etc.), even if the notions are (formally) mutually translatable. Similarly, Gödel numbering will not convince us that proving theorems within a formal theory T is really performing calculations on certain codes (even if there is such a number-theoretic counterpart). In general, we can encode finite structures as natural numbers, but we have a strong feeling that these structures are something different than their codes.

This simple example shows that we are not allowed to draw ontological morals in an automatic way just from the existence of an interpretation. The objects of the two theories differ in their ontological nature; this is an intuition which cannot be expressed formally but nevertheless sounds plausible.

---

[9] T_1 and T_2 have also non-standard models, of course, but the intended models for T_1 and T_2 are N and V_ω.

Mathematical Practice Objection

The "Mathematical Natural Ontological Attitude"[10] is rather non-reductive: in mathematics we investigate mathematical objects *per se*. It is quite unintuitive to claim that for example natural numbers or topological spaces, really are very complicated sets and that when studying, for example, the geometric properties of 4-dimensional differentiable manifolds, we are really studying complicated sets living somewhere within the set-theoretic universe.[11] If we accepted this point of view, we should – in principle – regard every mathematician to be a set theorist. I think that most mathematicians would reject this point of view and treat the set-theoretic reduction of, for example, real numbers to Dedekind cuts (which are pairs of sets of rational numbers, which are equivalence classes, which are sets…) as an auxiliary formal construction. A differential geometer studies objects which have a geometric nature, not a set-theoretic nature. The development of mathematics was possible without the presence of sophisticated set-theoretic notions, and probabilistic, algebraic, analytic or geometric intuitions stand on firm ground without making use of the notion of an inner model or collapsing cardinals within a forcing extension. In mathematical practice, set-theoretic notions are used at the naïve level – we do not exploit sophisticated, abstract, inherently set-theoretical constructs in their full generality in ordinary mathematics (cf. Simpson's remarks in footnote 5). Moreover, mathematicians do not use set-theoretic language, they rather express their ideas, definitions and proofs in a kind of "mathematical vernacular". They consider set theory to be rather a logician's tool for a formal reconstruction (formal idealization) of mathematics – not as mathematics proper. A functional analyst would probably not be interested in the fact that copies of his Hilbert or Banach spaces (e.g. C[0,1] or $L^2[0,1]$ or …) live within several models of ZFC. His object of study is Banach spaces *per se*, not their set-theoretic representations.

A very similar problem arises when we discuss the reconstruction of real numbers within Z_2 (i.e. in the context of reverse mathematics). The encodings are carried out using sparse formal resources, so they are quite tedious and not entirely intuitive. Assume we accept these reductions/encodings as legitimate: are then the real numbers really ZFC-sets or rather Z_2-constructs? This problem cannot be simply explained away within some kind of "if-thenism", according to which the only relevant issue is proving theorems and not looking for the "essences"

---

10 The phrase is borrowed from Fine (1986) and is modified.
11 Consider again the example of natural numbers and hereditarily finite sets. If we are interested in HF sets, we would **not** accept the thesis that we are really investigating number-theoretic facts. The question of whether A is the power set of B is a set-theoretic question, even if it might be encoded in a "number-theoretic disguise".

of mathematical objects (real numbers, in this case). The ontological questions concerning the nature of the objects of study are legitimate. I think that mathematical practice supports the intuition of a rich universe of objects of various kinds, rather than the intuition of a uniform, set-theoretic universe. So, the "natural ontological attitude" supports the view that they are objects *per se*, and that their set-theoretical representation is just a perhaps artificial encoding within a given formal framework.

The Problem of a Conceptual Background

A naïve formulation of PIE is "if P is a consistent definition of an object, then this object exists".[12] Of course, every definition is formulated within a certain conceptual environment. Mathematicians do not usually work in formalized theories, so we have to make clear exactly what is meant by the statement "P is a consistent definition formulated within a conceptual environment E". The question might also be stated differently: under what assumptions do we recognize the definition P as a legitimate (intended) definition of a certain mathematical object? Consider a toy example: a number p is recognized as prime only within (some version of) number theory NT. The notion of being prime does not make any sense in isolation. We have to be able to first define the natural number structure and then prove some facts concerning natural numbers.[13]

So, the claim concerning the existence of the P-object involves a claim concerning the existence of a mathematical domain, where the object defined by P exists. If we claim, for example, that an uncountable set exists, we commit ourselves to the existence of at least such a portion of the mathematical world in which the notion of uncountability makes sense. So, the question is how large should this portion of the mathematical universe be? What is the "quantum of mathematical existence", so to speak?

Of course, there is no general answer as it depends on the context, but the conceptual (and consequently — ontological) environment must be "self-contained"

---

12 A different formulation: if P consistently defines a property, it is exemplified.
13 In the language of arithmetic, the formula defining a prime number is simple, e.g. $prime(p) \Leftrightarrow \neg(\exists n, m \in \mathbb{N})(n \neq 1 \land m \neq 1 \land n \neq p \land m \neq p \land n \cdot m = p)$, but outside of number theory its meaning is certainly not the intended one. The formula only says that there is a certain operation · and a certain object 1, and that there are no objects $x, y \neq 1$, $x, y \neq p$, such that performing · on them yields the result p; however, this formula alone does not capture our intended notion of primeness as we have to understand the notion of multiplication (and other number-theoretic notions) first.

– it has to provide enough conceptual richness to make the considerations (concerning in particular the P-object in question) interesting and fruitful. Examples of such interesting environments are, for example, number theory, set theory, the theory of real numbers, etc. But, for example, group theory or the theory of partial orders are not interesting and rich enough to stand on their own (their axioms are meant rather to be definitions of a certain type of objects, not to be fundamental truths). Speaking metaphorically, mathematical conceptions are "Gestalt-like" – they appear to us as certain wholes. Usually these conceptions are open-ended in the sense that they can be developed, enriched, and combined with other theories, but they need to have a certain "critical mass" nonetheless, and PIE can only be reasonably applied to such "self-contained" conceptual systems.

PIE *versus* Completeness Theorems

The claim concerning the existence of all possible structures (implementations of coherent theories) is an informal, intuitive claim, and it is natural to ask whether it has a formal counterpart (paraphrase). The considerations in the previous paragraph make clear that PIE states the existence of mathematical domains, being ontological correlates of (self-contained) mathematical conceptions.

If we accepted set-theoretic reductionism, a natural formal counterpart of PIE would be the completeness theorem (provided we work in first-order logic). So, the vague, intuitive thesis that coherent mathematical conceptions have their implementations (ontological correlates) would be paraphrased as the claim that consistent theories have models (that are sets).

We could therefore identify the pre-theoretic notion of coherence with consistency, and also identify the notion of "being interpreted in an extra-linguistic domain" with "having a model". An important methodological advantage of this proposal is certainly the availability of model theory (with its powerful metamathematical tools). The choice of the first-order framework eliminates the possible tensions between the syntactic and semantic notion of consistency. Interpreting PIE as a completeness theorem provides us with powerful tools for a good formal treatment of the informal notions of "entailing", "being described by", "being a structure for", "being an ontological correlate of", etc. Instead of the vague notion of an ontological correlate for a mathematical conception, we have the precise notion of a model for a formal theory.

However, the philosophical costs of such an interpretation are high. Treating the completeness theorem as the adequate formal counterpart of PIE rests on the following assumptions:

1. The set-theoretic representation of mathematical notions is faithful and justified.
2. The notion of first-order consistency is an adequate counterpart of the intuitive notion of coherence.
3. The notion of a set-theoretic structure (model) is an adequate formal paraphrase of the metaphysical notion of the implementation (correlate) of a mathematical conception.

All these assumptions are problematic:

(re 1) This has been already discussed in section 10.

(re 2) This assumption rests on the much-disputed first-order thesis. First-order logic is certainly distinguished for many reasons, one of which is Lindström theorems, which identify first-order logic as the strongest logic that fulfills some natural conditions. Nevertheless, it is not clear whether first-order logic is the only "Genuine Logic", or just one of many possible choices (cf. e.g. the discussion in [Shapiro, 1991, 2014] or [Sher, 1991]).

(re 3) Identifying set-theoretic models with ontological correlates (counterparts) of mathematical theories seems to be the most problematic assumption from a philosophical point of view. The metaphysical intuition concerning the existence of structures S_K being realizations of mathematical conceptions K (provided we accept such a pre-theoretical intuition as legitimate) need not be satisfied just by exhibiting a formal model of given theory T_K.[14] The model might be artificial and its construction tricky. We expect something more: the existence of **genuine** mathematical (e.g. number theoretic) structures. Formal models are not (at least – not always) what mathematicians have in mind when they speak of the mathematical domains described by their theories.

I think therefore that completeness theorems do not capture the ontological intuitions underlying PIE as there are deep problems related to the relationship between set theory and (the rest of) mathematics. There is also a question concerning the interpretation of PIE as an internal problem of set theory. This problem is discussed in the next paragraph.

---

[14] "S_K satisfies K" translates into "$M_K \models T_K$" – where T_K is the formal paraphrase of the conception K, and S_K is the (intuitively conceived) structure for K, whereas M_K is a formal model for T_K.

PIE and Set Theory

What is the interpretation of PIE in the context of set theory? It cannot simply mean that all possible sets exist – this problem has been discussed already in Section 10. Apart from many other reasons, there is a trivial one: the existence of one object can be equivalent to the non-existence of another even if both the existence claims treated separately are consistent. So, PIE should rather be understood as the claim that all conceptions of set theory (or all possible understandings of the concept of set) have ontological implementations. However, some problems arise immediately.

If we restrict our considerations to first-order logic, the completeness theorem holds. And if we accept the interpretation of PIE as a completeness theorem, this simply means that all conceptions of set (e.g. all consistent extensions of ZFC) have models. This is compatible with the (informal) claim that all these models live within the set-theoretic universe V. In particular, we have (many) models of, for example, ZFC+CH, ZFC+V=L, ZFC+$c=\aleph_2$ and also of ZF+AD, ZFC+¬Con(ZFC), etc. It might be claimed that in this case PIE is fulfilled to the maximum (possible) extent: not only do ontological correlates of these theories exist, but also there are plenty of them (as there are many models for theories).

However, this is not what the adherent of PIE aims at. After all, many of the models are rather artificial and will be considered to be purely formal constructions – in a sense, artefacts of first-order logic (which is unable to characterize structures categorically). We are not willing to accept these artificial models as "Genuine Ontological Correlates". The idea behind PIE applied to set theory is rather that there are many genuine set-theoretic universes and that each (consistent/coherent) set-theoretic conception T has its "metaphysical implementation" in the form of a genuine universe V_T. Of course, "genuine" is not a formal notion – it is an expression of a metaphysical intuition. In particular, the claim that both ZFC+V=L and ZFC+MC have implementations does not only mean that there are models M_1 and M_2 for these theories. Being an ontological maximalist, a PIE-ist would rather postulate the existence of many universes *per se* where these two conceptions are implemented. Knowledge about relative consistency and the existence of models can be helpful and motivating but is not a proper ontological thesis. The idea of many set-theoretical universes living happily side by side in some giant "Meta-environment" is somehow vague, but it corresponds better to the ontological intuition underlying PIE than the claim concerning the existence of models living within one set-theoretic universe V.

Maximality considerations in set theory have found their expression in various forms.[15] The investigations into the theory of large cardinals are an example: we consider the universe to be open-ended so that the existence of larger and larger objects is postulated. PIE seems to be quite coherent with the recent multiverse/hyperuniverse programs in set theory, but this is an issue worthy of separate investigation (and, therefore, will not be discussed here).[16]

PIE and Realism

PIE looks attractive to the mathematical realist, but closer examination reveals some fundamental difficulties (at least for some versions of the realistic stance). An important feature of mathematical realism is the belief that mathematical claims have truth values, i.e. that mathematical questions have definite answers.[17] In particular, the following sentences have definite truth values (in fact, all these sentences (except CH) are theorems of standard mathematics):
- 19 is a prime number;
- there are infinitely many prime numbers;
- there are no positive natural numbers x,y,z,n ($n > 2$) such that $x^n + y^n = z^n$;
- any function continuous on the interval [0,1] is uniformly continuous;
- not all subsets of the interval [0,1] are Lebesgue-measurable;
- there is a continuous function $f : \mathbb{R} \to \mathbb{R}$ which is not differentiable;
- the second derivative of e^x is e^x;
- fundamental groups of homeomorphic topological spaces are isomorphic;
- CH;
- (...and many other sentences...)

---

15 The methodological aspects of such maximality principles within set theory were the subject of (Maddy, 1997). She analysed two principles – MAXIMY and UNIFY – where the idea behind MAXIMIZE is to provide as many isomorphic types as possible.
16 The results of Hamkins, S.-D. Friedman, Woodin can be mentioned here. The results are partially formal and have a deep philosophical interpretation. They involve the problem of justifying axioms for set theory and affect our understanding of the notion of truth in set theory (therefore also in mathematics), so they are important for the realism-antirealism debate (cf. (Arrigoni & Friedman, 2013), (Antos et al., 2015), (Hamkins, 2012) for details on these programs).
17 There is a distinction between "realism in ontology" and the (weaker) "realism in truth value", where this second stance is based on the claim that all sentences have a truth value. This fits Dummett's characterization of realism.

If a mathematical claim has a proof, then this proof warrants the possession of a truth value; however, according to the realist having a truth-value transcends possessing a proof.[18] Some sentences are independent of the commonly accepted axioms (CH is an obvious example), so they have no proof (nor their negation, of course). The claim that they nevertheless have a definite truth value is a fundamental philosophical declaration. Gödel, in particular, claimed that mathematics is a "body of mathematical propositions which hold in an absolute sense, without any further hypothesis" (Gödel, 1951/1995b, p. 305) (i.e. mathematics must not be interpreted in an "if-thenism" manner). Of course, due to the independence results there can be no single (recursive) mathematical theory which would encompass all mathematical truths; nevertheless, it is possible to look for new axioms which could enable us to settle the open problems. In Gödel's opinion, mathematics is inexhaustible and the formation of new axioms has no end: "the very formulation of axioms up to a certain stage gives rise to the next axiom" (Gödel, 1951/1995b, p. 307). The task of the set theorist is not only to prove new theorems, but also to look for arguments to justify new, plausible axioms ("Gödel's program"). In particular, CH has a definite truth value and our mission is to find it. Interestingly, Gödel also tried to find reliable axioms that determine the actual size of the continuum (Gödel 1970/1995a, 1970/1995c).

The truth-makers for these independent sentences cannot be proofs (as they do not exist), so they are true due to what happens in the abstract realm of mathematical objects. Even if we are not able to provide answers to all mathematical questions concerning this universe, due to a lack of ingenuity or even to some fundamental incompleteness phenomena, we believe that these questions are well posed. Importantly, these questions cannot be reduced to meta-mathematical problems concerning provability (i.e. "what follows from what"). The continuum hypothesis is a genuine question concerning sets, not a question about the provability of CH within ZFC (the answer to the last question is well-known and negative). If we ask whether, for example, there are infinitely many pairs of twin prime numbers (p, $p + 2$) (i.e. whether the Prime Number Hypothesis holds), we want to know – being good realists – whether this sentence is really true or false (even if it happened by some strange accident that it is independent from PA or even from ZFC).[19]

---

18 We might accept a reductive, deflationary interpretation of truth and declare the truth-makers for mathematical sentences to be proofs: i.e. the proof is not only the vehicle to learn the truth, it is rather constitutive for being true, but this is certainly **not** the point of view of the realist.
19 There are sentences with an interesting mathematical content which are independent of ZFC, and there are also examples of such sentences which are even independent of ZFC+V=L (such examples were given in (Friedman, 1981, 1986)); in order to prove these sentences, we have to assume the existence of Mahlo cardinals. (The sentences from (Friedman, 1981) concern Borel

But this point of view seems to be incompatible with PIE. If all (consistent) theories have their implementations (i.e. some ontological correlates), then the question "which one of them is true?" becomes meaningless. We could only say that they are true in one part of the mathematical realm (a set-theoretic universe, perhaps), but false in another.[20] And even if we accept some form of the multiverse/hyperuniverse program and believe that there are certain "global truths" concerning the ensemble of all these universes, these truths will not concern, for example, real numbers, but rather the (very abstract and "high-level") relationships between various universes (the existence of embedding, isomorphic sub-models, forcing extensions etc.) So, even if we accept some notion of "hyperuniverse truth", we shall only be able to answer questions like CH in very few cases. There are many sentences φ that are independent of ZFC (CH being the generic example), such that both ZFC+φ, ZFC+$\neg\varphi$ seem to be reasonable candidates for "The Very Conception of Set".[21]

Also, independence phenomena are present not only within set theory. There are many examples of independent sentences which have mathematical content and even possibly some bearing on mathematical physics. For example, da Costa and Doria proved several results of this kind concerning ZFC:

1. There is an expression for a motion $m(t)$ on $\mathbb{R}^2$ such that the assertion '$m(t)$ is ergodic in $\mathbb{R}^2$', can neither be proved nor disproved in ZFC.
2. There is an expression for a differentiable dynamical system v such that the statement 'v has a Smale horseshoe' can neither be proved nor disproved in ZFC.
3. There is a flow X such that the statement 'X is a Bernoullian flow' can neither be proved nor disproved in ZFC.
4. If φ is a ZFC predicate that gives a nontrivial characterization for chaos in a dynamical system, then there is an expression Z for a dynamical system such

---

functions from the Hilbert cube into the unit interval.) Friedman claims that "these propositions provide examples of interesting theorems whose proofs necessarily involve the outer limits of what is commonly accepted as valid principles of mathematical reasoning" (Friedman, 1981, p. 209).

20 But if it is so, we are faced with a very uncomfortable situation as all the ordinary mathematical truths (there are infinitely many prime numbers ...; every bounded set of real numbers has an infimum ...; the letters "I" and "O" are not homeomorphic ... etc.) cease to be truths concerning the mathematical universe – they become just "local" truths.

21 There are no compelling reasons to accept either CH or its negation, and there is an ongoing discussion. For example, in (Woodin, 1999, 2001) subtle arguments in favour of $\neg$CH are formulated. So, if the set-theoretic community finally accepted Woodin's point of view, then the problem of the true value of the continuum would be solved (in the same sense in which the problem of AC has been solved). But CH is only one of many such independent sentences, and there is no guarantee that other problems of this kind could be solved (or "solved") as well.

that neither $\varphi(Z)$, nor $\neg\varphi(Z)$ can be proved in the formal theory (da Costa & Doria, 1996, p. 112).

They also give examples concerning applied mathematics and claim forcefully that "Gödel incompleteness is no outlandish phenomenon; it is an essential part of the way we conceive mathematics" (da Costa & Doria, 1994, p. 188).

We might redefine the primarily robust notion of mathematical truth (of the "one-universe Platonist") into a kind of "supervenient truth": a sentence α is true simpliciter iff it is true in all universes of the multiverse. So, if, for example, in every universe there are infinitely many prime numbers, then this would be an absolute number-theoretic truth[22], but there is no reason to suppose that this will apply to **every** mathematical hypothesis of interest because this notion of truth will depend of the form of the multiverse. If it is rich (in the sense that many different set theories have their universes "to live in"), then the class of sentences with definite truth values may happen to be just ZFC, so many interesting sentences remain without a truth value. Even if we restrict our attention to "reasonable" universes, we cannot expect that this will provide answers to all interesting mathematical questions.

Some questions will therefore have "multiverse-supervenient" answers, but others will not. And then, "φ is true" can only mean "φ is a consequence of T".[23] So, a metaphysical claim is "explained away" as a metamathematical claim. There can be no universal theory describing the mathematical universe and full knowledge is not really possible.[24] This is quite uncomfortable for the realist, who is usually convinced that there is an objective theory describing the mathematical world. The seemingly "hyper-realistic" PIE violates the basic intuitions of mathematical realism.

Let me comment briefly on the status of PIE from the point of view of two important realistic positions concerning mathematics, i.e. Gödel's and Quine's.[25]

---

[22] I assume that this is such a basic fact that it should be reflected in any reasonable conception of set (of course, it is true within ZFC and its extensions).
[23] If T is first-order, this is (due to completeness) equivalent to saying that φ has a proof within T. In a general case, the notion of consequence can be understood in various ways.
[24] Importantly, it is not impossible for practical reasons – it is impossible for very fundamental reasons.
[25] Let me mention Balaguer's FBP (Full Blooded Platonism; Balaguer, 1998) here. The fundamental claim of FBP is that all logically possible mathematical objects exist. Balaguer considers consistency to be a primitive concept that is not defined in set-theoretical terms (he adopts some of the ideas of Field (1991)). Interestingly, according to Balaguer, there are really no good arguments either for or against mathematical realism, and FBP is just a hypothetical stance: if you are going to be a mathematical realist, you should choose FBP!

Gödel's realism. Gödel's version of mathematical realism is not compatible with PIE. In a very brief formulated version, the main theses of his stance are:
1. There exists an objective mathematical universe.
2. It has a set-theoretic character (the basic mathematical notion is the notion of set).
3. We obtain knowledge about the truth of axioms thanks to mathematical intuition.
4. In particular, we should (and can) try to elucidate the notion of set by finding new, plausible axioms (extending the axioms of ZFC).
5. Set-theoretical sentences have truth values (even if we cannot discover them).

This is not the multiverse view in any sense. Mathematical sentences have truth value. So, either there exists a bijection between $P(\omega)$ and $\aleph_1$ (i.e. CH is true), or there exists an uncountable subset of the real number line which is not equipollent to the real number line (i.e. CH is false). This is an absolute truth, not a truth relativized to sophisticated models for ZFC. Metaphysical truth is not reducible to its model theoretic counterpart.

If we accept Gödel's belief (which is shared by many contemporary set theorists) that investigations into the theory of large cardinals present the best unfolding of the concept of a set, then we will also believe that the true universe V is huge. False set theories (like ZFC+V=L, or even ZFC+¬Con(ZFC)) have their models (e.g. Henkin term models, or other models obtained by formal constructions) within this huge universe, but they do not deserve to have a universe of their own.

Quine's realism. PIE does not fit Quine's realism either. Quine's version of mathematical realism is justified by indispensability arguments, resting on the premise that mathematics is indispensable in science. The holistic view of ontology, combined with Quine's existence criterion ("to be is to be a value of a bound variable"), allows us to accept the existence of **exactly** those mathematical objects which are indispensable in scientific theories. But this class is far sparser than the class of objects which exist according to ZFC. We do not need $\aleph_{2000}$ in physics (not to mention even larger sets, whose existence is provable with ZFC, or large cardinals, whose existence is consistent with ZFC). So, in no way are we entitled to claim that all possible objects exist (regardless of the exact interpretation of this claim). Ockham's razor is a basic (meta-)ontological principle here: Quine claims explicitly that **not all** possible mathematical objects exist – only those which are really indispensable in science. Two passages are often quoted in this context:

> So much of mathematics as is wanted for use in empirical science is for me on a par with the rest of science. Transfinite ramifications are on the same footing insofar as they come of

a simplificatory rounding out, but anything further is on a par with uninterpreted systems. (Quine, 1984, p. 788)

I recognize indenumerable infinities only because they are forced on me by the simplest known systematizations of more welcome matters. Magnitudes in excess of such demands, e.g. Beth ω [the cardinal number of $V_\omega(N)$ and of $V_{\omega+\omega}$] or inaccessible numbers, I look upon only as mathematical recreation and without ontological rights. (Quine, 1986, p. 400)

To be exact, Quine does not speak of possible mathematical objects as he is quite suspicious about modal notions (the infamous "slum of possibles" from (Quine, 1953)).
So, PIE and Quine's version of mathematical realism are incompatible.

Concluding Remarks

PIE seems to be a natural expression of the ontological intuition of the richness of the mathematical world, and it also seems to provide an explanation for (at least some) theses of the mathematical realist. On closer inspection, however, it turns out to be difficult to reconcile with mathematical realism as it rests on strong assumptions (so it will be rejected by Quine's adherents), and – in a sense – it destroys the notion of objective mathematical truth (which makes it less attractive for the Gödelian Platonist). It seems to have a formal counterpart (completeness theorem), but its underlying ontological intuitions are very different. PIE is also difficult to reconcile with the view that mathematics is reducible to set theory, and – regardless of the plausibility of such a "reducibility claim" – it is not easy to interpret within the set-theoretic framework itself (the multiverse/hyperuniverse view seems to be closest to it).

All this does not mean, however, that PIE should be definitely rejected. It expresses some plausible intuitions, and it is natural to ask how it could be reformulated (interpreted) so as to regain as much of its ontological power as possible and be coherent with the argumentation in favour of mathematical realism. It is important to explain the rationale behind these *prima facie* plausible claims, which express a certain metaphysical attitude. It can also be viewed as a part of the project of investigating the problem of explanation within mathematics and the philosophy of mathematics. Both justifying axioms (i.e. finding philosophical motivation for them) and interpreting formal results from the philosophical point of view are part of this project.[26] I hope this paper is a step toward clarifying this issue.

Acknowledgment: The preparation of the final version of this paper was supported by an National Science Centre grant, number 2016/21/B/HS1/01955.

Bibliography

Antos, C., Friedman, Sy-D., Honzik, R., & Ternullo, C., (2015). Multiverse conceptions in set theory. *Synthese, 192*, 2463–2488.
Arrigoni, T., & Friedman, Sy-D. (2013). The hyperuniverse program. *Bulletin of Symbolic Logic, 19*(1), 77–96.
Balaguer, M. (1998). *Platonism and anti-platonism in mathematics*. New York, Oxford: Oxford University Press.
da Costa, N. C. A., & Doria, F. A. (1994). Suppes predicates an the construction of unsolvable problems in the axiomatized sciences. In P. Humpreys (Ed.), *Patric Suppes: scientific philosopher* (pp.151–193). Kluwer Academic Publishers.
da Costa, N. C. A., & Doria, F. A. (1996). Structures, Suppes predicates, and boolean-valued models in physics. In P. I. Bystrov & V. N. Sadovsky (Eds.), *Philosophical logic and logical philosophy* (pp. 91–118). Kluwer Academic Publishers.
Field, H. (1991). Metalogic and modality. *Philosophical Studies, 62*, 1–22.
Fine, A. (1986). The natural ontological attitude. In *The shaky game: Einstein, realism, and the quantum theory* (pp. 112–135). Chicago: Chicago University Press.
Friedman, H. (1981). On the necessary use of abstract set theory. *Advances in Mathematics, 41*, 209–280.
Friedman, H. (1986). Necessary uses of abstract set theory in finite mathematics. *Advances in Mathematics, 60*, 92–122.
Friedman, H. (2000). Normal mathematics will need new axioms. *The Bulletin of Symbolic Logic, 6*, 434–446.
Gödel, K. (1995). *Collected works: Vol.3* (S. Feferman et al., Eds.). Oxford University Press.
Gödel, K. (1970/1995a). A proof of Cantor's continuum hypothesis from a highly plausible axiom about orders of growth. In (Gödel, 1995, pp. 422–423).
Gödel, K. (1951/1995b). Some basic theorems on the foundations of mathematics and their implications. In (Gödel, 1995, pp. 304–323).
Gödel, K. (1970/1995c). Some considerations leading to the probable conclusion that the true power of the continuum is $\aleph_2$. In (Gödel, 1995, pp. 420-421).
Hamkins, J. D. (2012). The set-theoretic multiverse. *Review of Symbolic Logic 5*(3), 416–449.
Hellman, G. (1989). *Mathematics without numbers*. Oxford: Clarendon Press.
Maddy, P. (1997). *Naturalism in mathematics*. Oxford: Clarendon Press.

---

[26] PIE is a philosophical claim which can motivate new axioms. I think that the multiverse project can be viewed as being motivated by claims such as PIE.

Quine, W. V. (1953). On what there is. In *From a logical point of view* (pp. 1–19). Cambridge, Harvard University Press.
Quine, W. V. (1984). Review of Charles Parsons' Mathematics in philosophy. *The Journal of Philosophy, 81*, 783–794.
Quine, W. V. (1986). Reply to Charles Parsons. In I. Hahn, & P.A. Schilpp (Eds.) *The philosophy of W.V. Quine* (pp. 396–403). La Salle: Open Court.
Shapiro, S. (1991). *Foundations without foundationalism*. Oxford: Clarendon Press.
Shapiro, S. (2014). *Varieties of logic*. Oxford: Oxford University Press.
Sher, G. (1991). *The bounds of logic: a generalized viewpoint*. Cambridge, MA, MIT Press.
Simpson, S. G. (1999). *Subsystems of second order arithmetic*. Springer.
Woodin, H. (1999). *The axiom of determinacy, forcing axioms and the nonstationary ideal*. Berlin–New York: de Gruyter.
Woodin, H. (2001). The continuum hypothesis, Part I, II. *Notices of the AMS, 48*(6,7), 567–576, 681–690.

Rafal Urbaniak
Neologicism for Real(s) – Are We There Yet?

Abstract: I survey existing approaches to neologicist reconstructions of real numbers. Simons' early approach is a step towards such a reconstruction, but it's not a fully developed one. Shapiro's reconstruction merely mimics a set-theoretic one and for this reason is lacking (despite the fact that his reconstruction is technically correct). Hale's reconstruction is much more in the neologicist spirit, but philosophical worries related to Frege's constraint remain.

Keywords: neologicism, abstraction principles, reconstruction of real numbers, real number theory, Frege, Simons, Shapiro, Hale.

The goal of the game is to provide a philosophically compelling and technically adequate reconstruction of standard mathematical theories. One of the players is neologicism – the view that mathematical theories can be reconstructed as following from appropriate abstraction principles by higher-order logic. To be successful, it should provide foundations not only for natural number arithmetic, but also for other mathematical theories, such as real number theory (RNT) (its axioms will be specified in section 11). Finding appropriate abstraction principles is not simple: they should be strong enough to give the desired theory, not strong enough to prove undesired claims and (last but not least) provide a basis for a philosophically acceptable story about RNT and its applicability. This paper is not an exposition of Frege's own extensive views on real numbers. It is well known that Frege has not provided foundations for RNT in terms of abstraction principles. Instead, I survey the existing attempts of developing neologicist foundations of RNT (Simons, 1987; Shapiro, 2000; Hale, 2000, 2005), compare them, and discuss their plausibility.

Neologicism is briefly introduced in section 11. The axioms of RNT are described in section 11, and the standard construction of real numbers is described (with emphasis on the underlying abstraction principles) in section 11. In section 11 I discuss the non-arbitrary role of abstraction principles in contrast with a certain degree of arbitrariness involved in the set-theoretic reconstruction. Simons's perspective on Fregean RNT is presented in section 11 and discussed in section 11. I then move on to reconstructions provided by Shapiro (sections 11 and 11) and Hale (sections 11 and 11).

Rafal Urbaniak, 1. Centre for Logic and Philosophy of Science, Ghent University, Belgium; 2. Institute of Philosophy, Sociology and Journalism, University of Gdańsk, Poland.

https://doi.org/10.1515/9783110669411-011

Neologicism: The Main Gist

Thanks to neologicism (Wright, 1983; Zalta, 1983; Hale and Wright, 2001), the Abstraction Principles (APs) first introduced by Frege (1884, 1893, 1903), now enjoy a revived interest in the philosophy of mathematics. APs are expressions of the form:

$$f(\sigma) = f(\tau) \equiv \sigma R \tau,$$

where R is an equivalence relation, and f is a newly introduced operator (σ and τ can be first- or higher-order variables). Note that since APs are meant to be used in the foundations of mathematics, R cannot be understood set-theoretically, because neologicists cannot rely on set theory (which is only, hopefully, to be founded in appropriate abstraction principles) in their construction. Instead, the toolkit of second-order logic is assumed as logical and not dependent on set theory, and second order variables are rather taken to range over concepts rather than sets.

On the neologicist's approach, an AP is meant to fix the reference of abstract terms and explicate operation f, which, on the intended interpretation, assigns abstract objects to things that σ and τ range over.

A classic example of this strategy is the derivation of second-order Peano Arithmetic from Hume's Principle:

> The number of one concept is the same as the number of (HP)
> another concept if and only if those concepts are equinumerous.

More formally:

$$N(F) = N(G) \equiv F \approx G \qquad \text{(HP)}$$

where equinumerosity ($\approx$) is defined in pure second-order language, without employing the notion of a number (in terms of the existence of a 1-1 binary relation between the Fs and the Gs). By introducing this principle, we are supposed to: fix references of expressions such as "the number of F", determine an operation that assigns numbers to concepts, and explicate our sortal concept of a number. Adding a comprehension principle for concepts and Hume's Principle to second-order logic yields a consistent system, which allows for the derivation of second-order Peano Arithmetic (Boolos, 1987).

It should be clear why neologicism is tempting even if we put philosophical considerations aside: if we can replace a whole bunch of independent arithmetical axioms of Peano arithmetic with a single very intuitive principle that only gives us an identity criterion for numbers, then why not?

For some, neologicism is also tempting for philosophical reasons. Say you agree with neologicists that abstraction principles indeed are definitions introducing objects of new types by providing appropriate identity conditions. Then the fact that a mathematical theory can be reconstructed as a set of logical consequences of an abstraction principle (or a set thereof) shows that the mathematical knowledge involved is in fact logical, because it can be obtained by definitions and logical reasoning.

Neologicists, however, take a more philosophically involved stance towards mathematics (Hale, 1984). They accept the classical semantics of mathematical language, assuming that the singular terms of the language of mathematics purport to refer to mathematical objects, and its quantifiers purport to range over such objects. On this approach, for a mathematical sentence to be true, singular terms occurring in it must refer to mathematical objects and seemingly first-order quantification occurring in it has to range over such objects. Given that they also are convinced that most sentences accepted as mathematical theorems are true, they end up upholding platonism about mathematical objects. From this perspective, abstraction principles play an important epistemological role. For in response to the question how we, mundane beings, can have knowledge of abstract objects, such as natural numbers, the neologicists reply: we obtain it through the use of conceptually true abstraction principles which introduce identity criteria of such objects. For instance, (HP) is a conceptually true statement of the identity conditions of natural numbers, and indeed, from (HP) and (second-order) logic alone, we can obtain all the standard theorems of arithmetic. In this sense, neologicism is a systematic approach to mathematical platonism and its epistemological difficulties.

Another classic example of an abstraction principle is Basic Law V, employed by Frege to introduce the notion of an extension:

$$\{x \mid Fx\} = \{x \mid Gx\} \equiv \forall x \, (Fx \equiv Gx). \tag{BLV}$$

It says that the extension of F is the same as the extension of G just in case F and G apply to exactly the same objects. It is well known that (BLV), with second-order logic and full comprehension for concepts in the background, is inconsistent and leads to Russell's paradox. This is where one type of difficulties arises for the neologicism: it turns out that finding appropriate abstraction principles underlying mathematical theories other than natural number arithmetic is quite a challenge. In particular, no system of APs for set theory has been agreed on by the neologicist.

Neologicism also faces other challenges (see e.g. Fine, 2002). The neologicist would like to claim that the APs they use are conceptually true. However, how can this claim be justified? One cannot say that all APs are true. Within a sensible logical framework certain APs, such as BLV, lead to straightforward contradictions. Moreover, there is no consistency test for APs. What is worse, certain APs are separately consistent but mutually exclusive, and hence not all of them can be true. These and related problems give rise to the fairly open problem of finding sensible acceptability conditions for APs. All proposed solutions are quite complicated and there is no general agreement as to their plausibility and effectiveness. Moreover, explications of these solutions require quite a lot of set theory in the background. As a result, as Incurvati (2007) points out, even if there was a complicated acceptability criterion that neologicists would agree on, a person about to decide whether to accept given APs as the foundations of given mathematical theories would be in no position to know that those APs satisfy this criterion, because she wouldn't be able to rely on a mathematical theory (namely, set theory) to judge that.

More philosophical issues also arise. It is not clear that APs are capable of *fixing* the reference of abstract terms. Given a domain of non-abstract objects, no AP determines unambiguously the set of abstract objects that have to be added to this domain in order for the principle to hold. Say I introduce a new sort of objects, which I call "utopiec". I insist that these sorts of objects are correctly introduced by the following abstraction principle (the variables range over people):

x's utopiec is the same as y's utopiec iff x and y weigh the same.

Notice that the relation on the right-hand side is an equivalence relation. If the neologicist story is correct, the left side of any substitution of this principle entails the existence of an utopiec (just as the left-hand side of Hume's Principle can be used to prove the existence of numbers), whereas the right side does no such thing.

If the ontological commitments of two sides of an equivalence differ, how can it be conceptually true? Even if we accept the principle, how does it help in determining whether a given object is an utopiec? Even if we were able to identify the sort of objects in question, how would we be able to assign them to appropriate objects?

But the focus of this paper is not mainly about general philosophical concerns related to neologicism. Rather, we'd like to focus on the following, perhaps more technical, problem. Neologicists so far have conquered rather few hills and obtaining Peano Arithmetic is their main point of pride. No sufficiently strong replacement of (BLV) is on the horizon,[1] and it is not clear what standard mathematical theories can be regained by means of APs. The main concern of this paper is with one of them – I'd like to focus on neologicist attempts to obtain Real Number Theory from acceptable APs. But before we move on to the discussion of the existing approaches and our own developments, I first would like to describe our target theory in more detail.

---

[1] Although, a predicative cumulative approach is available, which results in a consistent theory containing Zermelo set theory with Choice (Urbaniak, 2010).

Real Numbers: Axiomatic Description

To fix our target, in this section we briefly survey a standard axiomatization of RNT.[2] Readers familiar with this material can safely skip this presentation.

A structure $\mathbb{R} = \langle \mathbb{F}, +, \times, \leq, 1, 0 \rangle$ is a real number structure iff it is a **complete ordered field**. For it to be a *field*, the following conditions have to be satisfied:

$$\forall x, y \in \mathbb{F}\ \exists z \in \mathbb{F}\ z = x + y \qquad \text{(Closure)}$$
$$\forall x, y \in \mathbb{F}\ \exists z \in \mathbb{F}\ z = x \times y$$
$$\forall x, y, z \in \mathbb{F}\ x + (y + z) = (x + y) + z \qquad \text{(Associativity)}$$
$$\forall x, y, z \in \mathbb{F}\ x \times (y \times z) = (x \times y) \times z$$
$$\forall x, y \in \mathbb{F}\ x + y = y + x \qquad \text{(Commutativity)}$$
$$\forall x, y \in \mathbb{F}\ x \times y = y \times x$$
$$\exists y \in \mathbb{F}\ \forall x \in \mathbb{F}\ x + y = x \qquad \text{(Identity)}$$
$$\exists y \in \mathbb{F}\ \forall x \in \mathbb{F}\ x \times y = x$$
$$\forall x \in \mathbb{F}\ \exists y \in \mathbb{F}\ x + y = 0 \qquad \text{(Inverses)}$$
$$\forall x \in \mathbb{F}\ [x \neq 0 \rightarrow \exists y \in \mathbb{F}\ x \times y = 1]$$
$$\forall x, y, z \in \mathbb{F}\ x \times (y + z) = (x \times y) + (x \times z) \qquad \text{(Distributivity)}$$
$$0 \neq 1 \qquad \text{(Neq)}$$

The first three types of conditions are straightforward. The fourth type requires the existence of identity elements which are (respectively) called 0 and 1. The next type of conditions requires that addition and multiplication have their inverses. The last two requirements are straightforward.

For a field to be ordered, $\leq$ has to be reflexive, antisymmetric, transitive and total:

$$\forall x \in \mathbb{F}\ x \leq x$$
$$\forall x, y \in \mathbb{F}\ [x \leq y \wedge y \leq x \rightarrow x = y]$$
$$\forall x, y, z \in \mathbb{F}\ [x \leq y \wedge y \leq z \rightarrow x \leq z]$$
$$\forall x, y \in \mathbb{F}\ [x \leq y \vee y \leq x]$$

---

[2] Other axiomatizations, for instance ones employing < instead of ≤, are available. This is just one out of many.

Moreover, this order should have the following properties:

$\forall x, y, z \in \mathbb{F} \ [x \leq y \rightarrow x + z \leq y + z]$ (Preservation)

$\forall x, y \in \mathbb{F} \ [0 \leq x \wedge 0 \leq y \rightarrow 0 \leq x \times y]$

$\forall \emptyset \neq A \subseteq \mathbb{F} \ [\exists x \ x \in u.b.(A) \rightarrow \exists y \ y \in l.u.b.(A)]$ (Dedekind-completeness)

(Preservation) requires that addition be monotone and that multiplication of positive reals preserve the sign. For an ordered field to be (Dedekind-) complete, each non-empty subset of $\mathbb{F}$ which has an upper bound in $\mathbb{F}$ should have a least upper bound.[3] All these requirements taken together determine the structure of real numbers up to isomorphism.[4]

Note that the notion of completeness used here is Dedekind-completeness. It is somewhat different from (and not equivalent to) Cauchy completeness. Cauchy-complete fields which aren't Dedekind-complete are known. In fact, Cantor only showed his construction is Cauchy-complete.[5]

One remark before we move on. These axioms are used to *define* what it is to be a complete ordered field. This task is separate from showing that a complete ordered field exists. On the standard approach, this goal is achieved by using set theory in the background and defining a set-theoretic structure which is a complete ordered field and which can be proven within set theory to exist. Mathematicians sometimes are inclined to say that once such a structure is described and proven to exist, given that there is a standard and agreed way of achieving this, real numbers are defined as elements of this structure. This, however, is a bit too hasty from the philosophical perspective. Real numbers were talked about and used in mathematics prior to the development of set theory, and so what the standard approach recommends is at best a re-definition of real numbers, which boils down to indicating which set-theoretic objects behave (as far as mathematical theorems of real number theory are concerned) the same way we'd expect real numbers to behave. Since whether real numbers in fact *are* such sets can still be sensibly asked, I wouldn't want to call this move a definition. Instead, I'll call it a *(re-)construction of real numbers*.[6]

---

[3] An upper bound of a set S in $\mathbb{F}$ is an $x \in \mathbb{F}$ such that for any $y \in S$ we have $y \leq x$. A least upper bound of S in $\mathbb{F}$ is an upper bound x of S in $\mathbb{F}$ such that additionally, for any upper bound y of S in $\mathbb{F}$ we have $x \leq y$.

[4] The theorem in its basic formulation is due to Cantor. See (Jech, 2002, p. 38) for a modern statement and proof.

[5] I thank Piotr Błaszczyk for this observation.

[6] Note that this notion of construction has nothing to do with the notion of construction as used in early geometry.

Real Numbers: Standard Construction

Since neologicist approaches to RNT will be evaluated in comparison to the standard construction of real numbers in set theory, I present the standard construction before moving forward. Readers familiar with this well-known material are asked to either bear with me for a bit longer, or to skip this section and come back to it as needed.

There are multiple ways of constructing real numbers in set theory. In this section I'll describe the one that's most common and relied on in neologicist constructions. To reach the reconstruction of real numbers in set theory we need to proceed through: ordered pairs, natural numbers, integers and rational numbers.[7]

An **ordered pair** $\langle x, y \rangle$ is identified with the set $\{\{x\}, \{x, y\}\}$. Using this definition, it is possible to prove in standard set theory:

$$\langle x, y \rangle = \langle x', y' \rangle \equiv x = x' \wedge y = y' \qquad \text{(Pairs)}$$

where (Pairs) is usually used as a measuring stick to assess the adequacy of a given set-theoretic definition of ordered pairs.

This is by no means the only way to represent pairs in set theory. Equally well we could have agreed to use $\{\{y\}, \{x, y\}\}$, or, in fact, as far as the preservation of (Pairs) is concerned, we could take any dummy set S and use $\{\{x\}, \{x, y\}, S\}$. This multiplicity of reconstructions is inherited by any further reconstructions relying on a set theoretic notion of a pair.

One way to identify **natural-number**-like objects in set theory is to take:

$$0 = \emptyset \qquad \text{(Natural)}$$
$$1 = \{0\} = \{\emptyset\}$$
$$2 = \{0, 1\} = \{\emptyset, \{\emptyset\}\}$$
$$3 = \{0, 1, 2\} = \{\emptyset, \{\emptyset\}, \{\emptyset, \{\emptyset\}\}\}$$
$$\vdots$$

In general, the successor of a set x is $x \cup \{x\}$. A set is called inductive if it contains $\emptyset$ and is closed under the successor operator. The set of natural numbers is identified with the smallest inductive set. The ordering of natural numbers thus understood is just elementhood. Appropriate definitions of addition and multiplication provide

---

[7] Note, however, that not all these stages are necessary. The so-called Eudoxus construction (Arthan, 2004) proceeds directly from the integers to the real numbers bypassing the intermediate construction of the rational numbers. I thank Piotr Błaszczyk for pointing this out.

us with a set-theoretic model of natural numbers. (Again, this is by no means the only way to construct a model of natural number arithmetic in set theory.)[8]

Once set-theoretic natural numbers are in place, the general strategy is extended to **integers**. One way to go about this is to observe that every integer can be represented as $\langle m, n \rangle$ for $m, n \in \mathbb{N}$. The underlying intuition is that $\langle m, n \rangle$ corresponds to the integer $m - n$.

Representation of an integer by a pair of natural numbers isn't unique – after all there are different pairs of natural numbers $\langle k, l \rangle$, $\langle m, n \rangle$ such that $k - l = m - n$. The natural move is to "identify" all such pairs – that is, to represent an integer k as the set of all pairs of natural numbers $\langle m, n \rangle$ such that $m - n = k$. Ordering for integers satisfies the condition that $\langle m, n \rangle \leq \langle r, s \rangle$ just in case $m - n \leq r - s$.

A minor obstacle is that it's too early to do this within the system, because we haven't defined subtraction for natural numbers yet.[9] This obstacle, however, can be overcome by defining an appropriate relation on pairs of natural numbers. Whatever way of explicitly defining subtraction on pairs of natural numbers we take, the adequacy of the account is measured by whether

$$x - y = x' - y' \equiv x + y' = x' + y \qquad \text{(Difference)}$$

can be obtained. Thus, the set of integers, $\mathbb{Z}$, is identified with the set of all such equivalence classes of pairs of natural numbers. It is countable, because the set of pairs from a countable set is also countable. Again, this is not the only way to obtain from (set-theoretic counterparts of) natural numbers set-theoretic objects as proxies for integers.

Once integers are in place, we move on to **rationals**. The idea now is that each rational number k can be represented as $\frac{m}{n}$ for some integers m and n. The situation is analogous to the one with integers. Now the key observation here is that in the standard setting, assuming $y, y' \neq 0$, we have:

$$\frac{x}{y} = \frac{x'}{y'} \equiv x \times y' = x' \times y \qquad \text{(Ratio I)}$$

Again, we haven't officially defined division for integers yet, but multiple different set-theoretic representations of rationals and their division can be given (the multiplicity stemming for instance from various ways of set-theoretically representing pairs or natural numbers and re-defining related notions accordingly). As long as

---

8 Note another potential source of non-uniqueness: there are fields for which there is no unique linear ordering that agrees with addition. I thank Piotr Błaszczyk for this observation.
9 Of course, there is the natural number subtraction $\dot{-}$ which has the property that if $n > m$, $m \dot{-} n = 0$. But it clearly cannot do the job of yielding negative integers.

(Ratio I) follows, the reduction is usually considered successful. Just as in the case of (Difference), the right-hand side of (Ratio I) uses an equivalence relation (only this time between pairs of integers whose second elements differ from zero). The standard move is to identify each rational number with an appropriate equivalence class of pairs of integers and to introduce the ordering component-wise.

Next, one way to go is to define *Dedekind cuts* in terms of rationals. A Dedekind cut is a pair $\langle A, B \rangle$ of sets of rational numbers, such that $A, B \neq \emptyset$, A is closed downwards, B is closed upwards and A contains no greatest element.[10] Each cut is uniquely represented by either of its two elements, so it is customary to identify real numbers with non-universal and downward closed sets of rational numbers which contain no greatest elements. It can be proven that real numbers thus defined satisfy the axioms discussed in section 11.

The Sturdiness of Abstraction

One reason why we went through the standard construction is that refreshing it will help us trace the ghost of the standard construction in some neologicist approaches. More about this soon.

Another reason is more philosophical. The way we proceeded with the exposition emphasizes that it is abstraction principles such as (Pairs), (Difference) and (Ratio) that constitute natural constraints on otherwise somewhat arbitrary set-theoretic definitions. They are arbitrary in the sense that there are non-equivalent sets of definitions that yield the same result as far as the language of RNT reinterpreted in set theory is concerned, and there are no good reasons to prefer one way of reconstructing real numbers in set theory over another. This, I submit, provides some support for thinking about mathematical theories in terms of their underlying abstraction principles.

Indeed, what I take to be one of the nicest features of neologicism is the contrast between the arbitrariness of standard set-theoretic constructions described in Section 11 and the sturdiness of abstraction principles that need to be upheld. After all, there are infinitely many ways of constructing a set-theoretic complete ordered field.[11] Yet, it seems, all of them have to result in the same (up to reinterpretation) abstraction principles.

---

10 Writing this definition formally would not contribute to clarity.
11 Just to mention the most obvious options: instead of $\emptyset, \{\emptyset\}, \{\emptyset, \{\emptyset\}\}, \ldots$, one could identify natural numbers with $\emptyset, \{\emptyset\}, \{\{\emptyset\}\}, \ldots$ (there are infinitely many other ways of getting copies of natural numbers, really) and instead of using Dedekind cuts one could use Cauchy sequences.

As Benacerraf (1965) points out, set-theoretic "reductions" aren't a convincing guide to ontology. There are no good reasons to pick one of infinitely many possible "reductions" – and if there are no good reasons to think that one of many available set-theoretic stories is a better guide to ontology than the others, there seem to be no good reasons to think that any of them provides us with reliable ontological insights.

In contrast, on most approaches, neologicism takes each type of objects introduced by APs to be objects *sui generis*. While Frege himself attempted to define numbers as extensions, neologicists usually don't try to do that and take natural numbers to be objects of their own kind, introduced by their own AP, Hume's Principle.

Observe that even if one and the same function symbol could be equally well introduced by different APs, this would not fall prey to the "embarrassment of riches" objection which Benaceraff raised against set-theoretic reductions. For this would only mean that identity between new singular terms can be introduced in terms of various equivalent conditions, not that the reference of those new terms should shift with the change of the defining condition.

One minor comment before we move on. General concerns about the neologicist project aside, there is a particular price to pay for sui-generism. The situation is analogous to the one that *ante rem* structuralists have about identifying mathematical objects across structures. Assume natural numbers have been introduced by HP. Additionally, assume that some sensible APs have been used to introduce real numbers. It is highly unlikely that it will follow that, say, the natural number 1 is the real number 1. But this worry is beyond the scope of this paper – we have to have an actual neologicist construction of real numbers first. Let's proceed to this task.

Simons's Approach

One of the first papers devoted to the subject of obtaining real numbers via abstraction principles is (Simons, 1987). Following Frege,[12] Simons rightly notices that

---

Again, there are infinitely many other options. For instance, Dedekind cuts can be taken to be pairs of sets, first elements of pairs or second elements of pairs.

12 Note that this paper is by no means an attempt at an exposition of Frege's own extensive views on real numbers. For the purpose of this paper it is enough to observe that Frege has not developed RNT from abstraction principles. For Frege's own views, see (Dummett, 1995) and, of course, a large part of Frege (2013).

real numbers should be distinguished from magnitudes. The reason is, there is only one system of real numbers (or at least I hope so), while there are many kinds of magnitude, and even one and the same type of magnitude might be assigned different real numbers depending on the choice of unit. Rather, Frege suggests, real numbers should be identified with ratios of magnitudes.

Simons first notices that within Frege's framework one is allowed to use positive integers and infinite sums of negative powers of two. Then, every positive real number a may be expressed as $a = r + \sum_{k=1}^{\infty} [\frac{1}{2^{n_k}}]$ where $r \geq 0$, $n, k \in \mathbb{N}$ and $k \mapsto n_k$ is monotone increasing, which means that each real number can be associated with the corresponding pair composed of a positive integer and an infinite sequence of positive integers.[13] Addition and multiplication can be defined on such objects to come out right (although providing the right definitions, Simons works within informal mathematics, and doesn't prove that such definitions can be constructed by means of fairly limited neologicist resources).

Any particular rational number b determines the relation that holds between pairs representing rational numbers a, c such that $a + b = c$. Take all such relations as objects and you have a domain (the set of all such relations) that can go proxy for positive real numbers.

To define real numbers over the domain of magnitudes (for $b, d \neq 0$), Simons (1987, p. 40) suggests following Euclid's Definition 5 of Book V of *Elements*:[14]

$$\frac{a}{b} = \frac{c}{d} \equiv \forall n, m \in \mathbb{N} \; [a^n \lesseqqgtr b^m \text{ as } c^n \lesseqqgtr d^m] \qquad \text{(Ratio II)}$$

(Ratio II) says that the ratios of a to b and c to d are identical just in case taking corresponding arbitrary powers preserves the relations of being less than, identical, or lower than. (Simons, 1987) contains no precise formulation of an axiomatic system, and by the same token no proof of the axioms of RNT within the neologicist system.

Comments on Simons's Approach

Let us summarize the steps taken by Simons. First, rational numbers are built from natural numbers and their infinite sequences. Then, certain binary relations

---

[13] This is pretty much the construction mentioned by Frege in II.164 of (Frege, 2013).
[14] In fact, (Ratio II) doesn't follow Euclid's definition. The right formulation is (Ratio III), further on in this paper. Note that in (Simons, 2011, p. 8), Simons abandons (Ratio II) and uses the Euclidean definition. Thanks to Piotr Błaszczyk for noticing this.

between rationals thus understood yield the domain of magnitudes, which does have a structure of real numbers, but Simons wants to keep the domain of magnitudes and real numbers apart. So reals are introduced as ratios over this domain by means of an AP – (Ratio II). While this definitely is a serious step towards developing a neologicist approach to real numbers, are we there yet?

- The whole construction is developed in informal mathematics, without showing that all the moves really are available to the neologicist; no axiomatization is really given.
- The motivation for having two levels – of magnitudes and of real numbers – was that magnitudes can be adequately captured in terms of different choices of units by different real numbers. Simons's division into two levels: that of domains and that of ratios does not seem to preserve this property. There is no hint as to how different choice of units over a domain of magnitudes could result in the assignment of different real numbers to the same magnitudes.
- Another reason why the distinction between magnitudes and real numbers was made was that different magnitudes might be measured with the same real numbers. But here, there are no multiple domains of magnitudes; there is only one, composed of relations on pairs of natural numbers and infinite sequences, and it intuitively doesn't have too much to do (apart from, perhaps, structural similarity) with magnitudes as we normally understand them.

The key role in the reconstruction is played by constructing real numbers over some sets (of sets of sets...) of natural numbers. From the neologicist perspective, the problem with such approaches is that while they allow us to prove the existence of complete ordered fields, they do not seem to satisfy *Frege's constraint*:

> [...] a satisfactory foundation for any branch of mathematics should somehow also explain its basic concepts so that their applications are immediate. (Wright, 2000)

Indeed, the claim that an account of RNT should also somehow account for its applicability can already be found in Frege:

> At the same time, however, we avoid the emerging problems of the latter approaches, that either measurement does not feature at all, or that it features without any internal connection grounded in the nature of the number itself, but is merely tacked on externally, from which it follows that we would, strictly speaking, have to state specifically for each kind of magnitude how it should be measured, and how a number is thereby obtained. Any general criteria for where the numbers can be used as measuring numbers and what shape their application will then take, are here entirely lacking. (Frege, 2013, II.159, p. 157)

While intuitively real numbers should be properties of magnitudes (perhaps, relative to measurement method), Simons's strategy, doesn't cast much light on how real number theory is used to handle real magnitudes in the real world. This stands in stark contrast to (HP), which explicitly tells us how to associate numbers with predicates which can apply to extra-mathematical objects.[15]

If you're a structuralist, this might not be a serious worry, because all you care about is proving the existence of an appropriate structure. But if you think that real numbers are properties of magnitudes (or something similar), it would be nice if your reconstruction of RNT captured this feature.

Before we move on to accounts which attempt to satisfy Frege's constraint, let's take a look at another structurally adequate account of APs resulting in complete ordered fields (Shapiro, 2000). It improves on Simons's approach, because Shapiro develops his formal constructions in more detail and within the intended theory.

Shapiro's Reconstruction

Shapiro starts with the observation that APs can operate on multiple variables taken in order. Come to think of it, even if one tries to introduce ordered tuples by means of APs, say by (Pairs) (mentioned on p. 187), such a principle already employs multiple variables taken in a certain order.

The first serious move is to introduce integers by (Difference). Then, Shapiro introduces quotients by means of:

$$Q(m, n) = Q(p, q) \equiv (n = 0 \wedge q = 0) \vee (n \neq 0 \wedge q \neq 0 \wedge m \times q = n \times p) \quad \text{(Q)}$$

where m, n, p, q are variables ranging over integers. Rational numbers are identified with quotients whose second arguments are not 0.

If P is a property of rationals and r a rational number, let's say that $P \leq r$ just in case $\forall x \, (Px \rightarrow x \leq r)$. Let's say that a property of rationals P is a cut-like property if it is bounded above and instantiated in rationals. The cut of a predicate (of rationals) is introduced by:

$$C(P) = C(Q) \equiv \forall r \, [P \leq r \equiv Q \leq r] \quad \text{(Cut)}$$

Real numbers are taken to be cuts of cut-like properties.

One thing that's rather clear is that Shapiro obtained a domain for a complete ordered field by means of abstraction principles which on the right-hand side have

---

[15] Interestingly, these discussions have some real impact outside of academia. See a Frege-inspired appeal for a change in German mathematical teaching curriculum towards a more applicability-oriented (in the relevant sense) account of real numbers (Griesel, 2007).

only logical vocabulary or operators introduced (ultimately) by logical abstraction. Next, he needs to define appropriate operations on this domain. Addition and multiplication for integers and rationals are defined by:

$$(a - b) + (c - d) = (a + c) - (b + d) \tag{1}$$
$$(a - b) \times (c - d) = (a \times c + b \times d) - (b \times c + a \times d) \tag{2}$$
$$Q(m, n) + Q(p, q) = Q(m \times q + p \times n, n \times q) \tag{3}$$
$$Q(m, n) \times Q(p, q) = Q(m \times p, n \times q) \tag{4}$$

In the first two definitions variables range over natural numbers and addition symbol on the right-hand side represents addition for natural numbers. In the second two definitions they range over integers, and the addition on the right-hand side is addition for integers.

As for addition and multiplication for real numbers, things are a bit more complicated (Also, some more details are needed to show that what we get is a complete ordered field). Say r ranges over rational numbers and P, Q are cut-like properties of rational numbers. Define:

$$C(P) < C(Q) \equiv C(P) \neq C(Q) \wedge \forall r \, (Q \leq r \rightarrow P \leq r)$$

We say that r is a ZERO rational number iff $r < 0$. Then take real number 0 to be $C(ZERO)$. It is the additive identity of the field.

The additive inverse of $C(P)$ is $C(-P)$, where $-Pr$ iff $P \leq -r$. (To prove the existence of inverses, the property of being bounded and instantiated are used.) Now, r is said to be a $P + Q$-rational number iff $\exists x, y \, (Px \wedge Qy \wedge r < x + y)$. Finally, define:

$$C(P) + C(Q) = C(P + Q)$$

Shapiro also defines multiplication, in a similar style, but the details are a bit tedious, so I will not describe them here. What matters, though, is that Shapiro proves that the operations thus introduced satisfy the usual axioms that a complete ordered field is supposed to satisfy.

So, now we at least should be convinced that the existence of a complete ordered field can be proven by means of abstraction principles (and the principles employed seem pretty safe).

Comments on Shapiro's Reconstruction

As far as Frege's constraint is concerned, Shapiro's way of introducing real numbers is rather unlike (HP), which captures Frege's intuition that natural numbers are second order properties. When we try to read off what real numbers are assigned to from (Cut), it seems that they are properties of cut-like properties of rational numbers, which doesn't provide us with much guidance when it comes to understanding the applicability of real number theory to extra-mathematical reality.

This goes against intuitions many people have about what real numbers are naturally assigned to. For instance, Hempel (1952, p. 63) takes quantitative concepts to be functions assigning real numbers to objects from a given (extra-mathematical) domain and Carnap (1966, p. 62) says that measurement is an assignment of numbers to a body or process. The underlying intuition is that rather than being properties of cut-like predicates over rationals, real numbers are properties of magnitudes or even objects themselves (fair enough, relative to a certain way of measuring them, but still). This is a feature that Shapiro's construction doesn't capture. Again, for a structuralist this is perhaps not an essential problem. But if you think APs should not only result in the appropriate structures but also provide us with a better understanding of their applicability, you might wish for a more philosophically enlightening way of introducing reals within the neologicist program.

Wright (2000), even though not a structuralist, bites the bullet. Following Heck (1997), he distinguishes between two tasks an AP can fulfill: providing an axiomatic basis for a mathematical theory, and providing a philosophical insight into the nature of the objects involved. While (HP) should do both, it is not clear to Wright that APs for real numbers should achieve both tasks as well.[16]

For Wright it seems sensible to require from an abstraction principle introducing a mathematical theory to account for its applicability only if indeed the core concepts of that theory can be communicated by explaining their applications:

> Frege's Constraint is justified, it seems to me, when – and I am tempted to say, only when – we are concerned to reconstruct a branch of mathematics at least some – if only a very basic core – of whose distinctive concepts can be communicated just by explaining their empirical applications. (Wright, 2000, p. 328)

Crucially, this is where the contrast between natural numbers and real numbers, sufficient for dropping Frege's Constraint for real numbers, arises for Wright. While we can fairly easily assign natural numbers to groups of objects, he insists, there is no clear way of achieving something analogous for assigning real numbers to quantities:

> The immediate obstacle is, briefly, that it is simply not the case that the distinctive concepts of real analysis can be grounded in their applications after the fashion in which, at least in

---

16 Thanks to Bob Hale for pressing me on my criticism of Wright's approach.

> principle, arithmetical concepts and simple geometrical concepts can. For instance, while the cardinal number of a group can be empirically determined, and the application of at least small cardinals schematized in thought [...], no real number can ever be given as the measure of any particular empirically given quantity. There is simply no such thing as determining a real value of a quantity by measurement or indeed by any other empirical procedure – any set of measurements we take will be finite, and even in the best case there will be no empirical distinction between their convergence upon a particular real value as opposed to uncountably many others sufficiently close to but distinct from it. (Wright, 2000, p. 328)

Instead, Wright suggests that a neologicist can drop the "arguably tendentious presupposition" (p. 325) that there has to be more to the real numbers than any broadly structuralist view of them can accommodate. Rather, such a request should be discounted by saying that the best that can be done is providing a theory which appropriately characterizes the relevant class of structures. Further requests of an explanation of applicability of real number theory are to be answered with whatever structuralist story about the applicability of mathematical theories in general one prefers.

For one thing, the suggested contrast between the ways of assigning natural numbers and of assigning real numbers to a part of reality, is overstated. On one hand, natural numbers are not assigned to objects directly, but rather to objects relative to a concept: one army might be five divisions and twenty regiments, etc. On the other hand, in pretty much any practical context we do have an agreed upon method of assigning a real number to a measurement. Of course, this is relative to a type of quantity, measurement methods, the choice of a unit, and the choice of a precision level. But if we really didn't have an agreed on method of assigning real numbers whenever a measurement is made, real number theory would be much less useful than it is.

More importantly, going structuralist about real number theory doesn't square very well with neologicist general methodology for recognizing objects. After all, neologicists claim that natural numbers are *sui generis* objects because singular terms that seem to refer to them occur in true mathematical statements. However a neologicist fills in the details of the argument (crucially, one would have to provide a sensible syntactic criterion of being a singular term), it seems that the argument would apply equally well to real numbers. After all, at least apparently, singular terms that seem to refer to real numbers, do occur in true mathematical statements.

For this reason, I think it is still worthwhile to search for APs for real numbers that *can* address Frege's Constraint. In search for such APs, let's take look at Hale's approach.

Hale's Reconstruction

Hale's approach quite closely follows Euclid's axioms when it comes to quantities, and his construction of reals is very similar to Weber's *Lehrbuch der Algebra. Einleitung* (1895).[17]

Hale (2000, 2005) distinguishes between entities standing in various quantitative relations to one another and quantities as such: abstract objects introduced by abstraction on a quantitative equivalence class (such as: "the length of a = the length of b iff a is as long as b").

Mere availability of phrases like "more ϕ than" or "as ϕ as" doesn't make ϕ a quantity in the relevant sense. For the phrases make sense even for such predicates as *elegant* or *impressive*, which do not admit any clear numerical measurement.[18]

What makes a domain quantitative, according to Hale, is the availability of some combination operation which preserves ordering, so that $a \oplus b$ is more ϕ than a and than b. Accordingly, a *minimal Q-domain* is a collection Q with operation $\oplus$ which commutes, associates, and satisfies the trichotomy condition, according to which exactly one of $\exists c\ a = b \oplus c, \exists c\ b = a \oplus c$ and $a = b$ holds (order $a < b$ is then defined by the second of these).[19]

To exclude infinity and infinitely small quantities Hale moves to *normal Q-domains*, which are minimal Q-domains additionally satisfying the comparability condition: $\forall a, b\ \exists m\ (ma > b)$ where m ranges over positive integers (multiplication is defined inductively).

---

[17] This has been observed by Piotr Błaszczyk. See his (Błaszczyk, 2015) for a commentary on Weber's approach and (Błaszczyk, 2013) for an interesting survey of early axiomatizations of quantities.

[18] Focusing on domains of magnitudes is already a strategy proposed by Frege, who found attempts to define quantities explicitly unhelpful: "Instead of asking which properties an object must have in order to be a magnitude, one needs to ask: how must a concept be constituted in order for its extension to be a domain of magnitudes?" (Frege, 2013, II.161, p. 158). Notice also that the notion of a quantitative domain has been reinvented independently of Frege and prior to the birth of neologicism (Kirsch, 1970).

[19] Note that (Hale, 2000, p. 181) literally uses the word *collection*. This motivates the question whether Hale's project can even get started, given that he has no neologicist story about collections. But let's not dwell on this, assuming that the notion of collection as used in the project is rather deflationary, and we could be talking about a concept instead.

For normal domains Q, Q', one can compare ratios of quantities belonging to them using the following principle:

$$\forall a, b \in Q, c, d \in Q' \ a : b = c : d \equiv \quad \text{(Ratio III)}$$
$$\equiv \forall n, m \in \mathbb{N} \ [ma \lesseqgtr nb \text{ just as } mc \lesseqgtr nd]$$

Magnitudes are said to be in the same ratio, the first to the second and the third to the fourth, when, if any equimultiples whatever are taken of the first and third, and any equimultiples whatever of the second and fourth, the former equimultiples alike exceed, are alike equal to, or alike fall short of, the latter equimultiples respectively taken in corresponding order.

To ensure that it is always possible to look at ratios with common denominators and that there is no smallest quantity, Hale introduces *full Q-domains*. Such a domain is a normal Q-domain for which $\forall a, b, c \in Q \ \exists q \in Q \ a : b = q : c$.

Still, a full Q-domain can be countable. To ensure that it's more like reals, the following move is needed. Say that $S \subseteq Q$ is bounded above by b iff $\forall a \in S \ a \leq b$. $b \in Q$ is a least upper bound of $S \subseteq Q$ iff b bounds Q above and $\forall c$ (c bounds S above $\to b \leq c$). Then, a full Q-domain is complete iff every bounded above and non-empty set $S \subseteq Q$ has a least upper bound.

Now, ratios over complete Q-domains are order-complete and isomorphic to each other. So, *if there is a complete Q-domain, we can introduce positive real numbers*. To obtain negative real numbers it is enough to introduce (Difference) where variables range over positive reals.

The challenge now is to prove the existence of complete Q-domains. There is no prior guarantee that the physical world comprises real-valued quantities. Hale's way out is to prove the existence of a positive quantitative domain built over natural numbers. To do that, he takes ratios over positive natural numbers to obtain a full domain. Then he takes cut-like properties over the domain of such ratios and proves that the domain thus obtained is complete.

Hale's Reconstruction and Frege's Constraint

In (Hale, 2000) not much is said on how the reconstruction relates to Frege's constraint; the issue is somewhat more emphasized in (Hale, 2005, p. 27):

> There is no denying the importance of providing for the application of the real numbers in a satisfactory account. But it is not obviously correct to insist – as Frege appears to do – upon a strong application constraint which demands that the definition of the real numbers should explicitly provide, at least in general terms, for their application. A weaker constraint would be just that the reals be defined in a way that allows for an explanation of their applicability.

This weaker constraint is surely irresistible, but it is very weak. Given that the reals, taken as a whole, constitute a structure of a certain kind, it will anyway be possible to explain empirical applications by drawing attention to appropriate mappings from the elements of the empirical domain and the reals – an isomorphism or partial isomorphism, as with standard representation theorems in measurement theory[20] – so it is hard to see how any otherwise satisfactory definition/construction of the reals could fail to satisfy it. The strong constraint, by contrast, is harder to satisfy – but for that very reason requires further defense. I believe it can be justified, but will not pursue the issue here. Even if it cannot be upheld, it is at least a welcome feature of a definition of reals as ratios of quantities that it makes their application in measurement integral to them.

So, there is at least one reading, let's call it the *weak reading* of Frege's constraint, on which an account of reals satisfies the constraint as soon as empirical applicability is explained by indicating a mapping from the empirical domain into the reals.

The weak reading is in the focus of an objection raised by Batitsky (2002). He argues that representation theorems provided by standard measurement theory, which connect elements of an empirical domain with real numbers and show how results about real numbers correspond to results about that empirical domain, are sufficient for an explanation of the applicability of real number theory.

Hale addresses Batitsky's concerns in (Hale, 2002), where – in line with the already quoted passage (Hale, 2005, p. 27) – he argues that being susceptible to such an explanation of applicability is not enough to satisfy Frege's constraint in the relevant sense (hopefully, to be specified). Hale's complaint is that the story provided by standard measurement theory is not a philosophical account of real numbers. Measurement theory takes real numbers as already given and explains how they can be used for measurement:

> [...] the explanation provided is external, in the sense that it makes no connection with the way the reals are defined (if they are indeed defined at all, rather than taken as primitive). And for these reasons, it cannot be what Frege had in mind, when he expressed the view that applications were essential and had to be provided for by an adequate definition – that is, whilst there remains a substantial question about what Frege's Constraint should be taken to require, it is at least clear that the availability of an explanation of the kind furnished by standard measurement theory cannot be enough to satisfy it. If it were, the Constraint would be entirely toothless. (Hale, 2002, p. 312)

Thus, Hale's problem with Batitsky's approach seems to be that Frege's constraint requires an account of real numbers to provide an explanation of their applicability,

---

[20] A representation theorem for a structure $\langle X, \leq_x, \oplus \rangle$ proves the existence of an $f : X \mapsto \mathbb{R}$ such that for all $x, y \in X$, (i) $x \leq_x y \Leftrightarrow f(x) \leq f(y)$, (ii) $f(x \oplus y) = f(x) + f(y)$, and (iii) for any f' satisfying (i) and (ii) there is a positive $r \in \mathbb{R}$ such that $f'(x) = r \times f(x)$.

whereas the explanation of applicability of real number theory provided by measurement theory instead of providing an account of real numbers as it proceeds, already assumes that real numbers have been already introduced.

Batitsky brings up – as the only alternative to the weak reading – another reading of the constraint, according to which the theory of real numbers

> [...] must explain this applicability [...] by defining real numbers as certain constructions from quantities. Such a definition would allow us to interpret statements about real numbers as general statements about all their applications in measurement. (Batitsky, 2002, 286)

Such a definition, Batitsky's argument continues, is certainly impossible, because domains of real quantities that we'd like to measure do not satisfy the conditions required for being a complete Q-domain (they often don't even satisfy the conditions for being a minimal Q-domain, because they aren't closed under composition).

Hale (2002, p. 309) concedes that much:

> There is, of course, a more-or-less obvious price to be paid for so characterizing complete quantitative domains that it is not only impossible to show that physical quantities of any kind – such as lengths, masses, etc., – constitute such a domain (or indeed, a full, normal or even minimal domain), but virtually certain that they do not do so.

The notion of a minimal domain is crucial here. For if real quantities constituted such a domain, we could simply take a ratio abstraction explicitly restricted to the elements of such a domain; if the minimality fails, no explanation of applicability *via* restricting the ratio abstraction is available (for more details see Hale, 2002, footnote 15).

There is, however – Hale claims – a more balanced reading of Frege's constraint as applying to this case, according to which, an account of real numbers should not be tied to any particular application (and so, should not build real numbers from some particular real quantities), and it shouldn't "so characterise the objects of the theory that the possibility of their application appears merely as a kind of happy accident, entirely unforeshadowed in their definition" (Hale, 2002, p. 307). Such an account should reveal the general principles underlying the use of real numbers in measuring. Given that real quantities can't be expected to have the same structure as the real numbers themselves, the account is bound to be less direct and more complicated than in the case of natural numbers and (HP).

The remaining important question is: How exactly does Hale's account of real numbers satisfy Frege's constraint as explicated in the course of the Batitsky-Hale exchange?

The straightforward explanation relying on the real quantities in question satisfying the appropriate structural requirements to underlie the construction of real numbers is not available. The proof of the existence of real numbers relies on the existence of a complete Q-domain. This existence is proven by mimicking the fairly standard construction, but that proof does nothing to explain the applicability of real numbers to real quantities. What else can be done?

> A proper development of this idea would require a much fuller discussion than I can give here, where the briefest of sketches of what I had in mind must suffice. [...] How, then, are empirical applications of the reals to such quantities to be explained? In a nutshell, and so inevitably prescinding from many important complications and qualifications, quite simply: by observing that whilst physical quantities of any given kind do not – and arguably could not – exhibit the abstract structure of even a minimal, much less a complete, quantitative domain, this does not mean that they cannot constitute a partial realisation of it. (Hale, 2002, p. 314)

To elucidate this brief suggestion, Hale considers the example of mass. Given a domain of real physical objects, we can use the equivalence relation of *being just as massive as* and introduce the masses of those objects (and only those objects) by the abstraction principle:

$$mass(x) = mass(y) \leftrightarrow x \text{ is just as massive as } y$$

Then, Hale insists, for some of those massive bodies x and y there might exist a composed massive body $x©y$, where © is the physical composition operation. If it doesn't exist, it sometimes might be brought into existence, although Hale admits that massive bodies aren't closed under ©. So, the domain of masses (with the addition operation $\oplus$, corresponding to ©) thus introduced will not even constitute a minimal Q-domain. Just as massive bodies aren't closed under composition, masses themselves aren't closed under addition. He continues:

> However, given any minimal (and hence any normal, full or complete) quantitative domain, Q, there exists a one-to-one mapping, ϕ, from masses into Q which is structure-preserving in the obviously relevant way. In particular, wherever masses m_1 and m_2 are such that $m_1 < m_2$, we have $\phi(m_i) < \phi(m_2)$, and where q_1 and q_2 are any elements of Q $\phi^{-1}(q_1)$ and $\phi^{-1}(q_2)$ both exist, $q_1 < q_2$ only if $\phi^{-1}(q_1) < \phi^{-1}(q_2)$. Further, wherever $m_1©m_2$ exists, $\phi(m_1) + \phi(m_2) = \phi(m_1©m_2)$. In virtue of this homomorphism, the (obviously incomplete) domain of masses may be viewed as partially realising the complete quantitative domain Q. And this relationship, I contend, constitutes the basic core of an explanation of the applicability of the reals in measurements of mass. (Hale, 2002, p. 314)

The problem with this approach is that if the existence of an order-preserving morphism (from something that isn't even a minimal quantitative domain into a minimal quantitative domain) is good enough to establish that the domain in question is a partial realisation of real numbers (and there is no other explication of what it means to be a partial realisation in Hale's paper), it's hard to see how this account fares better than Batitsky's measure-theoretic story. After all, the measure-theoretic explanation in fact consisted in showing that a morphism of a certain type holds, and calling the phenomenon partial realisation just seems to be a syntactic sugar that provides no deeper insight into the applicability of real number theory to real quantities than that provided by Batitsky's account.

A more satisfactory development of the explanation presented by Hale would require a more precise explication of the notion of partial realisation, an account of how it differs from the simple existence of a morphism, and how its use in the story makes a difference. The story would have to, moreover, be strongly connected to the way real numbers are introduced, to make it abundantly clear that the possibility of partial realizations between real domains and the domain of reals is not, in Hale's own words, merely a kind of happy accident, entirely unforeshadowed in the definition of real numbers. The account would also have to explain how developing the real numbers the way Hale developed them and explaining their utility in terms of partial realizations is different from (and better than) the standard structuralist story about the real number structure and the applicability of real number theory in virtue of the existence of an appropriate embedding. So far, such a development, I submit, is missing.

Before wrapping up, one more issue that might be worth considering. In the story about composition operations on real objects, a modal factor snuck in quite quickly, and it seems indispensable if we are to obtain anything close to potential minimal Q-domains starting from real quantities. After all, not all results of composition of massive bodies exist, and as soon as one wants to claim that they nevertheless are (in some, perhaps merely logical sense of the word) possible, or is willing to consider the set of results of possible composition operations on massive bodies, one needs to take modal logic seriously, and explicitly present and justify all the modal assumptions needed. So, it seems, a plausible story about the applicability of real number theory to domains of real quantities might also require some systematic approach to modalities employed. Again, so far, such a development is missing.

Acknowledgment: This research was funded by Research Foundation Flanders and Visiting Fellowship at The Long Room Hub of Trinity College Dublin. I would also like to thank Piotr Błaszczyk, Cezary Cieśliński, Michał Tomasz Godziszewski, Bob Hale, James Levine and Pawel Pawlowski for their comments on earlier versions of this paper. I would also like to express my gratitude to Nathan Wood for his patience in proofreading the manuscript.

Bibliography

Arthan, R. D. (2004). *The Eudoxus real numbers*. Retrieved from https://arxiv.org/abs/math/0405454v1
Batitsky, V. (2002). Some measurement-theoretic concerns about Hale's "Reals by abstraction". *Philosophia Mathematica, 10*(3), 286–303.
Benacerraf, P. (1965). What numbers could not be. *Philosophical Review, 74*, 47–73.
Błaszczyk, P. (2013). Nota o rozprawie Otto Höldera Die Axiome der Quantität und die Lehre vom Mass. *Annales Universitatis Paedagogicae Cracoviensis. Studia ad Didacticam Mathematicae Pertinentia, V*, 129–142.
Błaszczyk, P. (2015). Nota o Lehrbuch der Algebra. Einleitung Heinricha Webera. *Annales Universitatis Paedagogicae Cracoviensis. Studia ad Didacticam Mathematicae Pertinentia, VII*, 117–127.
Boolos, G. (1987). The consistency of Frege's foundations of arithmetic. In *On being and saying: Essays for Richard Cartwright*. MIT Press. Reprinted in *Logic, logic, and logic*, 1998, Harvard University Press.
Carnap, R. (1966). *Philosophical foundations of physics*. Basic Books.
Dummett, M. (1995). Frege's theory of real numbers. In W. Demopoulos (Ed.), *Frege's philosophy of mathematics* (pp. 388–404). Harvard University Press.
Fine, K. (2002). *The limits of abstraction*. Clarendon Press.
Frege, G. (1884). *Die Grundlagen der Arithmetik: eine logisch-mathematische Untersuchung über den Begriff der Zahl*. W. Koebner. English translation: L. Austin (Trans.) (1974), *The foundations of arithmetic: A logico-mathematical enquiry into the concept of number* (2nd Rev. ed.). Oxford: Blackwell.
Frege, G. (1893). *Grundgesetze der Arithmetik*, volume (band 1). Verlag Hermann Pohle. Partial translation: M. Furth (Trans.) (1964), *The basic laws of arithmetic*. Berkeley: University of California Press.
Frege, G. (1903). *Grundgesetze der Arithmetik*, volume (band 2). Jena: Verlag Hermann Pohle.
Frege, G. (2013). *Basic laws of arithmetic* (P. A. Ebert & M. Rossberg, Trans.). Oxford University Press.
Griesel, H. (2007). Reform of the construction of the number system with reference to Gottlob Frege. *ZDM Mathematics Education, 39*(1-2), 31–38.
Hale, B. (1984). Frege's platonism. *The Philosophical Quarterly, 34*, 225–241.
Hale, B. (2000). Reals by abstraction. *Philosophia Mathematica, 8*(2), 100–123.
Hale, B. (2002). Real numbers, quantities, and measurement. *Philosophia Mathematica, 10*(3), 304–323.
Hale, B. (2005). Real numbers and set theory – extending the neo-Fregean programme beyond arithmetic. *Synthese, 147*(1), 21–41.
Hale, B., & Wright, C. (2001). *The reason's proper study. Essays towards a neo-Fregean philosophy of mathematics* [Introduction] (pp. 1–27). Oxford University Press.
Hale, B., & Wright, C. (2005). Logicism in the twenty-first century. In S. Shapiro (Ed.), *The Oxford handbook of philosophy of mathematics and logic* (pp. 166–202). Oxford University Press.
Heck, R. (1997). Finitude and Hume's principle. *Journal of Philosophical Logic, 26*, 589–617.
Hempel, C. G. (1952). *Fundamentals of concept formation in empirical science*. University of Chicago Press.

Incurvati, L. (2007). On some consequences of the definitional unprovability of Hume's principle. In P. Joray (Ed.), *Contemporary perspectives on logicism and the foundations of mathematics*. CDRS.

Jech, T. (2002). *Set theory. The third millenium edition, revised and expanded*. Springer.

Kirsch, A. (1970). *Elementare Zahlen-und Größenbereiche*. Göttingen: Vandenhoeck und Ruprecht.

Shapiro, S. (2000). Frege meets Dedekind: A neologicist treatment of real analysis. *Notre Dame Journal of Formal Logic, 41*(4), 335–364.

Simons, P. (1987). Frege's theory of real numbers. *History and Philosophy of Logic, 8*(1), 25–44.

Simons, P. (2011). Euclid's context principle. *Hermathena, 191*, 5–24.

Urbaniak, R. (2010). Neologicist nominalism. *Studia Logica, 96*, 151–175.

Wright, C. (1983). *Frege's conception of numbers as objects*. Aberdeen University Press.

Wright, C. (2000). Neo-Fregean foundations for real analysis: Some reflections on Frege's constraint. *Notre Dame Journal of Formal Logic, 41*(4), 317–334.

Zalta, E. (1983). *Abstract objects: An introduction to axiomatic metaphysics*. D. Reidel.

Jacek Paśniczek
Possible Worlds and Situations: How Can They Meet Up?

Abstract: The main goal of this paper is to discuss the most general ontological features of possible worlds and situations and to compare the two categories of entity. Possible worlds and situations are formally characterised using the notions of consistency, completeness and deductive closure. Usually the notions appeal to a class of propositions – consequently, to a fixed language. In this paper, we apply an algebraic framework (based on the De Morgan lattice) to render the ontological structure of possible worlds and situations. Such a theoretical move will make ontological issues independent of any language.

Keywords: possible worlds, situations, non-standard possible worlds, (in)consistency, (in)completeness, De Morgan lattice.

Concepts of possible worlds and situations are certainly among the most widespread, exploited and even abused concepts in contemporary analytic philosophy (and beyond). Needless to say, it was in the realm of logic and formal semantics that the idea of possible worlds emerged for the first time, and where the concept of "worlds" continues to be successfully applied. Situations, which are undoubtedly very closely related to worlds, are mainly associated with "situation semantics" – i.e. a theory of semantic information.

Surprisingly enough, the concepts of possible worlds and situations are so general that they are almost devoid of any ontological content. All one can say about possible worlds is that they are possible (consistent) and complete – i.e. they are maximally consistent. This means that for any possible world and proposition A, either A or $\neg A$ – but not both – can be true in that world. As regards situations, it is assumed that they can be incomplete and partially ordered by a containment relation (part-whole relation). Sometimes situations are referred to as partial worlds (see, for example, Fox and Lappin, 2005, ch. 2.7). Every proposition that is true in a situation will also be true in any broader situation in the sense of the relation.[1] Situations are usually considered as parts of the real world or of merely possible worlds. As such, situations which are maximal with respect to the containment

---

[1] In terms of situation semantics, every infon supported by the former situation will also be supported by the latter one.

Jacek Paśniczek, Maria Curie-Skłodowska University, Lublin, Poland.

https://doi.org/10.1515/9783110669411-012

relation can be identified with possible worlds themselves. From this perspective only, possible worlds appear to be a kind of situation.

Despite its generality, the concept of possible worlds has turned out to be inadequate for some logical and philosophical applications. There has been a need to construct a semantics for non-classical logic, and most importantly for paraconsistent logic, and also to construct theories of fine-grained meaning for natural language. Thus, the consistency assumption came to be dropped, and *impossible* worlds appeared, in which inconsistent propositions can be true. And, since completeness is the dual for the notion of consistency, these "new" worlds were also allowed to be incomplete. Conspicuously, impossible worlds need not be deductively closed, and so the inconsistency does not lead to their over-determination: i.e. it does not render all impossible worlds identical.

There are a variety of names used for worlds which can be inconsistent: "impossible worlds",[2] "non-normal worlds" (cf. Kripke, 1965), "non-standard possible worlds" (cf. Rescher and Brandom, 1979), "non-classical worlds" (cf. Cresswell, 1972), or "impossible possible worlds" (cf. Sillari, 2007). Perhaps the term "impossible worlds" is used most often to refer to all those "worlds". There are numerous semantical and logical applications of formal frameworks which include the set of possible worlds together with the set of impossible worlds (e.g. paraconsistent and modal logics, intensional semantics etc.).

Usually, consistency and completeness are predicated of worlds and situations with respect to assumed truth conditions of formulas. Worlds and situations as such are devoid of any internal structure, which then renders them consistent or inconsistent, and complete or incomplete. The truth conditions also allow one to define the containment (inclusion) relation between worlds or situations: v contains w iff every proposition that is true in v is true in w. (In the case of possible worlds, this relation will be that of simple equivalence.)

Our objective in this paper is to define notions of consistency, completeness and containment for worlds, and for situations, treating these as concepts independent of language, so that they may be taken to pertain to the ontological nature of worlds and situations themselves. We have focused our attention on a very weak conception of worlds/situations, in the sense that while it allows for inconsistency and incompleteness, it is not as radical as many notions of impossible worlds (such as, for example, those coined by Rantala, 1982, pp. 106–115, and Priest, 2001 – see, especially, his N_4). In particular, we shall appeal to the idea of non-standard

---

[2] Presumably, Charles Morgan was the first scholar ever to have used this term (cf. Morgan, 1973, pp. 280–289).

possible worlds proposed by Rescher and Brandom (cf. Rescher and Brandom, 1979).

Introducing the Algebraic Approach

Let us consider an algebraic structure commonly referred to as the De Morgan lattice $\langle M, \leq, \sqcup, \sqcap, ^- \rangle$, where $\langle M, \leq, \sqcup, \sqcap \rangle$ is a distributive lattice. The operation $^-$ is called the De Morgan *negation* and it is characterised by the following conditions:

$$\bar{\bar{a}} = a, \quad \text{DM1}$$
$$a \leq b = \bar{b} \leq \bar{a}, \quad \text{DM2}$$
$$\overline{a \sqcap b} = \bar{a} \sqcup \bar{b}, \quad \text{DM3}$$
$$\overline{a \sqcup b} = \bar{a} \sqcap \bar{b}.^3 \quad \text{DM4}$$

It is worth emphasising that the equalities $a \sqcup \bar{a} = V$ and $a \sqcap \bar{a} = \emptyset$ will not generally hold, even if the lattice has the greatest and the least element, V and $\emptyset$ respectively. What is particularly important for us is that in the De Morgan lattice the following may hold true: $a \leq \bar{a}$, $\bar{a} \leq a$ and $a = \bar{a}$. The De Morgan negation occurring in the formulas can hardly be read as "not".[4] However, there will be a very important meaning associated with these three formulas. To see this, let us consider the following theorems of the De Morgan lattice, which are easy to prove:

$$a \leq \bar{a} \wedge b \leq a \supset b \leq \bar{b}, \tag{1}$$
$$a \leq \bar{a} \wedge b \leq \bar{b} \supset a \sqcap b \leq \overline{a \sqcap b}, \tag{2}$$
$$\bar{a} \leq a \wedge a < b \supset \neg b \leq \bar{b}, \tag{3}$$
$$\bar{a} \leq a \wedge a \leq b \supset \bar{b} \leq b, \tag{4}$$
$$\bar{a} \leq a \wedge \bar{b} \leq b \supset \overline{a \sqcup b} \leq a \sqcup b, \tag{5}$$
$$a \leq \bar{a} \wedge b < a \supset \neg \bar{b} \leq b. \tag{6}$$

Assume, provisionally, that $a \leq b$ means that a is contained within b, whereas $a < b$ means that a is properly contained within b, $a \leq \bar{a}$ means that a is consistent, $\bar{a} \leq a$

---

3 The De Morgan lattice is an algebraic structure that is frequently applied in logical investigations (e.g. many-valued and paraconsistent logics, first-degree entailment). However, our use of this algebraic structure here will not have much in common with the above-mentioned applications. It should be noted that **DM3** and **DM4** follow from **DM1** and **DM2**.

4 These are not true in Boolean algebra.

means that a is complete, and $a = \bar{a}$ means that a is maximally consistent. Then, the theorems can be read in the following way, respectively: (1) if a is consistent and b is contained within a, then a will be consistent; (2) if a and b are consistent, then their meet will be consistent as well; (3) if a is complete and is properly contained within b, then b will be inconsistent; (4) if a is complete and a is contained within b, then b will be complete as well; (5) if a and b are complete, then their join will be complete as well; (6) if a is consistent and b is properly contained within a, then b will be incomplete. As we can infer therefrom, the proposed reading of $a \leq \bar{a}$ and $\bar{a} \leq a$ perfectly matches the classical meaning of consistent and complete sets of formulas, as long as the symbols $\leq$, $\sqcap$, and $\sqcup$ are understood set-theoretically (as inclusion, intersection and union, respectively). So, let us define consistency and completeness in formal terms:

$$\mathbf{con}(a) \equiv_{df} a \leq \bar{a},$$
$$\mathbf{com}(a) \equiv_{df} \bar{a} \leq a.$$

What should be emphasised is that the notions of consistency and completeness involve both the De Morgan negation and the ordering relation. It is worth noting that $\mathbf{con}(a)$ iff $\mathbf{com}(\bar{a})$ and vice versa, i.e. $\mathbf{con}(\bar{a})$ iff $\mathbf{com}(a)$. So $\mathbf{con}$ and $\mathbf{com}$ are, in a sense, dual notions. Still, it is not easy to discern the proper significance of the De Morgan negation, and consequently we have no guarantee that $\mathbf{con}$ and $\mathbf{com}$ express our ordinary notions of consistency and completeness. Needless to say, nothing is decided about the ontological nature of the entities which are elements of M. To be more specific, they could be interpreted as sets, as objects in a very general sense (i.e. Meinongian objects[5]), or as worlds and situations. Obviously, we are mainly interested in the last of these interpretations. With this one in mind, we can understand $a \leq b$ thus: whatever is true in a is true in b; a proposition is true in $a \sqcup b$ iff it is true in a or in b; a proposition is true in $a \sqcap b$ iff it is true in a and in b. Even so, our algebraic approach, as it stands, explains little about the nature of consistency and completeness. In what follows, we shall enrich the De Morgan lattice with an additional operation which will enable us to introduce more advanced ontological notions. Before doing so, we shall first introduce and discuss the concept of non-standard possible worlds in the set-theoretic framework.

---

[5] See, for example: Paśniczek (2014a, 2014b, 2016).

Non-Standard Possible Worlds

As we mentioned earlier, the concept of non-standard possible worlds was introduced by Rescher and Brandom (1979). The concept is meant as a generalisation of the concept of (standard) possible worlds. Generally speaking, we can think of non-standard possible worlds – n-worlds, for short – as determined by sets of propositions that are believed to be true in them (i.e. not only maximally consistent sets of propositions, as in the case of possible worlds!). Thus, n-worlds can be inconsistent (for some proposition A, both A and $\neg A$ may be true in a given world[6]) and incomplete (for some proposition A, neither A nor $\neg A$ may be true in a given world). However, n-worlds are not entirely irregular, in the sense of being correlated with a quite arbitrary set of propositions, as happens in the case of strongly impossible worlds. Some properties of standard possible worlds – let us refer to them as p-worlds, for short – survive in n-worlds; these worlds remain deductively closed in a certain sense of that term while preserving logical laws.

Let us now develop a set-theoretic approach to the ontology of non-standard possible worlds. Let W be the set of all p-worlds. In the contemporary analytic tradition, propositions are treated as subsets of W. (Let us disregard, for now, certain obstacles to such a treatment of propositions.[7]) N-worlds can be thought of as correlated with subsets of $\mathcal{P}(W)$ (i.e. the power set of W). For the sake of simplicity, if there is no danger of ambiguity, we shall identify n-worlds with subsets of $\mathcal{P}(W)$ (i.e. members of $\mathcal{P}(\mathcal{P}(W))$).

The crucial semantic principle associated with the ontology of n-worlds will be the following:

(*) A proposition A obtains[8] in an n-world U (in short: $U \Vdash A$) iff there exists a set $X \in U$ such that $X \subseteq [A]$, where $[A]$ is the set of p-worlds in which A obtains.

In order to facilitate an understanding of the concept of n-worlds, let us consider some instances of these. Let $s, u, v, w \in W$, and let us assume that for any two distinct p-worlds, there will be a proposition which is held in one of them but not in the other. According to this principle (*), exactly the same propositions are true in p-world s and in the n-world $\{\{s\}\}$. Thus we can interpret p-worlds as some

---

[6] We are operating on the assumption here that for every proposition, there will be a proposition which is its negation.

[7] See, for instance, Dunn and Hardegree (2001, ch. 13).

[8] The expression "a proposition obtains in" is frequently used in Rescher and Brendom (1979). In the present paper, we shall, in many contexts, use the expression interchangeably with the more familiar expression "is true in". (This is mainly for stylistic reasons.)

n-worlds (in other words, we may embed p-worlds in n-worlds). It is worth noting that a proposition will be true in an n-world $\{\{s\}, \{w\}\}$ iff it is true in s or in w; consequently, this n-world is basically inconsistent.[9] A proposition is true in an n-world $\{\{s, w\}\}$ iff it is true both in s and in w; what then follows is that the n-world is basically incomplete.[10] A proposition is true in an n-world $\{\{s, u\}, \{v, w\}\}$ iff it is true in both s and u or both in v and w; consequently, this n-world is basically incomplete and inconsistent.[11] A proposition is true in the n-world $\{W\}$ iff it is true in every p-world; call that n-world the *universal* n-world. A proposition is true in the n-world $\mathcal{P}(W) \setminus \{\emptyset\}$ iff it is true in at least one p-world; let us call that n-world the *particular* n-world. Every proposition obtains in the n-world $\{\emptyset\}$ (the *full* n-world), and no proposition obtains in n-world $\emptyset$ (the *empty* n-world). The ontological status of the last two n-worlds is somewhat controversial as, by way of contrast with other n-worlds, in $\{\emptyset\}$ even negations of logical laws obtain, and in $\emptyset$ no logical law obtains.[12]

Rescher and Brandom also proposed an algebraic approach to n-worlds. According to this approach, n-worlds can be conceived of as built out of standard possible worlds – p-worlds – by means of the *join* and *meet* lattice operations. Thus, we start, for example, with p-worlds u and v, and we get the following n-worlds: $u \oplus v$ and $u \otimes v$ (note that the meet $u \otimes v$ is never "empty", because at the very least logical laws obtain in v and w). Moreover, we can repeatedly apply these operations. Generally, for any n-worlds U and V, $U \oplus V$ will be understood as the n-world in which everything that is true in U or in V is also true in this very n-world; $U \otimes V$ will be understood as the n-world in which everything that is true in U and in V is also true in this very n-world. These operations can be defined set-theoretically as follows:

$$U \oplus V = U \cup V,$$
$$U \otimes V = \{X \cup Y : X \in U \wedge Y \in V\}.$$

The algebraic structure $\langle \mathcal{P}(\mathcal{P}(W)), \oplus, \otimes \rangle$ turns out to be a distributive lattice. It should be stressed that every n-world U can be displayed in the following canonical form: $U = \{\{w_{ij} : j \in J_i\} : i \in I\}$, where w_{ij} is a p-world.[13]

---

[9] If a proposition A is true in s but not in w, then $\neg A$ is true in w. Thus, A and $\neg A$ will be true in $\{\{s\}, \{w\}\}$.

[10] If a proposition A is true in s, but not in w, then $\neg A$ is true in w. Thus, $\neg A$ is not true in s and A is not true in w. It follows that neither A nor $\neg A$ are true in $\{\{s, w\}\}$.

[11] See notes 9 and 10.

[12] Note that we are not stipulating non-emptiness for the set of p-worlds. If $W = \emptyset$, then there are only two n-worlds – the full world and the empty world.

[13] Generally, sets I and J_i may be empty.

The ordering of $\langle \mathcal{P}(\mathcal{P}(W)), \oplus, \otimes \rangle$ is defined as follows:

$$U \leq V \equiv_{df} (\forall X \in U)(\exists Y \in V)(Y \subseteq X).^{14}$$

If we enrich the structure with the following operation,

$$\overline{U} = \{X \subseteq W : \text{ for every } Y \in U, X \cap Y \neq \emptyset\}$$

then the operation $^-$ will turn out to be the De Morgan negation and, consequently, the structure $\langle \mathcal{P}(\mathcal{P}(W)), \oplus, \otimes, ^- \rangle$ will turn out to be a De Morgan lattice.[15] It is worth noting that, in particular, $\{\{v\}\} = \overline{\overline{\{\{v\}\}}}$, i.e. n-world $\{\{v\}\}$ identified with p-world v is consistent and complete.

Let us assume that $\vDash A$ means that A is logically valid. In that case, the following can be easily proved:

$$U \leq V \supset (U \Vdash A \supset V \Vdash A), \tag{7}$$

$$U \leq \overline{U} \supset (U \Vdash A \supset \neg U \supset \Vdash \neg A), \tag{8}$$

$$\overline{U} \leq U \supset (\neg U \Vdash A \supset U \Vdash \neg A), \tag{9}$$

$$\vDash A \supset (U \neq \emptyset \supset U \Vdash A), \tag{10}$$

$$\vDash (A \supset B) \supset (U \Vdash A \supset U \Vdash B), \tag{11}$$

$$\{W\} \Vdash A \equiv \vDash \Box A, \tag{12}$$

$$\mathcal{P}(W) \setminus \{\emptyset\} \Vdash A \equiv \vDash \Diamond A, \tag{13}$$

$$\{W\} \Vdash (A \supset B) \supset (U \Vdash A \supset U \Vdash B),^{16} \tag{14}$$

$$\{\emptyset\} \Vdash A, \text{ for any } A. \tag{15}$$

(However, $U \Vdash (A \supset B) \supset (U \Vdash A \supset U \Vdash B)$ does not hold!) It should be noted that according to (2), $U \leq \overline{U}$ entails consistency, while according to (3), $\overline{U} \leq U$ entails

---

14 E.g. $\{\{u, v, w\}\} \leq \{\{u, v\}, \{s, v, w\}\}$. By the way, the same kind of formal construction is adopted by Van Fraassen (1969, pp. 477–487).
15 Proofs thereof can be found in Paśniczek (2014a, 2016).
16 $U \Vdash A$ can be expressed using a simpler formula: UA where U is interpreted in UA as an n-world operator and can be read "in U it is true that". According to (6) and (7), the formulas $\{W\} \Vdash A$ and $\mathcal{P}(W) \setminus \{\emptyset\} \Vdash A$ can also be read using the classical S5 formulas $\Box A$ and $\Diamond A$ respectively. This way of interpreting n-worlds ontology enables us to create a modal logic of n-worlds. The logic, n-logic, is axiomatised by the following axioms and rules:

N1 Truth-functional tautologies,
N2 $\Box(A \supset B) \supset (\alpha A \supset \alpha B)$,
N3 $\Box A \supset A$,
N4 $\neg \Box \neg \alpha A \supset \Box \alpha A$,
N5 $\Diamond \neg A \supset \neg \Box A$,

completeness. It can be verified that U is consistent iff there are no disjoint sets $X, Y \in U$; U is complete iff for every X, Y such that $X \cup Y = W$, either $X \in U$ or $Y \in U$.

Yet another notion of consistency can be introduced. Since every p-world is consistent, every n-world contained within a p-world must be consistent, as well. Here, by "p-worlds" we mean n-worlds of the form $\{\{v\}\}$, where v is a p-world *per se*. (We observed earlier that exactly the same proposition is true in n-world $\{\{v\}\}$ as in p-world v.) This notion of consistency can be expressed by a simple condition:

$$\mathbf{scon}(U) \equiv_{df} \bigcap U \neq \emptyset.$$

scon is called *strong* consistency since $\bigcap U \neq \emptyset$ entailes $U \leq \overline{U}$, but not conversely; i.e.

$$\bigcap U \neq \emptyset \supset U \leq \overline{U} \tag{16}$$

One can see that the same propositions will be true in n-worlds U and V, if:

for every $X \in U \cup V$ there exists some $Y \in U \cap V$ such that $X \subseteq Y$. (**)

Since the condition (**) defines an equivalence relation,[17] and for every equivalence class there is a maximal element (with respect to $\subseteq$[18]), we can represent n-worlds only with the help of these maximal elements. If $U \in \mathcal{P}(\mathcal{P}(W))$, then

$$[[U]]_W = \{X \subseteq W : (\exists Y \in U)(Y \subseteq X)\}.^{19}$$

Now let us assume that n-worlds are represented by maximal elements: i.e. let $\mathbb{N}_W = \{[[U]]_W : U \in \mathcal{P}(\mathcal{P}(W))\}$ be the set of n-worlds. Then the structure $\langle \mathbb{N}_W, \subseteq$

MP   if $\vdash_n A \supset B$ and $\vdash_n A$, then $\vdash_n B$,
G   if $\vdash_n A$ then $\vdash_n \Box A$.

Here, α represents the n-world operator or classical modal operators. In particular, one should note that (8) correspond to **N2**. The semantics of n-logic is based on (*), and it should be emphasised that truth conditions will be recursive. Clearly, n-logic contains S5. For a complete description of n-logic, see Paśniczek (1998a,b).

[17] For example, $\{\{u,v\}, \{v,w\}, \{s,u,v\}\}$ and $\{\{u,v\}, \{v,w\}, \{s,v,w\}\}$ are n-worlds that are equivalent in this sense.

[18] The two n-worlds mentioned above are contained in the n-world $[[\{u,v\}, \{v,w\}]] = \{\{u,v\}, \{v,w\}, \{s,u,v\}, \{s,v,w\}, \{s,u,v,w\}\}$, which is maximal with respect to $\leq$.

[19] In most cases, providing that doing so does not lead to misunderstandings, we will drop the index referring to a particular set of p-worlds.

, $\cup, \cap, \bar{}\rangle$ will turn out to be a De Morgan lattice with $\subseteq$ as its partial ordering.[20] From now on, when discussing issues pertaining to n-worlds we shall refer to this structure.

As we remarked earlier, n-worlds are not deductively closed under *modus ponens*. In particular, if $[[U]] \models A$ and $[[U]] \models B$, it need not be the case that $[[U]] \models A \wedge B$. Only those n-worlds are closed in this sense for which $\cap[[U]] \in [[U]]$. This boils down to the following:

$$\mathbf{Clos}([[U]]) \equiv_{df} [[\{\cap U\}]] = [[U]].$$

Clos($[[U]]$) means that U is the principal filter generated by $\cap U$.

Situations as Non-Standard Possible Worlds

Non-standard possible worlds can be incomplete, inconsistent and partially ordered. Thus, according to our preliminary remarks, they can be interpreted as situations (and possibly inconsistent ones). Still, we may accept a stronger notion of situation, which imposes a requirement on n-worlds according to which these worlds must then form parts of p-worlds in the sense of ≤ (or $\subseteq$) in order to be recognised as situations.[21] Obviously, inconsistent n-worlds cannot be situations understood in this way. For let us suppose that U is a situation, $U \leq V$, and V is a p-world. Then V will be consistent and complete: $V = \overline{V}$. According to **DM2**, $U \leq V \leq \overline{U}$, and this means that U must be consistent.

Let us consider some examples. Let $W = \{u, v, w\}$. n-world $[[\{\{v\}\}]] = \{\{v\}, \{u, v\}, \{v, w\}, \{u, v, w\}\}$ is a p-world and, in particular, the n-worlds $[[\{\{u, v\}, \{v, w\}\}]] = \{\{u, v\}, \{v, w\}, \{u, v, w\}\}$ and $[[\{\{v, w\}\}]] = \{\{v, w\}, \{u, v, w\}\}$ are contained within $[[\{\{v\}\}]]$, and as such can be considered situations.[22] Nevertheless, whereas the first situation is contained within one p-world only, i.e. in $[[\{\{v\}\}]]$, the second is contained in two p-worlds, $[[\{\{v\}\}]]$ and $[[\{\{w\}\}]]$. So, generally, situations can form parts of only one possible world or many possible worlds. Besides, situations are not only contained within p-worlds, but also in one another: e.g. $[[\{\{v, w\}\}]]$ is contained in $[[\{\{u, v\}, \{v, w\}\}]]$. It is easy to see that n-world $[[U]]$ is a situation contained in a p-world $[[\{\{v\}\}]]$ if $v \in \cap[[U]]$. Generally, $[[U]]$ is a situation iff $\cap[[U]] \neq \emptyset$:

$$\mathbf{sit}([[U]]) \equiv_{df} \cap U \neq \emptyset.$$

---

20 $\cup, \cap$ are ordinary set-theoretical operations.
21 Of course, n-worlds which are a maximal with respect to the ordering will be p-worlds.
22 $[[\{\{u, v\}, \{v, w\}\}]] \subset [[\{\{v\}\}]]$ and $[[\{\{u, v\}\}]] \subset [[\{\{v\}\}]]$.

This means that situations are strongly consistent n-worlds. (By the way, note that $\bigcap[[U]] \neq \emptyset$ iff $\bigcap U \neq \emptyset$.)

Every p-world is maximally consistent. Still, one may wonder whether the converse holds true: i.e. whether every maximally consistent n-world is a p-world. Interestingly enough, the answer is "no". Consider n-world $[[\{\{u,v\},\{v,w\},\{u,w\}\}]] = \{\{u,v\},\{v,w\},\{u,w\},\{u,v,w\}\}$. It is complete and consistent, but it is not identical with neither $[[\{\{u\}\}]]$, $[[\{\{v\}\}]]$ nor $[[\{\{w\}\}]]$. The consistency and completeness of this n-world can be verified by noting that

$$\{\{u,v\},\{v,w\},\{u,w\},\{u,v,w\}\} = \overline{\{\{u,v\},\{v,w\},\{u,w\},\{u,v,w\}\}}$$

or by applying (*). Again, consistent and complete n-worlds need not be p-worlds. This result is a little bit surprising, since usually maximal consistency is considered a sufficient condition for being a possible world. However, within our formal framework this is simply not the case! So what are the necessary and sufficient conditions for constituting a p-world? We can give here two such (equivalent) conditions, which may also be treated as definitions of p-worlds:

$$\textbf{\textit{p}-world}(U) \equiv_{df} U = \overline{U} \wedge [[\{\bigcap U\}]] = [[U]], \tag{+}$$

$$\textbf{\textit{p}-world}(U) \equiv_{df} \textbf{scon}(U) \wedge \overline{U} \leq U. \tag{++}$$

Here, (+) says that U is a p-world iff it is maximally consistent and deductively closed. Meanwhile, (++) says that U is a p-world iff it is strongly consistent and (weakly) complete. An essential feature of such definitions is that they do not refer directly to the original p-worlds – i.e. members of W.

Even so, if we look at the n-worlds properly contained within the maximally consistent n-world $\{\{u,v\},\{v,w\},\{u,w\},\{u,v,w\}\}$, then each of them can be considered a situation, although $\bigcap\{\{u,v\},\{v,w\},\{u,w\},\{u,v,w\}\} = \emptyset$.[23] So is it enough for a situation to be contained within a maximally consistent n-world?

Yet another example shows that this is an insufficient condition. Let us suppose, now, that $W = \{u,v,w,z\}$, and take into account the following:

$$[[\{\{u,v\},\{v,w\},\{u,w\}\}]] = \{\{u,v\},\{v,w\},\{u,w\},\{u,v,w\},\{u,v,w\},$$
$$\{u,v,z\},\{u,w,z\},\{v,w,z\},\{u,v,w,z\}\}.$$

This n-world is consistent and complete, and it contains n-world $\{\{u,v,w\},\{u,v,w\},\{u,v,z\},\{u,w,z\},\{v,w,z\},\{u,v,w,z\}\}$, which is consistent but in-

---

[23] Note that $\bigcap[[\{\{u,v\}\}]\neq \emptyset$, $\bigcap[[\{\{v,w\}\}]]\neq \emptyset$, $\bigcap[[\{\{u,w\}\}]]\neq \emptyset$, $\bigcap[[\{\{u,v\},\{v,w\}\}]]\neq \emptyset$, $\bigcap[[\{\{u,v\},\{u,w\}\}]]\neq \emptyset$, $\bigcap[[\{\{v,w\},\{u,w\}\}]]\neq \emptyset$.

complete – and, moreover, this n-world cannot be considered a situation as it is not strongly consistent.[24]

Back to the De Morgan Lattice

One can have serious doubts concerning the set-theoretic interpretation of non-standard possible-worlds ontology. The reason is that this interpretation renders non-standard possible worlds ontologically dependent on standard possible worlds. Sets representing n-worlds are built on p-worlds, the latter being ur-elements of the former. Certainly, it is desirable to avoid a set-theoretic over-interpretation of n-world ontology, and to treat n-worlds and p-worlds as basically the same kind of entity.

Indeed, that is precisely why we are going to propose an algebraic approach to the ontology of n-worlds – one that extends our preliminary approach as outlined in Section 1. Hence, we shall return to the De Morgan lattice, where $\mathcal{M} = \langle M, \leq, \sqcup, \sqcap, \bar{\ } \rangle$, assuming that the lattice is complete. Our intended interpretation of M is to be the set of n-worlds. Let us bear in mind that we can define consistency and completeness in $\mathcal{M}$, although we cannot define p-worlds in the lattice. However, the notion of p-world is of great importance for the ontology of worlds and situations. So we must somehow enrich the structure $\langle M, \leq, \sqcup, \sqcap, \bar{\ } \rangle$ to strengthen its expressive power.[25]

First, we introduce a new one-argument operation – the closure operation f, defined in the following way:

$$a \leq f(a), \qquad \text{DM5}$$
$$f(a \sqcap b) = f(a) \sqcap f(b), \qquad \text{DM6}$$
$$f(f(a)) = f(a). \qquad \text{DM7}$$

This closure operation will be the counterpart of the set-theoretic operation $[[U]] \to [[\{\bigcap U\}]]$. The reader will recall that n-worlds closed under this operation are deductively closed.

Then we are able to define an *atom* of the lattice:

$$\mathcal{A}(a) \equiv_{df} a = \bar{a} \wedge f(a) = a.$$

---

[24] It should be mentioned that usually, situations (partial worlds) are identified with principal filters, while possible worlds are identified with ultrafilters, in a lattice. Here we shall not take into account the possibility of non-prime ultra-filters playing the role of p-worlds.
[25] For details, see Paśniczek (2016).

This is interpreted as a p-world. Let us compare this definition with the definition from the previous section: p-worlds are consistent, complete, and deductively closed (a more immediate motivation for adopting such an interpretation will be given shortly). The notion of strong consistency is defined here as follows:

$$\mathbf{scon}(a) \equiv_{df} \exists b (\mathcal{A}(b) \wedge a \leq b).$$

We can see that the meaning of this definition is the same as the meaning of the definition introduced in the previous section: an n-world will be consistent in the strong sense iff it is contained within a p-world.

The notion of strong completeness will be defined in an analogous way:

$$\mathbf{scom}(a) \equiv_{df} \exists b (\mathcal{A}(b) \wedge b \leq a).$$

Intuitively, an n-world will be complete in the strong sense iff it contains a p-world.

As before, we may identify the notion of strong consistency with the notion of a situation:

$$\mathbf{sit}(a) \equiv_{df} \exists b (\mathcal{A}(b) \wedge a \leq b).$$

Thus a is a situation if it is contained within an atom of $\mathcal{M}$ (p-world). It should be emphasised that the De Morgan lattice $\mathcal{M} = \langle M, \leq, \sqcup, \sqcap, \bar{} \rangle$ as it stands does not guarantee the existence of atoms – i.e. p-worlds. M may contain only two elements: the greatest one and the least one (where these will be counterparts of the full and empty n-worlds $\{\emptyset\}$ and $\emptyset$, respectively).

Nevertheless, the content of the lattice $\mathcal{M}$ will not exactly match the content of the lattice $\langle \mathbb{N}, \subseteq, \cup, \cap, \bar{} \rangle$. Even if M contains different n-worlds from the greatest and the least ones, still, there can be no p-worlds among them. It follows that these n-worlds need not be built on p-worlds as is the case in $\langle \mathbb{N}, \subseteq, \cup, \cap, \bar{} \rangle$. In order to secure structural equivalence between the two lattices, we must extend $\mathcal{M}$ to a lattice which is, in a sense, an atomic lattice. Let us therefore add to $\mathcal{M}$ the following two axioms:

$$\forall c\, (f(c) \leq a \supset f(c) \leq b) \supset a \leq b, \qquad \textbf{DM8}$$

$$f(b) \leq \overline{f(c)} \supset \exists a\, (\mathcal{A}(a) \wedge f(b) \leq a \wedge f(c) \leq \overline{a}). \qquad \textbf{DM9}$$

By applying **DM8** and **DM9** it can be proved that every element of M can be displayed in the following canonical form:

$$a = \bigsqcup_{i \in I} \bigsqcap_{j \in J_i} a_{ij},$$

where a_{ij} are atoms, i.e. p-worlds. (Compare this form with the canonical form of n-worlds in the set-theoretic framework.)[26] And this is why we are treating this lattice as an atomic lattice.

Non-emptiness of the set of atoms can be guaranteed if we adjoin one more axiom to $\mathcal{M}$:

$$\exists a\, (f(a) \leq \overline{f(a)}). \qquad \textbf{DM10}$$

From **DM9** and **DM10** we get the following:

$$\exists a\, \mathcal{A}(a).$$

Finally, let us formulate the representation theorem for De Morgan lattice $\mathcal{M}$:

Theorem. *Every De Morgan lattice $\mathcal{M}$ fulfilling the conditions of* **DM1–DM10** *will be isomorphic to a lattice $\langle \mathbb{N}_W, \subseteq, \cup, \cap, ^- \rangle$ for some $W \neq \emptyset$.*[27]

Thus, we see that although the set-theoretic approach to n-world ontology is structurally equivalent to the algebraic approach, the notion of set plays no essential role in establishing this ontology.

As has been demonstrated here, the ontology of (possible, impossible) worlds and situations can be effectively studied within a uniform category of n-worlds when examined in an algebraic setting. Such an approach reveals the most distinctive ontological features of worlds and situations without appealing to traditionally conceived possible worlds, situations, or propositions. The theorem just mentioned shows that n-world ontology can also be modelled within a set-theoretical framework. However, the algebraic approach to the ontology based on the De Morgan lattice is more general. According to the former, n-worlds are built out of sets of p-worlds. This means that these n-worlds are ontologically dependent on propositions understood as sets of p-worlds and, consequently, are dependent on p-worlds themselves. Since the De Morgan lattice need not be atomic, algebraically interpreted n-worlds can be treated as independent of both p-worlds and propositions.

Bibliography

Cresswell, M. J. (1972). Intensional logics and logical truth. *Journal of Philosophical Logic*, 1, 2–15.

---

[26] Roughly speaking, $\bigsqcap X$ is the result of applying the meet-operation to a possibly infinite subset of M; $\bigsqcup X$ is the result of applying the joint operation to a possibly infinite subset of M.
[27] For proof of this theorem, see Paśniczek (2016).

Dunn, J. M., & Hardegree, G. M. (2001). *Algebraic methods in philosophical logic*. Oxford: Clarendon Press.
Fox, C., & Lappin, S. (2005). *Foundations of intensional semantics*. Blackwell Publishing.
Kripke, S. (1965). Semantic analysis of modal logic II: Non-normal modal propositional calculi. In J. Addison, L. Henkin, & A. Tarski (Eds.), *The theory of models* (pp. 206–220). North Holland Publishing.
Morgan, C. (1973). Systems of modal logic for impossible worlds. *Inquiry, 16*, 280–289.
Paśniczek, J. (1998a). *The logic of intentional objects. A Meinongian version of classical logic*. Dordrecht–Boston–London: Kluwer Academic Publishers.
Paśniczek, J. (1998b). Beyond consistent and complete worlds. *Logique et Analyse*, 161–163.
Paśniczek, J. (2014a). *Predykacja. Elementy ontologii formalnej przedmiotów, własności i sytuacji*. Kraków: CCPress.
Paśniczek, J. (2014b). Toward a Meinongian calculus of names. In M. Antonelli, & M. David (Vol. Eds.), *Meinong Studies: Vol. 5. Logical, ontological and historical contributions on the philosophy of Alexius Meinong* (pp. 61–81). Berlin–Boston: De Gruyter.
Paśniczek, J. (2016). Meinongian predication. An algebraic approach. In P. Stalmaszczyk (Series Ed., Vol. Ed.), *Studies in Philosophy of Language and Linguistics: Vol. 7. Philosophy and logic of predication*. Peter Lang GmbH.
Priest, G. (2001). *An introduction to non-classical logic*. Cambridge: Cambridge University Press.
Rantala, V. (1982). Impossible worlds semantics and logical omniscience. *Acta Philosophica Fennica, 35*, 106–115.
Rescher, N., & Brandom, R. (1979). *The logic of inconsistency: a study in non-standard possible world semantics and ontology*. Oxford: Basil Blackwell.
Sillari, G. (2007). Quantified logic of awareness and impossible possible worlds. *Review of Symbolic Logic, 1*(4), 514–529.
Van Fraassen, B. (1969). Facts and tautological entailments. *The Journal of Philosophy, 66*(15), 477–487.

Marek Magdziak
The Ontologic of Actions

Abstract: The term *ontologic* was suggested by the Polish logician Jerzy Perzanowski, as a name for the theoretical or formal parts of ontology. The expression *ontologic of actions* refers to the formal logical study of ontological concepts that pertain to the domain of actions. Action is the intentional causing of change in the world. Such changes occur when certain states of affairs appear or disappear. The ontologic of actions should therefore be based on the ontologic of states of affairs. This paper provides a tentative formal logical study of ontological concepts pertaining to action, as these relate to the concept of a state of affairs and of the production or destruction of states of affairs through acting. It provides an axiomatic characterization of these concepts within the framework of a multi-modal propositional logic, and then presents a semantic analysis of them. The resulting deductive system also takes into account the concepts of the performance and the omission of an action. It thus aims to furnish a basic logical framework for addressing foundational ethical issues. The semantics amounts to a slight modification of the standard relational semantics for normal modal propositional logic.

Keywords: ontologic, action, state of affairs, performance, omission.

Introduction

From an ethical point of view, actions can be treated as the contents of evaluative concepts, the point being that we all sometimes describe certain actions as good or bad or right or wrong. However, states of affairs can also be regarded as good or bad. So not just actions, but also states of affairs, can make up the contents of evaluative concepts. This leads us to ask: how are evaluations of states of affairs related to evaluations of actions? This question concerns the relation between the rightness (or wrongness) of an action and the goodness or badness of the state of affairs produced or destroyed by that action. Moreover, an action can be evaluated in two different ways: on the one hand, it can be regarded as a good or bad action *per se*, but on the other, any particular performance or omission of that action can also as such be regarded as good or bad. The problem connects up with such matters as the relation between *prima facie* duties or obligations and the actual obligations or duties which, in certain given situations, may override them. This paper provides an ontological background for the exploration of such issues.

Marek Magdziak, Faculty of Social Sciences, Wrocław University, Poland.

https://doi.org/10.1515/9783110669411-013

For an action to exist means for it to be performed or omitted. So an action exists at a given state-description or at a given possible world if and only if it is performed or omitted at that world. In addition, an action can be possible, or not, at a given possible world. Hence there are at least three different basic *modi essendi* concerning actions: to be performed, to be omitted and to be possible. You might expect that if an action is performed, then it must be a possible action, and if an action is not a possible action, then it must be the case that it is omitted. However, an action is always performed or omitted by a particular agent. Therefore, that agent imparts existence to an action, performing or omitting that action. Action **a** can be a part of action **b** at a given possible world. In such a case, if **b** is performed at the possible world, then **a** is performed at the possible world, and if **a** is omitted at the possible world, then **b** is omitted at the possible world. Therefore, if action **a** is a part of action **b** and action **b** is a part of action **a** at a given possible world, then action **a** is performed at the possible world if and only if action **b** is performed at the possible world and action **a** is omitted at the possible world if and only if action **b** is omitted at the possible world. In such a case, actions **a** and **b** are equipollent at the possible world.

At a given possible world an action can at the same time produce some states of affairs and can destroy others. Moreover, an action is performed or omitted in order to produce some states of affairs or destroy others. Following von Wright, I assume that to act is to intentionally bring about or prevent a change in the world, while a change takes place when a state of affairs ceases to be or comes to be (von Wright, 1968, pp. 38–39). On the other hand, actions involve agents' distinguishing among alternative possibilities. For this reason, it seems reasonable to think that possible world semantics offers an appropriate theoretical framework for formulating an ontological account of actions. Possible world semantics is a theory that takes alternative possibilities as its basic primitive notion (see Stalnaker, 1980). From the standpoint of a given possible world, an action can be considered to be consistent with some possibilities and inconsistent with others. One can say that from the standpoint of that possible world, some possibilities are made possible and others impossible by the action in question.

Some Basic Insights

In order to elaborate on this idea, I shall study the logical interconnections between action-ontological concepts of performance, omission and being-a-part-of, as used when discussing actions as these relate to the concept of a state of affairs and of the production and destruction of states of affairs. Thus, I will study such contexts as **a** *produces* **B** and **a** *destroys* **B**, where **B** stands for a state of affairs and **a** stands for an action. Let me adopt an informal notation to express some basic insights. For action **a** and state of affairs **B**, let **a+B** mean that **a** produces **B** and let **a−B** mean that **a** destroys **B**. For action **a**, let Pa mean that **a** is performed, let Oa mean that **a** is omitted, and let Ma mean that **a** is a possible action. For actions **a** and **b**, let **a**⊂**b** mean that action **a** is a part of action **b**. The signs ∼, &, ∨ and → will then be used, respectively, as symbols for negation, conjunction, disjunction and material implication. In such contexts as Pa, Oa, Ma and **a**⊂**b**, the symbols **a** and **b** represent actions. In such contexts as **a+B** and **a−B**, the expressions **a+** and **a−** will also function as one-place modal operators.

It seems intuitively to be the case that if state of affairs **B** logically implies state of affairs **C**, and action **a** produces state **B**, then **a** also produces state **C**. It also intuitively seems to be the case that if state of affairs **B** logically implies state of affairs **C**, and action **a** destroys state **C**, then **a** also destroys state **B**.

It is quite clear that if action **a** produces state of affairs **B**, and **a** is performed, then **B** obtains:

(1) **a+B** → (Pa → **B**).

Thus, if the state of affairs produced by action **a** does not obtain, then action **a** is not performed:

(2) ∼**B** → (**a+B** → ∼Pa).

On the other hand, if action **a** destroys state of affairs **B**, and **B** obtains, then **a** is omitted:

(3) **a−B** → (**B** → Oa).

Thus, if action **a** is not omitted, and action **a** destroys state of affairs **B**, then state of affairs **B** does not obtain:

(4) ∼Oa → (**a−B** → ∼**B**).

Action **a** produces the state of affairs that **a** is performed:

(5) **a+**Pa,

and action **a** destroys the state of affairs that **a** is omitted:

(6) **a−**Oa.

Let **1** stand for an arbitrarily selected tautology of propositional logic, and let **0** stand for an arbitrarily selected counter-tautology of propositional logic. Of course, every action produces a tautological state of affairs:

(7) **a+1**,

and every action destroys a counter-tautological state of affairs:
(8) **a–0**.

Some action could be a possible action and some action could be an impossible action. Action **a**, we might say, is an impossible action if and only if **a** is not a possible action. However, the theory developed in the next two sections does not imply that there are some possible actions and some impossible actions. It can be shown, using counter-models, that no formula of the form **Ma** is a thesis, as is no formula of the form ∼**Ma**.

If there is a contradictory pair of states of affairs, such that each of them is produced by action **a**, then **a** is not a possible action:
(9) (**a+B** & **a+**∼**B**) → ∼**Ma**,

implying in turn that
(10) **Ma** → (**a+B** → ∼**a+**∼**B**).

Equally, if there is a contradictory pair of states of affairs, such that each of them is destroyed by action **a**, then **a** is not a possible action:
(11) (**a–B** & **a–**∼**B**) → ∼**Ma**,

implying in turn that
(12) **Ma** → (**a–B** → ∼**a–**∼**B**).

Furthermore, action **a** is an impossible action if and only if **a** produces a counter-tautological state of affairs:
(13) ∼**Ma** ≡ **a+0**.

Equally **a** is an impossible action if and only if **a** destroys a tautological state of affairs:
(14) ∼**Ma** ≡ **a–1**.

Also, if **a** is an impossible action, then **a** is omitted:
(15) ∼**Ma** → **Oa**.

And if **a** is performed, then **a** is a possible action:
(16) **Pa** → **Ma**.

At the same time, if action **a** is performed, then action **a** isn't omitted:
(17) **Pa** → ∼**Oa**.

Any action can be a part of another action. Action **a** is a part of action **b** if and only if action **b** produces the state of affairs that **a** is performed and destroys the state of affairs that **a** is omitted.
(18) **a**⊂**b** ≡ **b**+P**a** & **b**−O**a**.

The part-whole relation is a reflexive and transitive relation, thus:
(19) **a**⊂**a**,

and
(20) (**a**⊂**b** & **b**⊂**c**) → **a**⊂**c**.

The claim that this part-whole relation is a reflexive one is highly controversial. However, this property does appear quite attractive from a technical point of view. Let us assume that actions **a** and **b** are *equipollent* if and only if **a**⊂**b** and **b**⊂**a**. Due to the reflexivity and transitivity of the part-whole relation, the relation of equipollence is a reflexive, symmetric and transitive relation. In the following two sections, these ideas will be studied in two different ways: employing a syntactic approach in Section 3 and a semantic one in Section 4. The syntactic approach works at the level of a calculus and a theory construed as a set of all the relevant theses. The semantic approach deals with models and truth conditions in a model. With the help of these semantic notions, a set of all valid formulae will be singled out.

The Theory

Let me introduce a formal language. Its alphabet will be provided by the following:
(a) a denumerable set of propositional letters. I refer to these as $p_1, p_2, p_3, \ldots$, etc.,
(b) the symbols for logical connectives ∼ and &, for negation and conjunction, respectively,
(c) a denumerable set of action letters. I refer to these as $a_1, a_2, a_3, \ldots$, etc.,
(d) the specific symbols of the theory +, −, P, O, M and ⊂,
(e) symbols (and) for parentheses.

The letters **a, b, c,** ... will be used as metalogical variables to range over action letters. The action letters represent actions.

The set of well-formed sentential formulae will be the smallest set X satisfying the following conditions:

(For 1) Each propositional letter belongs to X,
(For 2) If x belongs to X and y belongs to X, then (x & y), ∼x and ∼y belong to X,
(For 3) If **a** is an action letter, then **P**a, **O**a and **M**a belong to X,
(For 4) If **a** and **b** are action letters, then **a**⊂**b** belongs to X,
(For 5) If x belongs to X and **a** is an action letter, then **a**+x and **a**−x belong to X.

The letters **A, B, C, ...** will be used as metalogical variables to range over well-formed sentential formulae. The formulae represent states of affairs.

Additional symbols are introduced by means of the following definitions:

(D1)   (**A** ∨ **B**) = ∼(∼**A** & ∼**B**),
(D2)   (**A** → **B**) = ∼(**A** & ∼**B**),
(D3)   (**A** ≡ **B**) = ∼(**A** & ∼**B**) & ∼(∼**A** & **B**).

Symbol **1** stands for an arbitrarily selected tautology of propositional logic; the symbol **0** stands for an arbitrarily selected counter-tautology of propositional logic.

Within the framework of the formal language, a simple calculus can be constructed. Let me construct the calculus on the following axiomatic basis.

The first group of axioms consists of all of the tautologies of classical propositional logic, with well-formed formulae substituted for the propositional letters.

The second group of axioms will be determined by the following schemata:

(A.1)   (**a**+**B** & **a**+**C**) → **a**+(**B** & **C**),
(A.2)   (**a**−**B** & **a**−**C**) → **a**−(**B** ∨ **C**),
(A.3)   **a**+**B** → (**P**a → **B**),
(A.4)   **a**−**B** → (**B** → **O**a),
(A.5)   **a**+**P**a,
(A.6)   **a**−**O**a,
(A.7)   **M**a → (**a**+**B** → ∼**a**+∼**B**),
(A.8)   ∼**a**−**B** → **M**a,
(A.9)   **a**−**B** → **a**+∼**B**,
(A.10)   **a**⊂**b** → (**a**+**C** → **b**+**C**),
(A.11)   **a**⊂**b** → (**a**−**C** → **b**−**C**),
(A.12)   **b**+**P**a & **b**−**O**a → **a**⊂**b**.

The rules of inference of the calculus wil be:
(R1)   The Rule of Detachment (*Modus Ponens*),
(R2)   A Rule of Monotonicity to the effect that if **B** → **C** is a provable formula, then **a**+**B** → **a**+**C** is also a provable formula,
(R3)   A Rule of Inverse Monotonicity to the effect that if **B** → **C** is a provable formula, then **a**−**C** → **a**−**B** is also a provable formula.

The axiom schemata and the inference rules capture some of the ontological content pertaining to actions discussed in the previous section, as well as some other intuitions concerning actions. Let me draw the reader's attention to (A.1), (A.2), (A.8), (A.9), (A.10) and (A.11). Axiom schemata (A.1) and (A.2) say that if action **a** produces state of affairs **A** and at the same time produces state of affairs **B**, then it also produces the conjunction of these states of affairs, and if action **a** destroys state of affairs **A** and at the same time destroys state of affairs **B**, then it also destroys the disjunction of these states of affairs. Due to (R2) and (R3), both of them could be strengthened to equivalence. According to (A.8), if there is a state of affairs **B** such that action **a** does not destroy **B**, then **a** is a possible action. According to (A.9), if action **a** destroys state of affairs **B**, then action **a** produces a state of affairs contradictory to state of affairs **B**. So, by (A.8), (A.9) and propositional logic, if there is a state of affairs **B** such that action **a** does not produce **B**, then **a** is a possible action. Due to (A.10) and (A.11), the state of affairs that action **a** is a part of action **b** implies that if action **a** produces state of affairs **C**, then action **b** produces state of affairs **C**, and if action **a** destroys state of affairs **C**, then action **b** destroys state of affairs **C**.

A proof is defined in the standard way as a finite sequence of formulae, such that each member either belongs to the axioms or is derived from earlier members of the sequence by *Modus Ponens*, the Rule of Monotonicity or the Rule of Inverse Monotonicity. A proof is said to be a proof of the last member in its sequence, and a thesis is a formula of which there is a proof. Among the theses are all of the formal counterparts of (1)–(20). In fact, the formal counterparts of (1), (3), (5), (6) and (10) are axioms. As for the remaining ones, as well as some other theses, let me state the following theorem. (Proofs of the theorems given are to be found in the appendix to this article, except in cases where such proofs are absolutely straightforward, in which case they have been omitted.)

Theorem 1. *The following expressions are thesis schemata of the calculus:*
(1.1) $a+(a+B \to B)$,
(1.2) $a-(a-B \,\&\, B)$,
(1.3) $a+1$,
(1.4) $a-0$,
(1.5) $Ma \to ((a+B \to \sim a+\sim B) \,\&\, (a-B \to \sim a-\sim B))$,
(1.6) $(a+B \,\&\, a+\sim B) \equiv \sim Ma$,
(1.7) $(a-B \,\&\, a-\sim B) \equiv \sim Ma$,
(1.8) $\sim Ma \equiv a+0$,
(1.9) $\sim Ma \equiv a-1$,
(1.10) $\sim Ma \equiv (a+B \,\&\, a-B)$,
(1.11) $Ma \equiv (a+B \to \sim a-B)$,

(1.12) $\sim Ma \to Oa$,
(1.13) $Pa \to \sim Oa$,
(1.14) $Pa \to Ma$,
(1.15) $a \subset b \equiv b+Pa \ \& \ b-Oa$,
(1.16) $(a \subset b \ \& \ b \subset a) \to (a+C \equiv b+C)$,
(1.17) $(a \subset b \ \& \ b \subset a) \to (a-C \equiv b-C)$,
(1.18) $(a \subset b \ \& \ b \subset a) \to (Pa \equiv Pb) \ \& \ (Oa \equiv Ob)$,
(1.19) $a \subset a$,
(1.20) $(a \subset b \ \& \ b \subset c) \to a \subset c$.

To prove the theorem, it is sufficient to show that the schemata (1.1)–(1.20) are thesis schemata.

Let me draw your attention to (1.1), (1.2), (1.10), (1.11) and (1.16)–(1.18), which capture some ontological intuitions about actions. Thesis schemata (1.1) and (1.2) say, respectively, that action **a** produces the state of affairs that if **a** produces state of affairs **B**, then **B** obtains, and that action **a** destroys the state of affairs that **a** destroys state of affairs **B** and **B** obtains. So, schemata (1.1) and (1.2), express a kind of principle of practical consistency of actions. According to (1.10) and (1.11), action **a** is an impossible action if and only if it produces and destroys the same state of affairs, and action **a** is a possible action if and only if it does not destroy the state of affairs it produces. Due to (1.16)–(1.18), if actions **a** and **b** are equipollent, then **a** produces state **C** if and only if **b** produces state **C**, **a** destroys state **C** if and only if **b** destroys state **C**, **a** is performed if and only if **b** is performed, and **a** is omitted if and only if **b** is omitted.

For any set of formulae X, I shall say that **A** is *deducible* from X if and only if there are formulae $B_1, B_2, \ldots, B_n$ belonging to X such that the formula $(B_1 \& B_2 \& \ldots \& B_n) \to A$ is a thesis. A set of formulae X is said to be *consistent* if and only if there is no formula **A**, such that **A** and $\sim$**A** are both deducible from X. Otherwise, X is said to be *inconsistent*. The definition implies that if any set of formulae X is inconsistent, then some finite subset of X is also inconsistent, and that if any set of formulae X is consistent and a formula **A** is not deducible from X, then the set $X \cup \{\sim A\}$ is also consistent. I take this theory to correspond to the class of all theses. Thus, the theory is the smallest set containing all of the axioms and closed with respect to *Modus Ponens*, the Rule of Monotonicity and the Rule of Inverse Monotonicity. Note that if the theory is inconsistent, then all sets of formulae are inconsistent. Fortunately, the following theorem holds.

Theorem 2. *The theory is consistent.*

A set of formulae is *complete* if and only if for any formula **A**, either **A** belongs to X or ∼**A** belongs to X. A set of formulae that is both consistent and complete is called a *maximal consistent* set of formulae. The theorem known as the Lindenbaum's Lemma, to the effect that any consistent set of formulae is a subset of a maximal consistent set of formulae, holds for the theory. Note that if X is a maximal consistent set of formulae, and **A** is deducible from X, then **A** belongs to X. It is easy to show that for every maximal consistent set of formulae X and for every formulae **A** and **B**, ∼**A** belongs to X if and only if **A** does not belong to X, and (**A**&**B**) belongs to X if and only if **A** belongs to X and **B** belongs to X.

Semantics

As was said, the action letters concatenated by the auxiliary symbol + or − behave like modalities, so they can be handled using possible world semantics. Thus, the semantics for the theory is a slight modification of the standard relational semantics for normal modal propositional logic. Possible worlds are also referred to by the term "worlds", "possibilities", "standpoints" or simply "points". The idea is roughly as follows. Usually, for any given action, there are points that are made possible by that action, and there are points that are made impossible by it. Moreover, the class of points that are possible due to an action, and the class of points that are impossible due to an action, can vary from different standpoints. Thus, for any given action, the set of possible worlds made possible by it and the set of possible worlds made impossible by it from the standpoint of a given possible world can be singled out. Then the action letters are to be interpreted as pairs of binary relations on the set of possible worlds. The intuition behind this modeling is that each action **a** is determined by a pair of binary relations <R_a, S_a>, where R_a correlates a possible world *w* with possible worlds made possible by action **a** at the possible world *w*, and S_a correlates a possible world *w* with possible worlds made impossible by action **a** at the possible world *w*. So, vR_aw means that at the possible world *w* action **a** makes possible a possible world *v*, and vS_aw means that at the possible world *w* action **a** makes impossible a possible world *v*.

Formally, I shall introduce the notion of an action-model. An action-model *M* is to consist of a non-empty set of possible worlds *W*, an infinite sequence P_1, P_2, P_3, ..., of subsets of *W*, here abbreviated as P_i, an infinite sequence R_1, R_2, R_3, ..., of binary relations on *W*, here abbreviated as R_i, and an infinite sequence S_1, S_2, S_3, ..., of binary relations on *W*, here abbreviated as S_i. Thus, I define an action-ontological model as a structure $M = \langle W, P_i, R_i, S_i \rangle$ satisfying the following additional conditions:

(C1) for each natural number k, and for any v and w belonging to W, if vR_kw, then vR_kv;

(C2) for each natural number k, and for any v and w belonging to W, if vS_kv, then vS_kw;

(C3) for each natural number k, and for any v and w belonging to W, if vR_kw, then it is not the case that vS_kw;

(C4) for each natural number k, and for any w belonging to W, if there is no v such that vR_kw, then for any v belonging to W, vS_kw;

(C5) for any natural numbers k and l, if for any v and w belonging to W, vR_kw implies that vR_lv, then for any v and w belonging to W, vR_kw implies that vR_lw;

(C6) for any natural numbers k and l, if for any v and w belonging to W, vS_kv implies that vS_lw, then for any v and w belonging to W, vS_kw implies that vS_lw.

From a technical point of view, conditions (C1), (C2), (C3) and (C4) are, respectively, semantic counterparts of the axiom schemata (A5), (A6), (A9) and (A8), and conditions (C5) and (C6) are semantic counterparts of the axiom schema (A12), and they are necessary for the completeness result.

Condition (C1) reflects the conviction that if an action makes a possible world v possible from the standpoint of given possible world w, then the action makes world v possible also from the standpoint of world v itself. Thus, if possibility v is made possible by an action in point w, then possibility v is made possible by the action in point v itself. So the transition from w to v does not change the quality of possibility v as a possibility made possible by a. Condition (C2) reflects the conviction that if an action does not make a possible world v impossible from the standpoint of given possible world w, then the action does not make world v impossible from the standpoint of world v itself either. Thus, if possibility v is not made impossible by an action in point w, then possibility v is not made impossible by the action in point v itself either. So the transition from w to v does not change the quality of possibility v as a possibility that is not made impossible by a.

The assumption stated in conditions (C2) and (C3) is to be referred to as the *first consistency principle*.

According to condition (C3), if an action makes a possible world v possible from the standpoint of given possible world w, then the action does not make world v impossible from the standpoint of world w. According to condition (C4), if an action makes no possible world possible from the standpoint of given possible world w, then the action makes all worlds impossible from the standpoint of world w.

Condition (C5) reflects the conviction that if any possible world v which is made possible by action a from the standpoint of given possible world w is a possible world made possible by action **b** from the standpoint of **v** itself, then any possible world **v** which is made possible by action **a** from the standpoint of possible world **w** is also a possible world made possible by action **b** from the standpoint of **w**. Thus, due to the idea that a possible world **v** that is made possible by action **b** from the standpoint of **v** itself is a possible world at which action **b** is performed, if point **v**'s being a point made possible by action **a** in the standpoint **w** implies that **v** is a point at which **b** is performed, then point **v**'s being a point made possible by action **a** in the standpoint **w** implies that **v** is a point made possible by action **b** in the standpoint **w**. If an action is performed in a world, then the action makes possible that world.

Condition (C6) reflects the conviction that if any possible world **v** which is made impossible by action **a** from the standpoint of **v** itself is a possible world made impossible by action **b** from the standpoint of any possible world **w**, then any possible world **v** which is made impossible by action **a** from the standpoint of possible world **w** is also a possible world made impossible by action **b** from the standpoint of **w**. Thus, due to the idea that a possible world **v** that is made impossible by action **a** from the standpoint of **v** itself is a possible world at which action **a** is omitted, if point **v**'s being a point at which **a** is omitted implies that **v** is a point made impossible by action **b** from the standpoint of any **w**, then point **v**'s being a point made impossible by action **a** in the standpoint **w** implies that **v** is a point made impossible by action **b** in the standpoint **w**. If an action makes a world impossible, then the action is omitted in that world.

The assumption stated in conditions (C5) and (C6) is to be referred to as the *second consistency principle*.

For any action-model M, and for any action letter **a**, there is a unique pair of binary relations $<R_a, S_a>$ which corresponds to **a** in model M. For each natural number k, relations R_k and S_k correspond to action letter a_k. For any possible world **w**, let $[w]^{R_a}$ be the set of possible worlds such that $[w]^{R_a} = \{v: vR_aw\}$ and let $[w]^{S_a}$ be the set of possible worlds such that $[w]^{R_a} = \{v: vS_aw\}$. Set $[w]^{R_a}$ will be called the *extension of action* **a** at possible world **w**, and set $[w]^{S_a}$ will be called the *anti-extension of action* **a** at possible world **w**. The extension of action **a** at possible world **w** contains the possibilities made possible by action **a** according to the standpoint **w**, and the anti-extension of action **a** at the possible world **w**, contains the possibilities made impossible by action **a** according to the standpoint **w**. Let me reformulate the conditions (C3) and (C4) in terms of the extension and anti-extension of an action.

(C3*)   for any action **a**, $[w]^{R_a} \cap [w]^{S_a} = \emptyset$;
(C4*)   for any action **a**, if $[w]^{R_a} = \emptyset$, then $[w]^{S_a} = W$.

Due to (C3*), (C4*) could be strengthened to

(C4**) for any action **a**, $[w]^{Ra} = \emptyset$ if and only if $[w]^{Sa} = W$.

As regards possible worlds in the action-model, I state the truth conditions for formulae according to their forms. I use $w \models^M A$ to mean that formula **A** is true at possible world w in action-model M. The truth conditions are as follows:

1. $w \models^M p_k$ if and only if w belongs to P_k for $k = 1, 2, 3, \ldots$.
2. $w \models^M \sim A$ if and only if not $w \models^M A$.
3. $w \models^M A \& B$ if and only if both $w \models^M A$ and $w \models^M B$.
4. $w \models^M A+B$ if and only if for any possible world v, if $v \in [w]^{Ra}$ then $v \models^M B$.
5. $w \models^M A-B$ if and only if for any possible world v, if $v \models^M B$ then $v \in [w]^{Sa}$.
6. $w \models^M Pa$ if and only if $w \in [w]^{Ra}$.
7. $w \models^M Oa$ if and only if $w \in [w]^{Sa}$.
8. $w \models^M Ma$ if and only if $[w]^{Ra} \neq \emptyset$.
9. $w \models^M a \subset b$ if and only if $[w]^{Rb} \subseteq [w]^{Ra}$ and $[w]^{Sa} \subseteq [w]^{Sb}$.

Clause (1) states what the true value of each propositional letter is to be at each possible world. It reflects the stipulation that in action-model M, a propositional letter p_k is true at a possible world w just in the case w is a member of the set P_k. Clauses (2) and (3) are simply repetitions of the usual propositional truth clauses. Due to definitions (D1)–(D4), they yield the classical truth tables for standard propositional connectives.

Clause (4) formulates my interpretation of the production of a state of affairs by an action: **a**+**B** is true at possible world w if and only if **B** is true at all possible worlds made possible by action **a** at possible world w, or an action produces a state of affairs at a possible world w if and only if the state of affairs obtains in any possible world belonging to the extension of action **a** at possible world w.

Clause (5) formulates my interpretation of the destruction of a state of affairs by an action: **a**−**B** is true at possible world w if and only if **B** is true only at possible worlds made impossible by action **a** at possible world w, or an action destroys a state of affairs at a possible world w if and only if the state of affairs obtains only in possible worlds belonging to the anti-extension of action **a** at possible world w.

Clauses (6), (7), (8) and (9) reflect some ideas about the action-ontological concepts of performance, omission, possibility and part-whole relation, as these relate to actions. According to (6), action **a** is performed at point w if and only if, from standpoint w, w itself is made possible by action **a**. According to (7), action **a** is omitted at point w if and only if, from standpoint w, w itself is made impossible by action **a**. Clause (8) states that action **a** is a possible action at point w if and only if the class of possibilities made possible by action **a** according to standpoint w is nonempty. The content of (9) is that action **a** is a part of action **b** at point w if and only if the class of possibilities made possible by action **b** according to standpoint

w is included in the class of possibilities made possible by action **a** according to standpoint **w**, and the class of possibilities made impossible by action **a** according to standpoint **w** is included in the class of possibilities made impossible by action **b** according to standpoint **w**.

Given my definition of an action-ontological model, I shall state the following theorem.

Theorem 3. *There are structures which are action-ontological models.*

For any given formula **A**, let $|\mathbf{A}|^M$ be the set of possible worlds that verify formula **A** in action-model M. Thus, $|\mathbf{A}|^M = \{v: v \models^M \mathbf{A}\}$. I shall call this the *propositional content of formula* **A** in action-model M. The propositional content of a formula in a model could be interpreted as the semantic counterpart of the state of affairs which corresponds to the formula in the action-model. Let me reformulate clauses (4) and (5) in terms of the extension and anti-extension of an action and the propositional content of a formula.

(4*) $w \models^M \mathbf{a}+\mathbf{B}$ if and only if $[w]^{R_a} \subseteq |\mathbf{B}|^M$,
(5*) $w \models^M \mathbf{a}-\mathbf{B}$ if and only if $|\mathbf{B}|^M \subseteq [w]^{S_a}$.

According to (4*), action **a** produces state of affairs **B** at point **w** if and only if the extension of action **a** at possible world **w** is included in state of affairs **B**. According to (5*), action **a** destroys state of affairs **B** at point **w** if and only if state of affairs **B** is included in the anti-extension of action **a** at possible world **w**.

A formula true at every possible world in some action-model M is said to be *valid in the action-model M*, while a formula valid in every action-model is just said to be *valid*. I write $\models^M \mathbf{A}$ to mean that formula **A** is valid in action-model M, and $\models \mathbf{A}$ to mean that formula **A** is a valid formula *simpliciter*.

As theorem 3 says, there are structures that are action-models. Let $W^\star$ be the set of maximal consistent sets of formulae. Due to theorem 2, $W^\star$ is a non-empty set. Let $P_i^\star$ be the infinite sequence of subsets of $W^\star$, such that for each natural number k, $P_k^\star$ is the set of maximal consistent sets of formulae containing the propositional letter p_k. Let $R_i^\star$ be the infinite sequence of binary relations on $W^\star$, such that for each natural number k, and for any **v** and **w** belonging to $W^\star$, $vR_k^\star w$ if and only if $\{\mathbf{C}: a_k+\mathbf{C} \in w\} \subseteq v$. Let $S_i^\star$ be the infinite sequence of binary relations on $W^\star$, such that for each natural number k, and for any **v** and **w** belonging to $W^\star$, $vS_k^\star w$ if and only if $\{\sim\mathbf{C}: a_k-\mathbf{C} \in w\} \not\subseteq v$.

Theorem 4. *The structure $M^\star = \langle W^\star, P_i^\star, R_i^\star, S_i^\star \rangle$ is an action-model.*

In order to show that the structure $M^* = \langle W^*, P_i^*, R_i^*, S_i^* \rangle$ is an action-model, it is sufficient to prove that the structure satisfies the conditions (C1)–(C6). I shall call this structure the *canonical model*.

Let me draw the reader's attention to the next theorem, which holds for the canonical model.

Theorem 5. *For any formula A and for any $w \in W^*$, $w \models^{M^*} A$ if and only if $A \in w$.*

I shall call this *the fundamental theorem for the canonical model*. The proof of the theorem is, of course, arrived at by induction, taking the construction of formulae as its point of reference.

Completeness

In the previous two sections, the formulae were studied in two different ways: a syntactic one in section 3 and a semantic one in section 4. A fruitful blend of the two approaches results in the following completeness theorem:

Theorem 6. *A formula is a thesis of the theory if and only if it is a valid formula.*

Thus, the syntactic and semantic approaches depict the same class of logically true formulae.

Some Comments

As was stated at the outset, actions can be treated as the contents of evaluative concepts. This approach is essential, for example, to ethical reasoning – the point being that we all sometimes describe certain things as good or bad. Actions, especially, can be regarded as good or bad. However, from an ethical point of view, states of affairs can also be regarded as good or bad. So not just actions, but also states of affairs, can make up the contents of evaluative concepts. The problem then arises of how evaluations of states of affairs are related to evaluations of actions.

Let us enrich the language of the theory with two one-place operators G and B, which are to serve as the formal counterparts of the evaluative concepts *it is good that* ... and *it is bad that* .... So, the expression G(**A**) means that it is good that **A**, and the expression B(**A**) means that it is bad that **A**, where **A** stands for a state of affairs. Thus, G(**A**) and B(**A**) are the formal counterparts of evaluations of states of affairs in the theory. However, in the enriched language of the theory there are also the expressions G(P**a**), G(O**a**), B(P**a**) and B(O**a**), where **a** stands for an action. The expression G(P**a**) means that it is good to perform **a**, the expression G(O**a**) means that it is good to omit **a**, the expression B(P**a**) means that it is bad to perform **a**, and the expression B(O**a**) means that it is bad to omit **a**. Hence, G(P**a**), G(O**a**), B(P**a**) and B(O**a**) are the formal counterparts of evaluations of actions in the theory. The theory can therefore help us to clarify the problem of how evaluations of states of affairs are related to evaluations of actions, but of course cannot settle it. From the point of view of the enriched language of the theory, one might, for example, consider the following formulae:

(F1)   **a**+**B** & G(**B**) → G(P**a**),
(F2)   **a**+**B** & G(**B**) → B(O**a**),
(F3)   **a**+**B** & G(**B**) → ∼B(P**a**),
(F4)   **a**+**B** & G(**B**) → ∼G(O**a**),
(F5)   **a**−**B** & G(**B**) → ∼G(P**a**),
(F6)   **a**−**B** & G(**B**) → B(O**a**),
(F7)   **a**−**B** & G(**B**) → ∼B(P**a**),
(F8)   **a**−**B** & G(**B**) → G(O**a**),
(F9)   **a**+**B** & B(**B**) → B(P**a**),
(F10) **a**+**B** & B(**B**) → G(O**a**),
(F11) **a**+**B** & B(**B**) → ∼G(P**a**),
(F12) **a**+**B** & B(**B**) → ∼B(O**a**),
(F13) **a**−**B** & B(**B**) → ∼B(P**a**),
(F14) **a**−**B** & B(**B**) → B(O**a**),
(F15) **a**−**B** & B(**B**) → ∼B(P**a**),
(F16) **a**−**B** & B(**B**) → B(O**a**).

Each formula reflects some intuition about the relation between evaluations of states of affairs and evaluations of actions. Now one may ask which formulae should be considered to count as axioms of ethical reasoning. As was stated earlier, the aim of the theory is to clarify the problem, not to settle it.

Another problem which arises is that of how evaluations of the performing of an action are related to evaluations of the omitting of that action. The problem concerns the relation between the formulae G(P**a**), G(O**a**), B(P**a**) and B(O**a**). From the point of view of the enriched language of the theory, we might, for instance, consider the following formulae:

(F'1) G(P**a**) → B(O**a**),
(F'2) G(P**a**) → ∼G(O**a**),
(F'3) B(P**a**) → G(O**a**),

(F'4) B(Pa) → ∼B(Oa),
(F'5) G(Oa) → B(Pa),
(F'6) G(Oa) → ∼G(Pa),
(F'7) B(Oa) → G(Pa),
(F'8) B(Oa) → ∼B(Pa).

Again, each formula reflects some intuition about the relation between evaluations of the performing of an action and evaluations of the omitting of that action. Once again, we can ask which formulae should be considered to count as axioms of ethical reasoning? Again, as was stated earlier, the aim of the theory is to clarify the problem, not to settle it.

In the language of the theory enriched by the two one-place operators G and B that are to count as the formal counterparts of the evaluative concepts *it is good that ...* and *it is bad that ...*, formulae such as G(Pa), B(Pa), G(Oa) and B(Oa) will be the formal counterparts of particular evaluations. For example, formula G(Pa) is to be read as saying that performing action **a** is good at some particular state description. To convey the thought that performing (or omitting) action **a** is in general good (or bad), the language of the theory will have to be enriched by the one-place operator for necessity □. Then, for example, the formula □G(Pa) can be be read as saying that it is necessarily (i.e. in general) good to perform **a**.

Conclusion

The axiomatic system for the ontologic of actions presented in the present article has been based on three assumptions. First of all, it is assumed that actions should be characterized in terms of their relations to states of affairs, because each action is aimed at producing a certain state of affairs or destroying a certain state of affairs, and thus at causing a change in the world. Secondly, it is assumed that an action takes place only if it is performed or omitted. Therefore, strictly speaking, it is the performance of an action or omission of an action that causes a change in the world. Thirdly, it is held that the language of the ontologic of actions should also take into account both possible and impossible actions, as well as the fact that an action can be part of another more complex action.

This system has been conceived as furnishing a formal-logical basis for what are more or less formalized ethical considerations. As an example, let us consider ethical evaluations: i.e. judgments of the form *'It's good when...'* and *'It's bad when...'*. Actions are, of course, subject to ethical evaluation, but let us note that in the strict sense it is potentially good or bad to perform or omit an action. The fact that an action is good may therefore mean that it is good when the action has been done, and the fact that an action is bad may mean that it is bad when the action has been done. The fact that an action is good can, however, be understood in a slightly different, stronger sense: namely, that it is good when the action has been done and bad when the action has been omitted. Similarly, the fact that an action is bad can be understood in a slightly different sense: namely, that it is bad when the action has been done and good when the action has been omitted. This is indirectly related to the question of the logical relationships between different types of evaluation of actions. One may ask whether, if doing a certain action is good, then it is bad to omit it, or whether, if omitting a certain action is bad, then it is good to do it, or whether, if doing a certain action is bad, then it is good to omit it, and so on. Further questions of this sort can be formulated, as is illustrated by the formulae (F'1)-(F'8) given earlier.

The expressions '*It is good when…*' and '*It is bad when…*' refer to states of affairs. For ethical evaluations, the states of affairs where an action has been performed or omitted are precisely those of key importance. The subjects of evaluation are then the actions. But these expressions can also refer to any other state of affairs. Therefore, there is no need to have separate evaluative expressions for actions and for states of affairs. This allows a precise distinction to be made between evaluations of actions and evaluations of the states of affairs that these actions produce or destroy. The ethics-related literature has repeatedly discussed various versions of the question of whether, if some action produces a good state of affairs, the performance of that action counts as good. However, one can also ask whether, if a certain action produces a bad state of affairs, the performance of that action is bad. Once again, further questions of this sort can be formulated, as is illustrated by the previously given formulae (F1)-(F16). To sum up, the examples presented show that the theory outlined in the present article allows for a much more precise approach to dealing with sophisticated ethical issues than did the previously developed logical tools.

Appendix: Proofs of Theorems

Proof of Theorem 1.
To prove the theorem it is sufficient to show that the schemata (1.1)–(1.20) are thesis schemata (where proofs are easy, they are omitted).

(1.1) **a+(a+B → B)**: From (A3), (A5) and (R2) by propositional logic.
(1.2) **a−(a−B & B)**: From (A4), (A6) and (R3) by propositional logic.

(1.3) **a+1**: By propositional logic P**a** → **1** is a thesis schema. Then, due to (R2), **a**+P**a** → **a+1** is a thesis schema. However, from (A5), **a**+P**a** is a thesis schema. Therefore, by (R1), **a+1** is a thesis schema.

(1.4) **a−0**: By propositional logic **0** → O**a** is a thesis schema. Then, due to (R3), **a**−O**a** → **a−0** is a thesis schema. However, from (A6), **a**−O**a** is a thesis schema. Therefore, by (R1), **a−0** is a thesis schema.

(1.5) M**a** → ((**a+B** → ∼**a+**∼**B**) & (**a−B** → ∼**a−**∼**B**)): By (A7), M**a** → ((**a+** ∼**B** → ∼**a+B**) is a thesis schema. However, by (A9), **a−B** → **a+**∼**B** is a thesis schema and, by transposition, ∼**a+B** → ∼**a−**∼**B** is a thesis schema. Then, by propositional logic, M**a** → (**a−B** → ∼**a−**∼**B**) is a thesis schema. Therefore, due to (A7) and propositional logic, M**a** → ((**a+B** → ∼**a+**∼**B**) & (**a−B** → ∼**a−**∼**B**)) is a thesis schema.

(1.6) (**a+B** & **a+**∼**B**) ≡ ∼M**a**: By (A7) and propositional logic, (**a+B** & **a+** ∼**B**) → ∼M**a** is a thesis schema. However, by (A8), (A9) and propositional logic, ∼M**a** → **a+B** is a thesis schema and therefore ∼M**a** → (**a+B** & **a+** ∼**B**) is a thesis schema. Hence, by propositional logic, (**a+B** & **a+** ∼**B**) ≡ ∼M**a** is a thesis schema.

(1.7) (**a−B** & **a−**∼**B**) ≡ ∼M**a**: Due to (1.5) and propositional logic, (**a−B** & **a−**∼**B**) → ∼M**a** is a thesis schema. However, by (A8) and propositional logic, ∼M**a** → **a−B** is a thesis schema and therefore ∼M**a** → (**a−B** & **a−**∼**B**) is a thesis schema. Hence, by propositional logic, (**a−B** & **a−**∼**B**) ≡ ∼M**a** is a thesis schema.

(1.8) ∼M**a** ≡ **a+0**: Due to (1.6), (**a+B** & **a+**∼**B**) ≡ ∼M**a** is a thesis schema. However, by (A1), (R2) and propositional logic, (**a+B** & **a+**∼**B**) ≡ **a+**(**B** & ∼**B**) is a thesis schema. So, by propositional logic, ∼M**a** ≡ **a+0** is a thesis schema.

(1.9) ∼M**a** ≡ **a−1**: Due to (1.7), (**a−B** & **a−**∼**B**) ≡ ∼M**a** is a thesis schema. However, by (A2), (R3) and propositional logic, (**a−B** & **a−**∼**B**) ≡ **a−**(**B** ∨ ∼**B**) is a thesis schema. So, by propositional logic, ∼M**a** ≡ **a−1** is a thesis schema.

(1.10) ∼M**a** ≡ (**a+B** & **a−B**): Due to (1.7), (**a−B** & **a−**∼**B**) ≡ ∼M**a** is a thesis schema. However, by (A9), **a−**∼**B** → **a+B** is a thesis schema and, by propositional logic, ∼M**a** → (**a+B** & **a−B**) is a thesis schema. On the other hand, by (A7) and propositional logic, (**a+B** & **a+** ∼**B**) → ∼M**a** is a thesis schema. However, by (A9), **a−B** → **a+**∼**B** is a thesis schema and, by propositional logic, (**a+B** & **a−B**) → ∼M**a** is a thesis schema. Therefore ∼M**a** ≡ (**a+B** & **a−B**) is a thesis schema.

(1.11) M**a** ≡ (**a+B** → ∼**a−B**): From (1.10) by propositional logic.

(1.12) ∼M**a** → O**a**: From (A8), by propositional logic, ∼M**a** → **a−B** is a thesis schema. However, due to (A4), by propositional logic, **a−1** → O**a** is a thesis schema. Hence, by propositional logic, ∼M**a** → O**a** is a thesis schema.

(1.13) P**a** → ∼O**a**: By (A3), P**a** → (**a+**∼O**a** → ∼O**a**) is a thesis schema. On the other hand, by (A9), **a−**O**a** → **a+**∼O**a** is a thesis schema, then by propositional logic P**a** → (**a−**O**a** → ∼O**a**) is a thesis schema. However, due to (A6), **a−**O**a** is a thesis schema. Therefore, by propositional logic, P**a** → ∼O**a** is a thesis schema.

(1.14) **Pa → Ma**: By (A3), **Pa → (a+B → B)** is a thesis schema. Then, by propositional logic, **Pa → (∼B → ∼a+B)** is a thesis schema and **Pa → (∼0 → ∼a+0)** is also a thesis schema (**0** stands for arbitrary chosen counter-tautology of propositional logic). Thus, by propositional logic, **Pa → ∼a+0** is a thesis schema. But, by (A8), (A9) and propositional logic, **∼a+0 → Ma** is a thesis schema and, by propositional logic **Pa → Ma** is a thesis schema.

(1.15) **a⊂b ≡ b+Pa & b−Oa**: From (A10), (A11) and propositional logic, **a⊂b → ((a+Pa → b+Pa) & (a−Oa → b−Oa))** is a thesis schema. However, due to (A5) and (A6), **a+Pa** is a thesis schema and **a−Oa** is a thesis schema. Then, by propositional logic, **a⊂b → b+Pa & b−Oa** is a thesis schema. On the other hand, by (A12), **b+Pa & b−Oa → a⊂b** is a thesis schema. Therefore **a⊂b ≡ b+Pa & b−Oa** is a thesis schema.

(1.16) **(a⊂b & b⊂a) → (a+C ≡ b+C)**: By (A10), (A11) and propositional logic, **(a⊂b & b⊂a) → (a+C → b+C)** is a thesis schema and **(a⊂b & b⊂a) → (b+C → a+C)** is a thesis schema. Then, by propositional logic, **(a⊂b & b⊂a) → (a+C ≡ b+C)** is a thesis schema.

(1.17) **(a⊂b & b⊂a) → (a−C ≡ b−C)**: By (A10), (A11) and propositional logic, **(a⊂b & b⊂a) → (a−C → b−C)** is a thesis schema and **(a⊂b & b⊂a) → (b−C → a−C)** is a thesis schema. Then, by propositional logic, **(a⊂b & b⊂a) → (a−C ≡ b−C)** is a thesis schema.

(1.18) **(a⊂b & b⊂a) → (Pa ≡ Pb) & (Oa ≡ Ob)**: By (A10), **a⊂b → (a+Pa → b+Pa)** is a thesis schema and, by (A11), **a⊂b → (a−Oa → b−Oa)** is a thesis schema. So, by (A5), (A6) and propositional logic, **a⊂b → b+Pa** is a thesis schema and **a⊂b → b−Oa** is a thesis schema. However, by (A3), **b+Pa → (Pb → Pa)** is a thesis schema and, by (A4), **b−Oa → (Oa → Ob)** is a thesis schema. Hence, by propositional logic, **a⊂b → (Pb → Pa)** is a thesis schema and **a⊂b → (Oa → Ob)** is a thesis schema.

On the other hand, by (A10), **b⊂a → (b+Pb → a+Pb)** is a thesis schema and, by (A11), **b⊂a → (b−Ob → a−Ob)** is a thesis schema. So, by (A5), (A6) and propositional logic, **b⊂a → a+Pb** is a thesis schema and **b⊂a → a−Ob** is a thesis schema. However, by (A3), **a+Pb → (Pa → Pb)** is a thesis schema and, by (A4), **a−Ob → (Ob → Oa)** is a thesis schema. Hence, by propositional logic, **b⊂a → (Pa → Pb)** is a thesis schema and **b⊂a → (Ob → Oa)** is a thesis schema.

Hence, by propositional logic, **(a⊂b & b⊂a) → (Pa ≡ Pb) & (Oa ≡ Ob)** is a thesis schema.

(1.19) **a⊂a**: From (A5), (A6) and (A12) by propositional logic.

(1.20) **(a⊂b & b⊂c) → a⊂c**: By (A10) and propositional logic, **(a⊂b & b⊂c) → (a+D → c+D)** is a thesis schema and, by (A11) and propositional logic, **(a⊂b & b⊂c) → (a−D → c−D)** is a thesis schema. Therefore **(a⊂b & b⊂c) → (a+Pa → c+Pa)** is a thesis schema and **(a⊂b & b⊂c) → (a−Oa → c−Oa)** is a thesis schema. However, by (A5), **a**+P**a** is a thesis schema and, by (A6), **a**−O**a** is a thesis schema. Hence (**a**⊂**b** & **b**⊂**c**) → **c**+P**a** is a thesis schema and (**a**⊂**b** & **b**⊂**c**) → **c**−O**a** is a thesis schema. Therefore, by (A12) and propositional logic, (**a**⊂**b** & **b**⊂**c**) → **a**⊂**c** is a thesis schema. □

Proof of Theorem 2.
To prove that the theory is consistent, let me take advantage of the standard language of classical propositional logic. The alphabet is given by a denumerable set Q of propositional letters: I refer to these as $q_1, q_2, q_3, \ldots$, etc., the symbols for the logical connectives ∼ and & for negation and conjunction, respectively, and the parentheses (and). The set of well-formed sentential formulae is defined inductively in the standard manner. The symbols ∨, → and ≡ are introduced by definitions in the standard way. Let me define the mapping T, from the language of the theory to the language of classical propositional calculus, as follows:

(T1) $T(p_k) = q_{2k}$,
(T2) $T(a_k) = q_{2k+1}$,
(T3) T(∼**A**) = ∼T(**A**),
(T4) T(**A** & **B**) = T(**A**) & T(**B**),
(T5) T(**a**+**B**) = T(**a**) → T(**B**),
(T6) T(**a**−**B**) = T(**B**) → ∼T(**a**),
(T7) T(P**a**) = T(**a**),
(T8) T(O**a**) = ∼T(**a**),
(T9) T(M**a**) = T(**a**),
(T10) T(**a**⊂**b**) = T(**b**) → T(**a**).

Thus, for every formula **A**, there is a unique formula T(**A**) in the language of classical propositional logic. Let me call it the PC-transform of **A**. It is easy to show that the PC-transform of every thesis of the theory is a tautology of classical propositional calculus. It follows that for every formula **A**, **A** and ∼**A** are not both theses, for if they were, T(**A**) and ∼T(**A**) would both be tautologies of classical propositional calculus, which is impossible. Thus, by definition given the theory, for every formula **A**, **A** and ∼**A** are not both deducible from the theory. □

Proof of Theorem 3.
The proof of the next theorem is in fact a proof of theorem 3. □

Proof of Theorem 4.
In order to show that the structure $M^\star = \langle W^\star, P_i^\star, R_i^\star, S_i^\star \rangle$ is an action-model, it is sufficient to prove that the structure satisfies the conditions (C1)–(C6).

(C1) For each natural number k, and for any ***v*** and ***w*** belonging to W, if ***v***$R_k^\star$***w***, then ***v***$R_k^\star$***v***. Assume that ***v***$R_k^\star$***w***. Thus, {**C**: a_k+**C** ∈ ***w***} ⊆ ***v*** and, by (A5), Pa_k belongs to v. Hence, by (A3), any formula depicted by schema (a_k+**C** → **C**) also belongs to v. Thus, for any formula **C**, if a_k+**C** belongs to v, then **C** belongs to v. Hence, {**C**: a_k+**C** ∈ v} ⊆ v and vR_k^*v.

(C2) For each natural number k, and for any v and w belonging to W, if vS_k^*v, then vS_kw. Assume that it is not the case that vS_k^*w. Thus, {∼**C**: a_k−**C** ∈ w} ⊆ v and, by (A6), ∼Oa_k belongs to v. Hence, by (A4), and propositional logic, any formula depicted by schema (a_k−**C** → ∼**C**) also belongs to v. Thus, for any formula **C**, if a_k−**C** belongs to v, then ∼**C** belongs to v. Hence, {∼**C**: a_k−**C** ∈ v} ⊆ v and it is not the case that vS_k^*v.

(C3) For each natural number k, and for any v and w belonging to W, if vR_k^*w, then it is not the case that vS_k^*w. Assume that vS_k^*w, so {∼**C**: a_k−**C** ∈ w} ⊄ v. Thus, for some **D**, a_k−**D** ∈ w and ∼**D** ∉ v, so **D** ∈ v. However, due to (A9), a_k+∼**D** ∈ w. Hence, {**C**: a_k+**C** ∈ w} ⊄ v, and it is not the case that vR_k^*w.

(C4) For each natural number k, and for any w belonging to W, if there is no v such that vR_k^*w, then for any v belonging to W, vS_k^*w. Assume that there is a v such that it is not the case that vS_k^*w. Thus, there is a v such that {∼**C**: a_k−**C** ∈ w} ⊆ v. Hence, {∼**C**: a_k−**C** ∈ w} is consistent. Thus, there is a formula **D** such, that a_k−**D** ∉ w and there is a formula **D** such, that ∼a_k−**D** ∈ w. Therefore, by (A8) Ma_k ∈ w and, by (A7) (a_k+**1** → ∼a_k+**0**) ∈ w. Hence, due to (1.3), ∼a_k+**0** ∈ w. Therefore, by (A1) and (R2), {**C**: a_k+**C** ∈ w} is consistent and there is a v such, that {**C**: a_k+**C** ∈ w} ⊆ v. So, there is a v such, that vR_k^*w.

(C5) For any natural numbers k and l, if for any v and w belonging to W, vR_k^*w implies that vR_l^*v, then for any v and w belonging to W, vR_k^*w implies that vR_l^*w. Assume that for any v and w, vR_k^*w implies that vR_l^*v. Thus, for any v and w, if {**C**: a_k+**C** ∈ w} ⊆ v, then {**C**: a_l+**C** ∈ v} ⊆ v. However, due to (A5), for any v, if {**C**: a_l+**C** ∈ v} ⊆ v, then Pa_l belongs to v. Thus, for any v, if {**C**: a_k+**C** ∈ w} ⊆ v, then Pa_l belongs to v. Hence, no maximal consistent set of formulae includes the set {**C**: a_k+**C** ∈ w} ∪ {∼Pa_l}, and therefore this set is inconsistent. Thus, there is a finite subset of this set {**C**$_1$, **C**$_2$, **C**$_3$,..., **C**$_k$, ∼Pa_l}, which is inconsistent, and therefore formula (**C**$_1$ & **C**$_2$ & **C**$_3$ & ...& **C**$_k$) → Pa_l is a thesis. Hence, (**C**$_1$ & **C**$_2$ & **C**$_3$ & ...& **C**$_k$) → Pa_l belongs to w and, by (A1), (R1) and (R2), (a_k+**C**$_1$ & a_k+**C**$_2$ & a_k+**C**$_3$ & ...& a_k+**C**$_k$) → a_k+Pa_l also belongs to w. However, formulae a_k+**C**$_1$, a_k+**C**$_2$, a_k+**C**$_3$, ..., and a_k+**C**$_k$ belong to w, and by (R1), a_k+ Pa_l also belongs to w. Thus, if vR_k^*w, then {**C**: a_k+**C** ∈ w} ⊆ v and Pa_l belongs to v. So, by (A3) and propositional logic, any formula depicted by schema (a_l+**C** → **C**) also belongs to v. Hence, {**C**: a_l+**C** ∈ w} ⊆ v, and therefore vR_l^*w.

(C6) For any natural numbers k and l, if for any v and w belonging to W, vS_k^*v, implies that vS_l^*w then for any v and w belonging to W, vS_k^*w implies that vS_l^*w. Assume that for any v and w, vS_k^*v implies that vS_l^*w. Thus, for any v and w, if {∼**C**: a_l−**C** ∈ w} ⊆ v, then {∼**C**: a_k−**C** ∈ v} ⊆ v. However, due to

(A6), for any v, if $\{\sim C: a_k-C \in v\} \subseteq v$, then $\sim Oa_k$ belongs to v. Thus, for any v, if $\{\sim C: a_l-C \in w\} \subseteq v$, then $\sim Oa_k$ belongs to v. Hence, no maximal consistent set of formulae includes the set $\{\sim C: a_l-C \in w\} \cup Oa_k$, and therefore this set is inconsistent. Thus, there is a finite subset of this set $\{\sim C1, \sim C2, \sim C3,\ldots, \sim C_k, Oa_k\}$, which is inconsistent, and therefore formula $Oa_k \to (C_1 \vee C_2 \vee C_3 \vee \ldots \vee C_k)$ is a thesis. Hence, $Oa_k \to (C_1 \vee C_2 \vee C_3 \vee \ldots \vee C_k)$ belongs to w and, by (A2), (R1) and (R3), $(a_l-C_1 \& a_l-C_2 \& a_l-C_3 \& \ldots \& a_l-C_k) \to a_l-Oa_k$ also belongs to w. However, formulae a_l-C_1, a_l-C_2, a_l-C_3, $\ldots$, and a_l-C_k belong to w, and by (R1), a_l-Oa_k also belongs to w.

Thus, if it is not the case that $vS_l{*}w$, then $\{\sim C: a_l-C \in w\} \subseteq v$ and $\sim Oa_k$ belongs to v. So, by (A4) and propositional logic, any formula depicted by schema $(a_k-C \to \sim C)$ also belongs to v. Hence, $\{\sim C: a_k-C \in w\} \subseteq v$, and it therefore is not the case that $vS_k{*}w$.

This completes the proof that the structure $M^\star = \langle W^\star, P_i^\star, R_i^\star, S_i^\star \rangle$ is an action-model. $\square$

Proof of Theorem 5.
The proof of the theorem is of course by induction on the construction of formulae. The definition of the canonical ontological model assures that for any propositional letter p_k, and for any $w \in W^\star$, $w \models^{M\star} p_k$ if and only if $p_k \in w$. In the case of the propositional connectives $\sim$ and &, you rely on the maximal consistency of each w to assure you that $A \in w$ if and only if it is not the case that $\sim A \in w$, and that $(A \& B) \in w$ if and only if $A \in w$ and $B \in w$.

In the case of action letter a, the induction hypothesis is that for any $w \in W^\star$, $w \models^{M\star} A$ if and only if $A \in w$.

At first, suppose $a+A$ belongs to w. Hence, if $vR_a{*}w$, then $\{C: a+C \in w\} \subseteq v$ and therefore $A \in v$. Thus, by the induction hypothesis, if $vR_a{*}w$, then $v \models^{M\star} A$. Hence, if $vR_a{*}w$, then $v \models^{M\star} A$ and therefore $w \models^{M\star} a+A$. Next, suppose that $w \models^{M\star} a+A$. Thus, if $vR_a{*}w$, then $v \models^{M\star} A$ and, by the definition of the canonical model and by the induction hypothesis, if $\{C: a+C \in w\} \subseteq v$, then $A \in v$. Hence, no maximal consistent set of formulae includes the set $\{C: a+C \in w\} \cup \sim A$, and therefore this set is inconsistent. Thus, there is a finite subset of this set $\{B_1, B_2, B_3, \ldots, B_k, \sim A\}$ which is inconsistent, and therefore formula $(B_1 \& B_2 \& B_3 \& \ldots \& B_k) \to A$ is a thesis. Hence, $(B_1 \& B_2 \& B_3 \& \ldots \& B_k) \to A$ belongs to w and, by (A1), (R1) and (R1), $(a+B_1 \& a+B_2 \& a+B_3 \& \ldots \& a+B_k) \to a+A$ also belongs to w. However formulae $a+B_1$, $a+B_2$, $a+B_3$, $\ldots$, and $a+B_k$ belong to w, and, by (R1), $a+A$ also belongs to w.

Now, suppose $a-A$ belongs to w. Hence, if it is not the case that $vS_a{*}w$, then $\{\sim C: a-C \in w\} \subseteq v$ and therefore $A \notin v$. Thus, by the induction hypothesis, if it is not the case that $vS_a{*}w$, then $v \not\models^{M\star} A$. Hence, if $v \models^{M\star} A$, then $vS_a{*}w$ and therefore $w \models^{M*}\mathbf{a}-\mathbf{A}$. Next, suppose that $w \models^{M*}\mathbf{a}-\mathbf{A}$. Thus, if it is not the case that $v S_a^* w$, then $v \not\models^{M*}\mathbf{A}$ and, by the definition of the canonical model and by the induction hypothesis, if $\{\sim\mathbf{C}: \mathbf{a}-\mathbf{C} \in w\} \subseteq v$, then $\mathbf{A} \notin v$. Hence, no maximal consistent set of formulae includes the set $\{\sim\mathbf{C}: \mathbf{a}-\mathbf{C} \in w\} \cup \mathbf{A}$, and therefore this set is inconsistent. Thus, there is a finite subset of this set $\{\sim\mathbf{B}_1, \sim\mathbf{B}_2, \sim\mathbf{B}_3,\ldots, \sim\mathbf{B}_k, \mathbf{A}\}$ which is inconsistent, and therefore formula $\mathbf{A} \to (\mathbf{B}_1 \vee \mathbf{B}_2 \vee \mathbf{B}_3 \vee \ldots \vee \mathbf{B}_k)$ is a thesis. Hence, $\mathbf{A} \to (\mathbf{B}_1 \vee \mathbf{B}_2 \vee \mathbf{B}_3 \vee \ldots \vee \mathbf{B}_k)$ belongs to w and, by (A2), (R1) and (R3), $(\mathbf{a}-\mathbf{B}_1 \,\&\, \mathbf{a}-\mathbf{B}_2 \,\&\, \mathbf{a}-\mathbf{B}_3 \,\&\, \ldots \,\&\, \mathbf{a}-\mathbf{B}_k) \to \mathbf{a}-\mathbf{A}$ also belongs to w. However formulae $\mathbf{a}-\mathbf{B}_1, \mathbf{a}-\mathbf{B}_2, \mathbf{a}-\mathbf{B}_2, \ldots$, and $\mathbf{a}-\mathbf{B}_k$ belong to w, and, by (R1), $\mathbf{a}-\mathbf{A}$ also belongs to w.

To prove that $w \models^{M*}\mathbf{Oa}$ if and only if $\mathbf{Oa} \in w$, at first suppose that $w \not\models^{M*}\mathbf{Oa}$. Hence, it is not the case that $w S_a^* w$ and, by the definition of M^*, $\{\sim\mathbf{C}: \mathbf{a}-\mathbf{C} \in w\} \subseteq w$. Thus, any formula depicted by schema $\mathbf{a}-\mathbf{C} \to \sim\mathbf{C}$ belongs to w. However, by (A6), $\mathbf{a}-\mathbf{Oa}$ belongs to w, and therefore $\sim\mathbf{Oa}$ also belongs to w. Hence, $\mathbf{Oa} \notin w$. Next, suppose $\mathbf{Oa} \notin w$. Thus, $\sim\mathbf{Oa} \in w$ and, by (A4), any formula depicted by schema $\mathbf{a}-\mathbf{C} \to \sim\mathbf{C}$ belongs to w. Hence, $\{\sim\mathbf{C}: \mathbf{a}-\mathbf{C} \in w\} \subseteq w$ and, by the definition of M^*, it is not the case that $w S_a w$. Therefore, $w \not\models^{M*}\mathbf{Oa}$.

To prove that $w \models^{M*}\mathbf{Pa}$ if and only if $\mathbf{Pa} \in w$, at first suppose that $w \models^{M*}\mathbf{Pa}$. Hence, $w R_a^* w$ and, by the definition of M^*, $\{\mathbf{C}: \mathbf{a}+\mathbf{C} \in w\} \subseteq w$. Thus, any formula depicted by schema $\mathbf{a}+\mathbf{C} \to \mathbf{C}$ belongs to w. However, by (A5), $\mathbf{a}+\mathbf{Pa}$ belongs to w, and therefore $\mathbf{Pa}$ also belongs to w. Hence, $\mathbf{Pa} \in w$. Next, suppose $\mathbf{Pa} \in w$. Thus, by (A3), any formula depicted by schema $\mathbf{a}+\mathbf{C} \to \mathbf{C}$ belongs to w. Hence, $\{\mathbf{C}: \mathbf{a}+\mathbf{C} \in w\} \subseteq w$ and, by the definition of M^*, $w R_a^* w$. Therefore, $w \models^{M*}\mathbf{Pa}$.

To prove that $w \models^{M*}\mathbf{Ma}$ if and only if $\mathbf{Ma} \in w$, at first suppose $w \models^{M*}\mathbf{Ma}$. Hence, $[w]^{R_a^*} \neq \emptyset$, and therefore there is a maximal consistent set of formulae v, such that $\{\mathbf{C}: \mathbf{a}+\mathbf{C} \in w\} \subseteq v$. Thus, $\{\mathbf{C}: \mathbf{a}+\mathbf{C} \in w\}$ is a consistent set, and therefore there is a formula $\mathbf{B}$, such that $\mathbf{B} \notin \{\mathbf{C}: \mathbf{a}+\mathbf{C} \in w\}$. Hence, there is a formula $\mathbf{B}$, such that $\mathbf{a}+\mathbf{B} \notin w$, and so $\sim\mathbf{a}+\mathbf{B} \in w$. Thus, by (A9) $\sim\mathbf{a}-\sim\mathbf{B}$ belongs to w and, due to (A8), $\mathbf{Ma} \in w$. Next, suppose $\mathbf{Ma} \in w$. Hence, by (A7), any formula depicted by schema $\mathbf{a}+\mathbf{B} \to \sim\mathbf{a}+\sim\mathbf{B}$ belongs to w. Therefore, due to 1.3, $\{\mathbf{C}: \mathbf{a}+\mathbf{C} \in w\}$ is a consistent set. Thus, there is a maximal consistent set of formulae v, such that $\{\mathbf{C}: \mathbf{a}+\mathbf{C} \in w\} \subseteq v$, and, by the definition of M^*, $[w]^{R_a^*} \neq \emptyset$. Therefore, $w \models^{M*}\mathbf{Ma}$.

To prove that $w \models^{M*}\mathbf{a}\subset\mathbf{b}$ if and only if $\mathbf{a}\subset\mathbf{b} \in w$, at first suppose $w \models^{M*}\mathbf{a}\subset\mathbf{b}$. Hence, $[w]^{R_b^*} \subseteq [w]^{R_a^*}$ and $[w]^{S_a^*} \subseteq [w]^{S_b^*}$, and by the definition of M^*, for any v, if $\{\mathbf{C}: \mathbf{b}+\mathbf{C} \in w\} \subseteq v$ then $\{\mathbf{C}: \mathbf{a}+\mathbf{C} \in w\} \subseteq v$ and for any v, if $\{\sim\mathbf{C}: \mathbf{b}-\mathbf{C} \in w\} \subseteq v$ then $\{\sim\mathbf{C}: \mathbf{a}-\mathbf{C} \in w\} \subseteq v$.

However, due to (A5), $\mathbf{a}+\mathbf{Pa}$ belongs to w, and therefore for any v, if $\{\mathbf{C}: \mathbf{b}+\mathbf{C} \in w\} \subseteq v$, then $\mathbf{Pa} \in v$. Hence, no maximal consistent set of formulae includes the set $\{\mathbf{C}: \mathbf{b}+\mathbf{C} \in w\} \cup \{\sim\mathbf{Pa}\}$, and, therefore, this set is inconsistent. Thus, there is a finite subset of this set $\{\mathbf{B}_1, \mathbf{B}_2, \mathbf{B}_3, \ldots, \mathbf{B}_k, \sim\mathbf{Pa}\}$ that is inconsistent, and therefore formula $(\mathbf{B}_1 \,\&\, \mathbf{B}_2 \,\&\, \mathbf{B}_3 \,\&\, \ldots \,\&\, \mathbf{B}_k) \to \mathbf{Pa}$ is a thesis. Hence,

(B_1&B_2&B_3&...&B_k) → Pa belongs to **w** and, by (A1), (R1) and (R2), (**b**+B_1 & **b**+B_2 & **b**+B_3 & ... & **b**+B_k) → **b**+Pa also belongs to **w**. However, formulae **b**+B_1, **b**+B_2, **b**+B_3, ..., and **b**+B_k belong to **w**, and by (R1), **b**+Pa also belongs to **w**.

Moreover, due to (A6), **a**−Oa belongs to **w**, and therefore for any **v**, if {∼C: **b**−C ∈ **w**} ⊆ **v**, then Oa ∉ **v**. Hence, no maximal consistent set of formulae includes the set {∼C: **b**−C ∈ **w**} ∪ {Oa}, and, therefore, this set is inconsistent. Thus, there is a finite subset of this set {∼B_1, ∼B_2, ∼B_3, ..., ∼B_k, Oa} that is inconsistent, and therefore formula Oa → (B_1∨B_2∨B_{k3}∨...∨B_k) is a thesis. Hence, Oa → (B_1∨B_2∨B_{k3}∨...∨B_k) belongs to **w** and, by (A2), (R1) and (R3), (**b**−B_1 & **b**−B_2 & **b**−B_3 & ... & **b**−B_k) → **b**−Oa also belongs to **w**. However, formulae **b**−B_1, **b**−B_2, **b**−B_3, ..., and **b**−B_k belong to **w**, and by (R1), **b**−Oa also belongs to **w**.

But due to (A12), any formula depicted by schema **b**+Ma & **b**−Oa → **a**⊂**b** belongs to **w** and therefore **a**⊂**b** ∈ **w**.

Next, suppose **a**⊂**b** belongs to **w**. Hence, by (A10), any formula depicted by schema **a**+C → **b**+C belongs to **w** and, by (A11), any formula depicted by schema **a**−C → **b**−C belongs to **w**. Therefore {C: **a**+C ∈ **w**} ⊆ {C: **b**+C ∈ **w**} and {∼C: **a**−C ∈ **w**} ⊆ {∼C: **b**−C ∈ **w**}. Thus, for any **v**, if {C: **b**+C ∈ **w**} ⊆ **v** then {C: **a**+C ∈ **w**} ⊆ **v** and for any **v**, if {∼C: **b**−C ∈ **w**} ⊆ **v** then {∼C: **a**−C ∈ **w**} ⊆ **v**. Therefore, $[w]^{R_b*} \subseteq [w]^{R_a*}$ and $[w]^{S_a*} \subseteq [w]^{S_b*}$. Thus, $w \models^{M*}$ **a**⊂**b**. This completes the proof of the theorem. □

Proof of Theorem 6.
In order to show that a formula is a thesis if and only if it is a valid formula, it is sufficient to prove that any thesis is a valid formula and that any formula that is not a thesis is not valid. The proof of the first implication requires the establishment of the validity of all axioms and the demonstration that the rules of inference (R1) and (R2) preserve validity. This can be easily done. To demonstrate that all axioms depicted by the axiom schemata (A5), (A6), (A9) and (A8) are valid, you should rely on conditions (C1), (C2), (C3) and (C4), as stated in the definition of the action-model. To demonstrate that all axioms depicted by the axiom schema (A12) are valid, you should rely on conditions (C5) and (C6), as stated in the definition of the action-model. To prove the converse implication, suppose a formula **A** is not a thesis. Then {∼**A**} is a consistent set. Thus, there is a maximal consistent set of formulae **v**, such that {∼**A**} ⊆ **v**. Hence, by the definition of $M*$, for some **w** ∈ $W*$, ∼**A** belongs to **w**, and **A** does not belong to **w**. Therefore, by theorem 5, for some **w** ∈ $W*$, $w \not\models^{M*}$ **A**, and so **A** is not a valid formula. □

Acknowledgment: The author wishes to thank Dr. Bartłomiej Skowron for his encouragement, and in particular Professor Marek Nasieniewski for his comments and suggestions. Several remarks that he made led to improvements in the text.

Bibliography

Perzanowski, J. (1990). Ontologies and ontologics. In E. Żarnecka-Biały (Ed.), *Logic counts* (pp. 23–42). Dordrecht: Kluwer.

Stalnaker, R. (1980). Logical semiotic. In E. Agazzi (Ed.) *Modern logic – a survey* (pp. 439–455). Dordrecht: Reidel.

von Wright, G. H. (1968). An essay in deontic logic and the general theory of action. *Acta Philosophica Fennica, 21*. Amsterdam: North Holland.

Michał Głowala
"Physical Intentionality" and the Thomistic Theory of Formal Objects

Abstract: In this paper I discuss the claim that the way in which powers are directed towards their manifestations resembles, in some crucial respects, the way in which intentional states or acts are directed towards their objects (the physical intentionality claim). My main thesis is that the scholastic notion of the formal object of a disposition or an act is indispensable for resolving the problem of "physical intentionality". In particular, (i) it is necessary for a sound analysis of the "marks of intentionality", (ii) it sheds considerable light on some parallels between the ontology of intentionality and the ontology of powers, and, finally, (iii) it allows one to demarcate the border between the physical and the mental.

Keywords: physical intentionality, metaphysics of powers, formal objects of powers, material objects of powers.

Introduction

Both intentionality and dispositionality pose numerous problems in ontology. Are these problems systematically interconnected? Some authors (like C. B. Martin, G. Molnar, A. Nes and D. Armstrong) think that dispositions are, in some respects, like mental states or acts: more precisely, that dispositions are *directed* towards their (possibly not occurring) manifestations in a way that resembles intentionality – the directedness of mental states or acts towards their (possibly not existing) objects; let us call this the physical intentionality claim.[1] Such a claim does not amount to the assertion that mental states or acts are essentially dispositional: what is at stake is just the alleged similarity between the directedness of mental states and the directedness of dispositions.

My aim here is to show that there is indeed some sort of deep similarity there, and that it can shed a great deal of light both on intentionality and on the ontology of powers. To grasp both the similarity itself and its limits, however, it is not enough to employ just the notion of intentionality; what is crucial is some more general

---

[1] See e.g. Martin and Pfeifer (1986), Place (1996), Armstrong (1999, pp. 29f), Molnar (2003, pp. 60–81), Nes (2008), Oderberg (2017).

Michał Głowala, Institute of Philosophy, Wrocław University, Poland.

https://doi.org/10.1515/9783110669411-014

notion of the *formal object* (*obiectum formale*) of a disposition or an act (which is one of the key notions of the scholastic ontology of powers and of intentionality). So my claim is that dispositionality and intentionality are two kindred cases of something more general: namely, possession of a formal object. In more general terms, I aim to show that certain key conceptual tools from that older approach to the ontology of powers can shed interesting light on some of the currently debated issues in that same area of inquiry. I will therefore be referring to certain scholastic texts and ideas whose relevance is crucial here, but my interest is strictly systematic rather than historical.

I shall proceed in three stages. First I will begin with a brief presentation of the problem and various sorts of putative solution to it (2). Then I shall introduce the notion of the formal object of a disposition or an act (3), and show some of its applications in the context of contemporary debates concerning physical intentionality (4). Finally, I shall briefly discuss three issues closely related to the topic of the intentionality of dispositions: the relationship between causal links and formal objects (5), the error of reifying formal objects both in the ontology of intentionality (including the ontology of action) and in the ontology of powers (6), and the border between the mental and the dispositional, construed at a general level in terms of formal objects (7).[2]

The Problem and Its Solutions

"The Marks of Intentionality" and Powers

Putting the matter in the briefest terms, the central thought here is that the directedness of a disposition towards its manifestations meets some of the conditions recognized (at least in some debates within analytic philosophy[3]) as being the

---

[2] I present a broader analysis of the issue of physical intentionality (and a number of other crucial issues concerning the relationship between powers and their manifestations) in terms of formal objects in my book (in Polish) *Możności i ich akty. Studium z tomizmu analitycznego* [*Powers and Their Manifestations. A Study in Analytical Thomism*], Wrocław 2016.

[3] I will not be discussing here an otherwise important and interesting question, which is whether the "marks of intentionality" of analytical philosophers offer a characterisation of intentionality that would be satisfactory for someone investigating the latter – e.g. from a phenomenological point of view; the answer to that is, at best, far from clear. The problem I am focussing on here concerns just the features included within the so-called "marks of intentionality", and not the relevance of "the marks of intentionality" for the issue of intentionality in general.

marks of intentionality. With regard to the latter, I would say that the following five are the most interesting:[4]

(i) *Essential directedness towards objects and manifestations.* Mental acts are directed towards their objects, and dispositions are directed towards their manifestations (or, as Armstrong puts it, they point to their manifestations). Moreover, directedness towards the object is essential for a mental act, and directedness towards its manifestation is essential for a disposition.

(ii) *Possible non-existence of objects and manifestations.* It is a widely discussed feature of intentionality that the directedness of intentional acts does not require the existence (at least with respect to what is, in *prima facie* terms, a fairly natural sense of "existence") of what a given act is directed towards; similarly, in the case of dispositions, their directedness does not require the existence of the manifestations. This shows, in particular, that either the directedness in question is not a typical relation requiring the existence of the relata, or that manifestations of powers that, in some sense, do not occur, do nevertheless exist in some sort of special way (for example, as some kind of Meinongian objects, or as Platonic ideas).

(iii) *Indeterminacy of objects and manifestations.* There is some sort of indeterminacy specific to the objects of intentional acts; for example, I may be looking just for *a* gift for someone (as opposed to this or that particular gift), or think about a mountain creek without thinking about any particular creek. Similarly, water-solubility is the disposition to be dissolved by water, not the disposition to be dissolved by this or that particular water, at this or that particular time, in this or that particular way.

(iv) *Aspectual shape.* According to Searle, each intentional act has some aspectual shape: i.e. the way the object is given in it. Nes claims that something similar holds for dispositions: for example, an electrically charged body has the power to attract other bodies *as electrically charged*, but not *as having gravitational mass*.[5] Another example could be the following: water dissolves sodium chloride as an ionic compound, but does not dissolve sugar in this way; things may be dissolved

---

[4] For more detailed (and critical) overviews of the marks of intentionality, see, e.g., Martin and Pfeifer (1986), Place (1996), Molnar (2003, pp. 63–66), Bird (2007, para. 5.7.1), and Oderberg (2017).

[5] Nes (2008, p. 207): "whenever an object is attracted, it is attracted in a certain way: gravitationally, electrically, magnetically, or perhaps in some other way. If a metal ball, say, gravitationally attracts a metal bar, say, it attracts it as having a certain mass. If the metal ball electrically attracts the metal bar, it attracts it as a thing with a certain charge. The ball attracts the bar here, in the different cases, as an exemplar of different features. It figures in the state of attraction under different aspects".

in many ways, and we characterise the way something is dissolved just by pointing to some aspect of it.

(v) *Intensionality*. It is widely agreed that intentional contexts are (at least often) intensional (although the very nature of intensionality poses various problems). Now it seems that ascriptions of manifestations also are intensional in the sense that various descriptions referring to the same manifestation are not interchangeable *salva veritate*. For example, as Molnar points out, from the fact that acid has the power to turn litmus paper red, and the fact that the colour of Post Office pillar boxes is red, it does not follow that acid has the power to turn litmus paper the colour of Post Office pillar boxes. Admittedly, this example poses a number of problems, discussed in Bird (2007), and I myself will address some of these in due course. Moreover, the general issue of the criteria of intensionality is a pretty complex one, and it is not possible to discuss it here at length. What seems clear, however, is that manifestation ascriptions *may* be read in an extensional way, so that the transition from "acid has the power to turn litmus paper red" to "acid has the power to turn litmus paper the colour of Post Office pillar boxes" is truth-preserving. The point is that it is also quite natural to read them in an intensional way, so that the transition in question is not truth-preserving, and acid does *not* have anything like the power to turn anything the colour of Post Office pillar boxes.

It seems that there is a parallel distinction to be made between intensional and extensional readings in the case of mental intentionality. In scholastic logic, for example, sentences like "*video papam*" or "*cognosco venientem*" were said to be read in an intensional way, while sentences like "*papam video*" or "*venientem cognosco*", having the reverse word order, were said to be read extensionally. So, if I otherwise know the man who happens to be approaching, but do not recognize him just as approaching, "*venientem cognosco*" would have counted as true, but "*cognosco venientem*" as false.[6] Similarly, Anscombe (1981a) points out that in some contexts it is quite natural to say "You can't have seen a lion, there wasn't any lion there to see" (p. 14), which shows that "seeing something" has been taken in an extensional way.

---

[6] See Nuchelmans (1980, pp. 57–59) and Geach (1980, pp. 131–132); Geach claims that the Latin statements with the object phrase preceding the verb should just be rendered in English with the existential quantifier.

Rejection and Acceptance of the Alleged Similarity

From a general standpoint, there are two ways in which one might reject the alleged similarity suggested in (i)–(v). The first is to reject it from the very outset, through appealing to some feature of intentionality. So, for example, one might claim that intentionality is something primarily given in the first-person perspective; or that intentionality is aboutness, whereas dispositions are not *about* their manifestations; or that intentionality involves representation, whereas dispositions do not represent their manifestations. Similarly, one might reject the similarity at the outset by claiming that it leads to animism or panpsychism (Mumford, 1999, pp. 221–222). I agree with Nes (2008, p. 207), however, that this is not a good way to solve the problem of the alleged similarity, because it "makes intentionality trivially sufficient for mentality", whereas the truth of claims associating intentionality with mentality is actually by no means trivial.

The other way (to be found in Alexander Bird (2007), Jan Hauska (2007) and, recently, David Oderberg (2017)) is to show, in a much more detailed manner, that the alleged similarity is only a superficial one – or, at least, that it is much weaker than it *prima facie* seems to be. This strategy, I think, is much more interesting and promising, and I will discuss some of its details in due course.

Some of the authors just mentioned claim, moreover, that there is in fact some similarity between the mental and the dispositional, although this similarity does not reflect any special feature of intentionality, but rather something more general. In particular, Bird claims that even if there are some important parallels between intentionality and dispositionality, they may have nothing to do with intentionality as such – instead reflecting, for instance, some more general requirements of logic (that are also revealed, for example, in the fact that the truth of statements like "Pegasus does not exist" is not trivial).[7] Oderberg (2016), on the other hand, claims in his recent paper that what is right in the physical intentionalist's arguments shows that both mental acts and dispositions display *finality*. Finality, according to him, is to be understood as "specific indifference" (the restriction of the manifestations to some range and, on the other hand, the indifference of the power to the circumstantial aspects of the manifestation). Moreover, there are fundamental differences between the finality of mental acts and the finality of physical dispositions, because the former, as opposed to the latter, involves freedom of abstraction.

---

[7] Bird (2007, p. 124): "if potencies require non-existent potential manifestations, that need have nothing to do with intentionality. [...] Rather it may reflect a much more general requirement of logic."

My claim is that there are some crucial parallels between intentionality and dispositionality, and that this in fact reflects something more general rather than anything specific to intentionality: namely, the fact that both dispositions and intentional states have *formal objects* – intentionality and dispositionality being two distinct cases of the possession of *formal objects*. I would like to show, however, that the parallels are much stronger than Bird claims they are: dispositions do have what is called the "marks of intentionality", but the latter should rather be called "marks of having a formal object". I agree with Oderberg that dispositions and mental states display two different brands of finality, yet I still hold that the notion of a formal object (*obiectum formale*) is required for a good treatment of certain details pertaining to finality – and, in particular, for a satisfactory resolution of certain problems posed by the "marks of intentionality".

Finally, it is worth noting that some authors arguing for the alleged similarity seem to suggest that it offers some prospect of naturalizing intentionality; some others, by contrast, think that it shows that dispositions are ontologically suspect in the way that mental acts are.[8] I do not embrace either of these standpoints: I just think that there is some deep interconnection between problems of intentionality and problems of dispositionality, and that this sheds important light both on the ontology of intentionality and on the ontology of powers. The notion of a formal object, I would say, is needed in order to grasp both the interconnection and the limits to the similarity.

The Notion of Formal Object and the Ambiguity of Object Ascriptions

A Preliminary Sketch of the Ambiguity

To introduce the scholastic concept of the formal object of a disposition or an act, let me begin with the distinction between the two uses of the word "object". This distinction has been discussed in some detail by Anscombe, who takes it to be a key to understanding some crucial issues in modern discussions of intentionality. In the sense that I am interested in here, "object" must always be accompanied by the genitive, as in "objects of thought" or "object of suspicion" or "object of contempt". An object in this sense is always an object *of* something. (This, I might

---

[8] According to Bird (2007, p. 121), John Heil suggests the former strategy; as for the latter, see Armstrong (1997, p. 79).

add, is an *old* sense of "object": one quite typical for scholastic philosophy). By contrast, in the sense which I am *not* interested in here, "object" does not require this sort of complement, as in "the most interesting objects in the British Museum" or "an unidentified flying object" (this being a more modern sense of "object"). My remarks concern the former, older sense of "object", and I will use the phrase "an object of something" to distinguish it from the latter one. (Anscombe uses "intentional object" for the old sense of "object", but since "intentional object" is used today in a radically different sort of way, I shall not employ her sense of "intentional object" here).[9]

Now the point is that the very phrase "an (the) object of *something*" is ambiguous, and the scholastic terms "*obiectum formale*" ("formal object") and "*obiectum materiale*" ("material object") were designed to mark the difference between these two senses. To see the ambiguity here, let us consider the question "What is the object of contempt?" On the one hand, we may be asking about something that happens to be the object of someone's contempt in a certain case: that would be a question about the *material* object of contempt. On the other hand, we may ask *what it is* for something to be the object of contempt: in other words, what is the general *form* of objects of contempt. That question concerns the *formal* object of contempt. One and the same thing may be the *material* object of both contempt and admiration, but by contrast, *formal* objects of contempt and admiration are necessarily distinct: what makes a thing the object of contempt is not what makes it the object of admiration.

A similar sort of ambiguity may be traced in reports of perception. Suppose you see a cart wheel as an oval: then the wheel you see is the material object of seeing, and the oval you see is its formal object. In particular, if the wheel is seen as an oval, part of what it is for it to be seen is to appear as an oval. (One might well wonder what it is for the wheel to appear just this way, and this is just a question concerning the details of the formal object of seeing in the case in question: a causal theory of perception would claim, for example, that this involves some sort of causal connection between the wheel and the seeing).

A parallel ambiguity is easy to trace in the case of "the object of dissolution" or "the object of attraction". What is the object of each of these? On the one hand, one may be asking here about, say, a particular sample of sodium chloride being dissolved, or a particular metal bar being attracted. On the other hand, this might

---

9 See Anscombe (1981a, p. 3): "objects on the other hand were formerly always objects *of* –. Objects of desire, objects of thought, are not objects in one common modern sense, not individual things, such as *the objects found in the accused man's pockets* [...] formerly, if something was called an object, that would have raised the question 'object of what?'". For the contemporary use of "intentional object", see Chrudzimski (2013).

be a question about what it is for the sample or the bar to be dissolved or attracted in a given case – in other words, about what the general form of the object of dissolution or attraction in that case is. Formal objects of dissolution or attraction are necessarily different, though one and the same thing may happen to be both dissolved and attracted: dissolution and attraction may happen to have one and the same material object. Moreover, there are various kinds of dissolution and attraction: sodium chloride, for example, is dissolved as an ionic compound through the dissolution of ionic bonds; the metal bar in Nes's example may be attracted gravitationally (as having mass) or electrically (as having electric charge). So, in this way, too, the question of "what it is to be the object of dissolution/attraction" turns out to be a non-trivial one.

Yet another example might prove useful here: suppose that George employs John as a gardener; there is a parallel ambiguity in the question of who is employed. On the one hand, one may be asking about the person who happens to be employed: that is, about the material object of employment. On the other hand, one may be asking whether John is employed as a gardener, solicitor, cook or someone else. This is a question about the formal object of employment, and the answer is typically given by the as-clause.

Finally, consider the "object of a disposition", or the "manifestation of a disposition".[10] On the one hand, something may happen to be a manifestation of a given disposition: for example, a destruction of ionic bonds in a given case may happen to be a manifestation of water-solubility. On the other hand, we may ask what the general form of the manifestation of a given sort of disposition is – in other words, what it is for a given change to be a manifestation of a given disposition. Again: one and the same thing may happen to be the manifestation of various dispositions – this is precisely what C.B. Martin and John Heil call mutual manifestation (for example, the dissolution of a sample of sodium chloride by water is the manifestation of both the solubility of NaCl and the relevant disposition of water).[11] On the other hand, the general form of the manifestation of the solubility is *not* the general form of the manifestation of the power of water: the former disposition is a disposition to be dissolved by water, whereas the latter is a disposition to dissolve. What makes the destruction of ionic bonds a manifestation of water-solubility is not what makes it a manifestation of the active power of water as a solvent. Here,

---

10 Here I am taking the object of a disposition to be what the disposition is a disposition to – and not, for example, the object that may be changed thanks to the disposition; so, for example, I am focusing on the sense in which the object of the power of a solvent is dissolving, and not something that gets dissolved.

11 See Martin (2008, p. 46), Heil (2005, p. 350) and Heil (2012, p. 127). This claim, of course, is rejected in many ways, and I discuss some of them in due course.

as in the former cases, it is natural to employ the as-clauses to express the identity and the distinctness of manifestations. One and the same change, namely the destruction of ionic bonds, as a change *undergone* by the particles of NaCl in virtue of some features of these bonds, is a manifestation of the water-solubility of NaCl; on the other hand, the same change, as a change *produced* by the particles of water in virtue of their polarisation, is a manifestation of the active power of water as a solvent of a certain type.

To sum up, we may say that what these examples show is a degree of ambiguity on the part of both object ascriptions and manifestation ascriptions. Statements of the form "the object/manifestation of ( ) is/was ( )" might be taken either in the material or in the formal sense, and what happens to be true in the material sense may be false in the formal one.

A parallel ambiguity is displayed by such contexts as "a power to ( )". On the one hand, we may ask whether anything has the power to dissolve that golden one-dollar coin; on the other, Molnar would say that "the power to dissolve that golden one-dollar coin" does not denote a "genuine intrinsic power" on the part of *aqua regia*. Dissolving a certain coin is clearly something like a material object of the power, and it may be asked what, precisely, makes that change a manifestation of the active power of *aqua regia* as a solvent (or what it is for the change in question to be a manifestation of that power). The change of the coin is a manifestation of that power *as* a certain sort of change produced by the solvent in a certain way, due to its properties; this gives (the outline of) the formal object of the power in question, and in a designation of a "genuine intrinsic power" something like its formal object must be given.

As the examples of employing John as a gardener, seeing a wheel as an oval, or dissolving sodium chloride as an ionic compound suggest, what the formal object (or at least part of the formal object) is can usually be expressed in a clause following "as" or "*qua*". An important thing to note here is that the clause "as ( )" is attached to the verb or predicate, and not to its object (the phrase giving the material object). The relationship between the formal and the material object seems to be a pretty complex matter (for example, if sodium chloride is being dissolved as an ionic compound, it has to be an ionic compound, whereas a wheel is seen as an oval, but is not an oval, and a person employed as a gardener need not be a gardener otherwise). Moreover, the as-clause often expresses something like a specific difference in respect of what is expressed by the predicate: for example, dissolving something as an ionic compound may be contrasted with other kinds of dissolving, and employing someone as a gardener may be contrasted with other kinds of employment. This reflects the scholastic idea that formal objects are essential for the *differentia specifica* of powers, states or acts having them.

In Latin, "object of something" ("*obiectum alicuius*") is derivative from a verb meaning "to be an object of something" ("*obiici alicui*"), and the formal object of something is defined just as the way in which something is an object ("*modus obiiciendi*") of it; to repeat, the formal object of x is what it is for a material object of x to be a material object of it, or what it is for x to have a material object. In other words, formal objects are just ways of having material objects: part of the formal object of x is the relevant sort of connection of x with its material object – the connection that makes something a material object of x. In many cases, the relevant sort of connection is a certain kind of causal connection. Paul Soncinas, a late-scholastic Thomist, says that a colour might happen to be an object (*obiici*) of sight, of thought, and of will, and what is distinct in these cases is just the way in which it is an object of a given power or activity.[12]

There is a closely related ambiguity that pertains to "having an object" or "having a manifestation". On the one hand, there is *actually* having an object or a manifestation: x actually has an object or a manifestation when there is something that happens to be an object or a manifestation of x. On the other hand, having an object or having a manifestation seems essential for that which has them: x has a given object or a given manifestation even if nothing actually happens to be an object or a manifestation of x (we may call this the essential sense of having an object or a manifestation). To see the relationship between this ambiguity in respect of "having an object/manifestation" on the one hand, and the distinction between material and formal objects on the other, we should first note that in the case of the essential sense of having an object or manifestation, we are referring to the formal object. At the same time, though, if we say that something is actually an object of x, both the formal and the material sense of "object" may be included: for example, a bell might happen to be both an object of seeing and an object of hearing at the same time – it is an object of seeing as something coloured (acting in a relevant way on the power of sight), and it is an object of hearing as something ringing (acting in a relevant way on the power of hearing). Similarly, one and the same change might happen to be a manifestation of two distinct powers at the same time: for example, a dissolution of some sample of NaCl by water is a manifestation of the solubility of NaCl (as a change that NaCl is undergoing due to some properties of its ionic bonds), and a manifestation of the relevant power

---

[12] Soncinas (1579, lib. VI, q. 8, p. 103a): "Dicitur enim obiectum, quia obiicitur potentiae et habitui; et ideo ubi sunt diversi modi se obiiciendi, ibi est diversa ratio obiecti, inquantum est obiectum"; "Nam color obiicitur visui, et intellectui, et voluntati a qua amari potest. Sed modus quo obiicitur visui est immutando ipsum per medium extrinsecum [...] Unde si volumus diffinire obiectum visus, inquantum est obiectum, dicemus quod est immutativum visus per medium extrinsecum etc".

of the solvent (as a change produced by that solvent due to the polarisation of its particles). What is one and the same here is just the material object of the powers in question, and what is expressed by the phrases following "as" are formal objects of the powers. Finally, if you say, in the usual sense of these words, that George employs a gardener, what is involved here is both the material object of employing (the person who is employed) and the formal object given by "a gardener".

Some Further Points Concerning the Ambiguity

This ambiguity pertaining to object or manifestation ascriptions is, I think, not difficult to detect. What is far from being trivial, however, is the application of the notion of formal object to some difficult issues within the debate on physical intentionality. For this reason, in the rest of this paper I shall be focusing just on these applications. Right now, however, we need to add four further introductory points concerning the distinction between formal and material objects.

(i) Formal objects are not just *types* that material objects are tokens of. A dissolved sample may instantiate various types, and the point is that most of them are not formal objects of dissolving. Similarly, what happens to be an object of contempt may well instantiate various types, but typically, most of them are not formal objects of contempt: they are not what makes these objects objects of contempt as such. An employed person may exemplify many skills, but typically, most of them are not the formal object of employment; they are not what makes that person employed in a given way. Moreover, it is quite possible that a person employed as a gardener is *not* a gardener in any sense that is not itself reducible to being employed as a gardener. Finally, if you see a wheel as an oval, "oval" is not a type exemplified by the wheel you see. Formal objects of x are more like *adverbial* characteristics of *acts*, or, more precisely, they are characteristics of the way in which something is an object of x. What is crucial for the formal object of x is not just the type that the material object of x instatiates or exemplifies, but rather the very sort of connection which is required for anything to be material object of x. Cardinal Cajetan, a late-scholastic Thomist, distinguishes *ratio obiecti ut est res* (the type of object insofar as it is a thing) and *ratio obiecti ut est obiectum* (the type of object insofar as it happens to be an object of something), asserting that the latter is the formal object in the strict sense (Cajetan, 1888, pars I, q. 1, art. 3, n. 3).

(ii) The connection between x and its *material* objects is typically a contingent one. Something just happens to be the object of contempt or dissolving, or a manifestation of a given disposition. Similarly, the employment relation between persons is a contingent matter. By contrast, the connection between x and its *formal* object is a necessary one. It is part of the nature of a disposition that it has a given form of manifestation (and, in particular, a certain sort of connection with its manifestation, or, in other words, a certain sort of connection that makes something a manifestation of it).

(iii) The relation between *x* and its material objects is a typical relation which requires the existence of both relata. By contrast, the relation between *x* and its formal object is, in a way, not a typical relation at all: the scholastics say it is a *transcendental* relationship. Transcendental relationships are not entities distinct from what is related, and hence basically do not require the existence of both relata. In many ways, transcendental relations are similar to what today are called formal relations (see e.g. Lowe, 2004). Hence formal objects – as opposed to "intentional objects" in the modern sense (see Chrudzimski, 2013) – are not introduced to provide a full-fledged relation in the case of apparently non-existent objects of acts or dispositions.

Tugby (2013, p. 460) remarks that the standard dispositionalists' reaction to the Meinongian issue – namely, claiming that "dispositional directedness must not consist in a genuine relation" – typically leaves their opponents unsatisfied. Admittedly, it is not enough to claim that dispositional directedness must not be a genuine relation, and similarly, in the case of the mental, it is not enough to claim that intentionality is not a genuine relation. I think, however, that focusing on the ambiguity of the phrase "the object of" that has just been outlined here paves the way for a better understanding of what these sorts of directedness or non-genuine relations *are*.

(iv) Material objects of *x*, and its formal object, are not two distinct sorts of object in the absolute sense. (In particular, material objects as considered here are not *material things*, but rather *objects in the material sense*.) More specifically, material and formal objects of employment are not two kinds of employed person, and material and formal objects of seeing are not two sorts of item one sees. In a sense, they are not two distinct kinds of object at all: what we have here is rather two radically different *senses* of "the object of ( )", these being the material and the formal one. In particular, formal objects should not be *reified* – that is, treated as some peculiar sort of *thing* (so, for example, formal objects of employment should not be treated as some peculiar sort of employed person). To repeat: formal objects are more like *adverbial* characteristics of the *act* or ways of having material objects. Nevertheless, in many cases there is a very strong temptation to reify formal objects – a topic I will return to in due course.

Formal Objects and the "Marks of Intentionality"

My main claim is that the "marks of intentionality" (i)–(v) discussed in section 2.1 are marks *of having formal objects* rather than marks of intentionality. Having a formal object is something more general than being intentional. Thus, dispositions have the "marks of intentionality" just by virtue of having formal objects. Intentional states, to be sure, have formal objects too, but these two cases of having formal objects are significantly different. I shall return to the topic of this difference in section 7 below, but first let us return once more to the "marks of intentionality".

(i) *Essential directedness*. The directedness of *x* towards its formal object, as opposed to its directedness towards its material object, is essential for *x*: for example, directedness towards ionic compounds is essential for a certain sort of dissolution ("directedness towards ( )" is ambiguous in a manner strictly parallel to the ambiguity of "the object of ( )"). Clearly it is *formal* directedness, as opposed to the material one, that is involved in the "marks of intentionality". As Mumford (1999, p. 221) rightly remarks, a falling rock might be directed towards the road below, but clearly there is much more involved in the directedness that is relevant in the case of the directedness of dispositions. Now the road below (as well as the larch or the grass below) are clearly material objects of the falling rock, or of its fall, but if the fall were to have a formal object, it might be something like "(things occupying) the place beneath on the route of the fall". The falling rock is directed towards the larch *as occupying the place beneath on the route of the fall*; a bird's flight, by contrast, might be directed towards the same larch as the place where its nest is located; and so on. In general, the formal object of the directedness of the fall of the rock is determined by the way in which the rock moves forward.

It may be useful to repeat here that clauses expressing formal objects typically attach to verbs or predicates (as in "( ) employs ( ) as a gardener" or "( ) dissolves ( ) as an ionic compound" or "( ) attracts ( ) as having gravitational mass"), and that these clauses often express the *differentia specifica* of what is expressed by the predicate.

(ii) *Possible non-existence of the object*. As Mumford rightly points out, "a falling rock cannot be directed towards a road which does not exist, whereas there can be fear of an intruder who does not exist" (Mumford, 1999, p. 221). The point is, however, that in a sense the falling rock (or the fall of the rock) may be said to be directed towards things occupying the place beneath on the route of the fall even if no object actually happens to occupy that place: if you consider the formal object of the fall, there is, in a sense, a possibility of non-existence on the part of the object; and the point is that in the case of "fear of an intruder", "an intruder" describes the *formal* object of fear, not a material one.

It has been pointed out that some mental states do not allow for the non-existence of the object: for example, knowing something by acquaintance (or *notitia intuitiva* in some late scholastic authors) excludes the possibility of the non-existence of what is thus known.[13] More generally, some things having formal objects do not allow for the non-existence of objects: for example, it seems that there is no dissolution if there is nothing actually being dissolved (no material object of dissolution).

In general, it would seem that "the possibility of the non-existence of the object" is a phrase that can carry many different senses. On the one hand, it seems clear that having an object or a manifestation in the essential sense does not depend on the existence of a material object, so there is some essential connection between x and its formal object that does not require the existence of material objects. In some cases, this means that a token of x may exist, even though it happens to be the case that there is no material object of x: for example, there is some sort of necessary connection between a particular trope of solubility and being dissolved, even if no dissolution actually takes place. In other cases, however, a token of x may only exist if it happens to be the case that there is something that is its material object: for example, a particular token of heating exists only if there is something that happens to be heated. Moreover, if I see something oval, there need not be anything oval I see, but one may wonder whether there must be something that I see (a material object of seeing) as oval. (This, however, is not the case when, for example, I imagine something oval or look for something oval). This wide variety, I think, is explained by the fact that the phrase expressing the formal object attaches primarily to the verb, so that it can stand in various relationships to the (material) object expressed by the object phrase.

(iii) *Indeterminacy of the object*. Formal objects are indeterminate in the sense that many details of material objects are not part of the formal object: for example, the history or the colour of a sample that happens to be the object of dissolution are not part of the form of the object of dissolution (of a given kind) – they are not part of what it is for something to be dissolved, and they do not make the sample an object of dissolution; and many properties of a given case of dissolution are not what makes it a manifestation of the power of the solvent. It is clear that formal objects of dispositions are typically indeterminate in this way and, at least *prima facie*, there is nothing mysterious about this sort of indeterminacy. In particular, this does not imply that there is anything indeterminate in the very manifestations (in the material sense) of powers.

---

[13] Oderberg (2017) claims that in general there is no possibility of non-existence of the object in the case of factives.

From this point of view, I think it is possible to defend the idea of indeterminacy in powers against Bird's charges. As Bird rightly remarks, a plurality of possible manifestations – for example, a plurality of ways in which a fragile vase might break – is not enough to establish any interesting parallel between the indeterminacy specific to intentionality and the indeterminacy of powers, since "an infinity of fully determinate objects is not the same as a single indeterminate one" (Bird, 2007, pp. 124–125). In terms of material and formal objects, each material object is fully determinate, and their plurality does not amount to any sort of interesting indeterminacy in the object. What is indeterminate, however, is the formal object of the power – not any of its material objects; no sort of indeterminacy is to be traced in material objects themselves.

Bird states that the indeterminacy specific to intentionality "reflects the descriptive character of much thought and, in particular, the fact that descriptions may be incomplete". I think there is, in fact, a close connection between indeterminacy and the incompleteness of descriptions: the point is, however, that there is a certain relevant kind of incompleteness of description that is specific to formal objects generally. To see this, let us note first that one and the same event might be a manifestation of various powers *under various descriptions*: hence, something is always a manifestation of a given power *under a certain description* relevant to the power, and the description in question is typically incomplete. So, for example, the breaking of a fragile vase might be an act of barbarism, but it is not a manifestation of the fragility of the vase under the description "an act of barbarism". (The way I am using "under a description" here forms a close parallel with its well-known use in Anscombe; it should especially be stressed that this use does not imply that the description in question is something that the agent *thinks of* [Anscombe, 1981b, pp. 209–210].) Now the point is that the description under which something is an object or a manifestation of a given power is just the description giving the formal object of that power, and the incompleteness of that description reflects only the indeterminacy of the formal object in question. So, to sum up, I agree with Bird that the indeterminacy of intentionality has something to do with the incompleteness of descriptions, but it seems to me that the relevant incompleteness of descriptions is not anything specific to thoughts: it is something more general, connected with having a formal object, because it is just the incompleteness of the descriptions *under which* something is a manifestation or an object of a given power or mental act.

(iv) *Aspectual shape*. The formal object of x (the description of which is the description *under which* something is the object of x) is clearly the aspectual shape particular to x.

(v) *Intensionality*. Let us note first that ascriptions of manifestations in the material sense might be read as extensional: if turning red happens to be the most unexpected change for observers, and acid has the power to turn litmus paper red, then acid has the power to produce the most unexpected change for observers. The power to turn the litmus paper red is manifested in the most unexpected change for observers. I think it is quite natural to read many manifestation ascriptions in just this extensional way. An extensional reading of "George employs a gardener" is also possible: if the gardener employed by George happens to be a murderer, then in some sense it is true that George employs a murderer.

The point is, however, that on a certain reading the transition in question is not truth-preserving. There is some important sense in which even if the person employed by George happens to be a murderer, George does *not* employ a murderer unless he employs that person *as* a murderer, and this is just the sense of "George employs a murderer" that involves the formal object of employing (the formal sense of "employing someone"). Similarly, in some important sense (namely the formal sense of "a power to ( )"), acid does *not* have a power to produce the most unexpected change for observers. This important sense is precisely the following: the colour change in question is not a manifestation of the relevant power of acid *as* the most unexpected change for observers, or, in other words, it is not a manifestation of that power *under the description* "the most unexpected change for observers". The description "the most unexpected change for observers" is *not* the description giving the *formal* object of the power in question (it might give the formal object of, say, a power of a perfect conjuror), and that explains the failure in respect of truth-preservation. There is, to my mind, a close parallel between the falsity of "George employs a murderer" and the falsity of "acid has the power to produce the most unexpected change for observers".

Here again, I think, it is possible to defend Molnar's idea of intensionality against Bird's charges. Bird agrees that there is a failure of truth-preservation in the case of Post Office boxes discussed by Molnar, but he argues that it has nothing to do with intensionality or referential opacity. According to Bird, in "the canonical characterization" of a power in terms of its manifestation, the descriptions of manifestations are used as expressions picking out (and referring to) natural or sparse properties (universals). Hence the point, according to Bird, is that in the case of the transition we do not have co-referring expressions but, instead, expressions referring to distinct sparse or natural properties (universals). That is why, according to Bird, in spite of the failure of truth-preservation there is no intensionality or referential opacity in the case in question (Bird, 2007, pp. 121–123).

Bird's standpoint seems to involve some sort of Platonic ontology of (sparse or natural) properties, but I am not going to discuss the general viability of Platonic ontologies of properties here. There are, however, three points that are relevant to the application of such ontologies in the case of powers and their alleged intentionality. The first is that one may wonder whether the strategy suggested by Bird might not be used to eliminate intensionality in (at least some) mental contexts. This would be a strategy for turning all the adverbial characteristics of mental acts – the characteristics of the way in which the object is given – into distinct objects of mental acts.

The second point is that even if Bird's proposal works for the essential sense of "having a manifestation" (as in "water-solubility manifests itself in being dissolved by water"), it does not seem to work in contexts like "the water-solubility of that piece of NaCl was manifested yesterday in NaCl's being dissolved by water".

The third point is that part of the formal object of x is just a sort of relevant connection with x that makes something a material object of x. For example, on the one hand, part of the formal object of the passive power of water-solubility (the power to be dissolved in water) is the connection between the dissolution of an ionic compound and the properties of that compound. (In other words, it is the way in which those properties of the compound are causally relevant to the dissolution). On the other hand, part of the formal object of the active power of water as a solvent is the connection between the same sort of dissolution and the properties of the molecules of water (in particular, the way in which their polarity is causally relevant to the dissolution). So it seems that for the canonical characterisation of a power, it is not enough to pick out some sparse property which is the manifestation type (for example, dissolution of an ionic compound): what must be embedded in the canonical characterisation is just the relevant sort of connection between the power and the manifestation type. In particular, one and the same sparse property might be a manifestation (in a material sense) of various powers. Here one might respond that the relevant sort of connection between the power and the manifestation is embedded in the very sparse universal which is the manifestation type, so that what we are picking out in the canonical characterisation of water-solubility is not just the dissolution of an ionic compound, but also some relevant sort of connection between the dissolution and the properties of the compound – in other words, that the sparse property in question is itself *causing a relevant sort of change in a relevant way*. It seems to me, however, that this does not solve the problem, because there are various possible connections between *causing a relevant sort of change in a relevant way* and a power: for example, something may be just a catalyser of *causing a relevant sort of change in a relevant way*. To my mind, the point is that if you make "in a relevant way" a feature of the object, there still remains the possibility of adding some adverbial characteristics pertaining to the connection between the manifestation type and a power. Or, from another point of view, the reason why formal objects cannot be just treated as manifestation types is, in a way, parallel to the reason why *formal* relations cannot be treated as just a special subset within relations generally.

Reification of Formal Objects: More Parallels between the Mental and the Dispositional

One of the main merits of the distinction between the formal and material senses of "the object of something" is that it helps reveal a certain general form of mistake that one encounters, I think, quite often, both in the ontology of intentionality and in the ontology of dispositions. In general terms, it is the mistake of treating formal objects of acts or dispositions as some peculiar (and usually privileged) kind of their material objects (or treating the formal sense of "the object of ( )" as a particular case of the material sense of "the object of ( )"): in other words, the mistake of *reifying* formal objects. To argue convincingly that this actually *is* mistaken (in various particular cases) is a sizeable task – one that I cannot hope to undertake in the present context. All I would like to show here is that there is a group of standpoints that may be treated as cases of the reification of formal objects.

An obvious example of such a mistake would be treating the gardener involved in "George employs John as a gardener" as some person distinct from John – a person employed in the most immediate and primitive way (so that it would be just the gardener, as opposed to John, that is *most properly* employed). In this case, of course, no one is likely to make such a mistake. In some philosophically more puzzling areas, however, the temptation to make an analogous move proves much more difficult to resist. Theories of sense data, construing these as the most immediate or primitive objects of sensation, seem to illustrate a similar step: here, formal objects of perception are taken to constitute a special sort of material object.[14] There seem to be many interesting similarities between the Thomistic theory of formal objects of perception and some contemporary versions of the adverbial theory of perception (although I cannot dwell on the details of the issue here). More generally, various theories of purely intentional objects, as well as the Meinongian ontology of objects, illustrate the same sort of move. Still more generally, theories that treat the directedness of mental acts or states as a relation involving the existence of both relata tend to treat objects in the formal sense as a kind of objects in the material sense. Finally, various versions of Platonism seem to rest on a shift from stating that something is an object of intellectual activity (in the formal sense

---

14 Anscombe (1981a, p. 11) says, that the position of Berkeley or of sense-data theorists "misconstrues intentional objects as material objects of sensation"; although my use of "formal object" and "material object" does not precisely correspond to her use of "intentional object" and "material object", her diagnosis, to my mind, concerns just what might be called the reification of formal objects.

of "object") to stating that it is an object of this activity in the material sense of "object", and hence that it exists as a correlate of the relevant mental acts.

There are also parallel examples in the ontology of dispositions. Armstrong's claim that dispositions require their non-existent manifestations as Meinongian objects in the way that mental acts do is a clear example of this parallel. Tugby's claim (2013) that a Platonic ontology of properties offers the best solution to the problem of directedness towards non-existent manifestations offers another example of the same strategy. There are, however, also less obvious examples: one of the most interesting is to be found in Molnar's distinction between the effects that cooperating powers produce together and proper manifestations of cooperating powers (which are *contributions* that powers make to their common effect). The claim that an effect of a power is *not* its manifestation (or that the proper manifestation of a power is something distinct from that effect) is, I think, clearly parallel to the claim that the whiteness of a given square on a chessboard is *not* the object of perception: in both cases, one focuses on *formal* objects and then treats them as a sort of material object – in fact, the only proper or immediate objects. In this context Molnar reifies the *contributions* of concurring powers: he does not treat the contribution of a given power to a common effect as *the way in which* the power in question (as opposed to other concurring powers) produces the effect, but rather as *something which* the power (as opposed to other concurring powers) contributes to the common effect.[15] This standpoint is closely related to excessive realism about component forces, so one may wonder whether realism about component forces is itself somehow related to the issue of formal objects.

There remains one further interesting parallel between the ontology of powers and the ontology of action, to which the distinction between formal and material objects proves relevant. To begin with the former: Lowe claims that the manifestation of water-solubility is being dissolved by water ("a causal state of affairs") as opposed to dissolution construed in an intransitive sense (as a sort of change); he thinks that taking dissolution in the intransitive sense (the change) to be the manifestation of a power is a sort of misunderstanding that stems from the ambiguity of the transitive and intransitive senses of "dissolution". The change itself, Lowe claims, is *not* a manifestation of the power: what is the manifestation is *causing* that change in some relevant way (2011, pp. 22–26). There is, according to Lowe, a similar misunderstanding in the ontology of action, stemming from the ambiguity of the transitive and intransitive senses of, say, "movement of a hand": it is not the change itself that is the action, but rather *causing that change* (2008, pp. 148–150).

---

[15] Molnar (2003, pp. 194–198) and McKitrick (2010); I say much more about this issue in the context of the scholastic metaphysics of powers in Głowala (2018).

If Lowe is right in his claim concerning manifestations of powers, then the assertion of Aristotle that there is a common manifestation of various powers (which is the very change produced thanks to the active power and undergone thanks to the passive one) rests on a mistake. To my mind, however, the distinction between a manifestation in the formal and in the material sense calls out to be introduced here: the dissolution in the sense of a change is a common manifestation (in the material sense) of concurring powers, and Aristotle is right in claiming that there is, in some sense (namely, the material one) a common manifestation of distinct powers. On the other hand, part of the formal object of each of these powers is a sort of causal connection between that common manifestation and the relevant power, and these sorts of causal connection are different for different powers.[16] If this is right, then Lowe's claim that the change itself *is not* a manifestation of any of these powers would be parallel to claiming that *only* formal objects of perception are objects of perception at all, and this, again, seems to amount to some kind of reification of formal objects.[17] Something parallel may be said, I think, about Lowe's claim concerning actions: causing a change is the action in the *formal* sense, whereas (*pace* Lowe) the change itself is the action (although precisely in the *material* sense), and in general actions in the formal sense should not be treated like special sorts of action in the material sense. There is, however, more to be said about actions in this context, and I shall return to this issue at the end of section 7 below.

Bird (2007, p. 126) claims that the philosophical problems concerning intentionality cannot be recreated within the ontology of dispositions. It seems to me, however, that the distinction between formal and material objects does reveal some striking parallels between the problems in the ontology of intentionality (including the ontology of intentional action) and in the ontology of powers.

---

[16] For an overview of the issue of common manifestation of distinct powers in the context of Aristotelian metaphysics, see Marmodoro (2007) and Coope (2007); I discuss the issue in the context of the scholastic metaphysics of powers in Głowala (2015).

[17] It might also be worth noting that from this perspective, the claim of Martin and Heil that there are no distinct manifestations of distinct powers (and hence no mind-independent asymmetry between active and passive powers), but only a common one (which is their "mutual manifestation" seen from various epistemic perspectives), is comparable to a kind of immediate realism in the case of sensation, according to which when I see the wheel as an oval, what I see is only the wheel, and in no way the oval. Anscombe says that such a position "allows only *material* objects of sensation" (Anscombe, 1981a, p. 11–12), although, to repeat, my use of "material object" does not precisely correspond to hers.

Formal Objects and Causal Links

Jan Hauska offers the following defence against Armstrong's charge regarding the Meinongian commitments of dispositionalism: he claims that the directedness of a power towards its possible manifestation is a robust causal link occurring in a possible world where the power is manifested: so a power that is not manifested is not actually directed towards a non-existent manifestation in a mysterious way resembling intentionality, but rather, the power *would be* directed towards a manifestation in some counterfactual circumstances, and then the directedness in question would be just a robust causal link.[18] The defence is based on the assumption (embraced by Armstrong himself) that the main reason to introduce the mysterious relation to non-existent manifestations is that dispositions *necessitate* their manifestations. A consequence of Hauska's defence of dispositionalism is the claim that the directedness of dispositions, being just counterfactual, is basically different from the alleged *actual* directedness of intentional states (Hauska, 2007, pp. 59–60).

I think there clearly is a robust causal link between a power and its manifestation in the *material* sense, and it is clear that it is this link that is crucial for the relation of necessitation (if powers are to necessitate their manifestations); it is clear that this robust causal link does *not* occur when the disposition is not manifested. The point is, however, that a particular manifestation in the material sense is a manifestation of a given disposition only *under a certain description*. This, I think, is a lesson to be drawn from the debates on mutual manifestation or common manifestations of distinct powers. Now, the very link between the disposition and what is expressed in the relevant description (the formal object of the disposition) is *not* itself a causal link, and it is precisely *this* non-causal link that poses the problem of non-existing manifestations. Admittedly, what is expressed in the description in question is *inter alia* a certain sort of causal link; the point is, though, that the very link between the disposition and the relevant sort of causal link is *not* itself a causal link.

On the other hand, this also shows what is wrong with Armstrong's charge levelled against dispositionalism: it is not true that dispositions "in some way point toward" their non-existent manifestations in the material sense. On the one hand,

---

18 Hauska (2007, p. 59): "since the manifestation of an undisplayed disposition does not exist, the disposition is not related to it. But they would stand in a full-blooded causal relation were the disposition's activating conditions to hold. Thus, instead of postulating a mysterious relation to a nonexistent manifestation, dispositionalism defends a robust causal link in counterfactual situations". I am grateful to Jan Hauska for a discussion concerning these issues.

they have robust causal links with their existent manifestations in the material sense; on the other, they are directed towards their manifestations in the formal sense, even if their manifestations in the material sense do not occur.

Formal Objects and the Mental

The final question I would like to address here is the issue of the borderline between the mental and the physical: can it be drawn in terms of formal objects? Putting my cards on the table, I think the core of the right answer is the following: all physical powers have *fixed* formal objects of their manifestations; in other words, there is no *real change* of their formal objects (no real change of the powers which would be a change of their formal object). Admittedly, there is some indeterminacy in their formal objects, and all their material objects (or manifestations in the material sense) are fully determinate, but the point is that the determination here in the case of a given particular manifestation is not any internal change of the power or its formal object, but rather the change of external circumstances. For example, the power of a solvent is not essentially directed towards NaCl (as opposed to, say, KCl); but if the solvent happens to dissolve NaCl (and not KCl), this involves no internal change of the directedness (in the formal sense) of the power of the solvent; it is just a matter of some external circumstance that on a given occasion the power is manifested in the dissolution of NaCl (as opposed to KCl). By contrast, mental powers allow for *changes* in the very formal objects of their actions: for example, I may see something as green or oval, but the details of the way I see it (and in particular the aspect under which I see it) are not fixed by the very nature of my power to see. In other words, it is not the case that all details pertaining to what it is for the wheel to be seen by me are fixed by the very nature of my sight. On the other hand, the details of the way I see it are not just external circumstances of my seeing: in particular, the aspect under which I see something is part of the very formal object of my seeing. So there is no directedness in the formal sense towards NaCl in the power of the solvent, but there is a directedness in the formal sense towards a given aspect of what I see in my sight. Similarly, I can think about one and the same thing either in gravitational or moral terms, and these are not fixed by the very nature of my power of reason – although they are not just external circumstances of its manifestation; the aspect in which I think about something is not determined by the very nature of my power of reason (and it is not just an external circumstance of my thinking). In some scholastic texts, changes to the formal objects of acts of powers were called *intentional changes*, just because that which is changed in them is the intrinsic directedness of powers.[19]

There is a further crucial difference between perception and thought here. In the case of sight, changes of formal object are basically produced by the external circumstances of one's seeing: the fact that I see something as a green oval basically depends on some external circumstances, so the way in which one sees something (along with the aspect under which one sees it) is basically determined by such external circumstances of seeing. We can change it, basically, by changing the external circumstances of seeing. By contrast, in the case of reason, the way I think about something and the aspect I focus on is basically up to me: I can choose between different ways of thinking about something. In particular, I can think about John either as a gardener or as a murderer, or even as a gardener who happens to be a murderer. Similarly, in the context of willing, I basically choose the way in which I will something.

Now let us return for a moment to the question (discussed in section 5 above) of whether it is the case that a change such as a movement of my hand is itself an action, or, as Lowe claims, that the action consists in the very *causing* of the change in question. I have claimed in section 5 that causing the change is the action in the formal sense, whereas the change itself (for example, the movement of my hand) is the action in the material sense, and that in general actions in the formal sense should not be treated as some special kind of actions in the material sense. The point is, however, that in the case of actions of mental agents what is involved here is a change of action in the formal sense; moreover, the change of action in the formal sense is itself a special sort of action. So here we have some reason to treat causing my hand to move as a sort of action of mine, distinct from the movement of my hand itself. Causing my hand to move should, in my opinion, be identified with *trying* to move my hand, which both McCann (1975) and Lowe claim are in some sense special actions of rational agents.

This, of course, is just a sketch of an answer, rather than an argument for it. The answer being put forward is, I believe, to be found in Aquinas and in various late-scholastic Thomists (although my interest in it here is purely systematic rather than historical, and I am not offering here any arguments for the historical accuracy of my claim). In many respects it is similar to the solution recently proposed by Oderberg, who states that "both powers and mental intentionality display specific indifference, but there is a difference in the indifference" (2017, p. 32). In the case

---

19 Toledo (1615, lib. II, q. 34): "[species] intentionales dicuntur, quia immutant intentionem sensuum, et movent sensus ad intendendum rebus".

of physical powers, the indifference is determined by the essence of the power, whereas in the case of mental powers, it is founded in "the freedom of abstraction".

Nevertheless, I would also say that the answer presented in outline here differs in at least four respects from the proposal of Oderberg.

The first difference concerns the determinability of powers: Oderberg says that the water-solubility of sodium chloride is a determinable power that may have various *modes* corresponding to various water samples, times and places of dissolution (2017, pp. 9 and 31). What seems absent in Oderberg's account is the difference between two sorts of determination of a determinable power that is crucial for my account: on the one hand, determinations of circumstance, which do not constitute an internal change of the power or its formal object, and on the other, such internal changes of a power itself, which do amount to a change in respect of the formal object of the act. On my account, the water-solubility of sodium chloride is, in a crucial sense, not determinable at all, and displays no indifference, just because the determination of the circumstances of its particular manifestation does not involve any change with respect to the formal object of the power itself.

The second difference concerns identity of thoughts. Oderberg claims that if Fred thinks about flight characteristics of birds, there are infinitely many ways "in which the very same kind of thought (defined by its object) can be had". He also claims that "Fred's thought about birds is the same kind of thought, whether it is about their flight characteristics, their reproductive characteristics, or both, or neither", and offers some arguments in favour of this identity (2017, pp. 32–33). On my account, what is crucial for the identity of the thought is its *formal* object (and not just its subject matter), so it is at least debatable whether all the thoughts Fred has about birds are of the very same kind. Hence the freedom displayed in thinking about birds in this or that way does not consist in the fact that one and the same kind of thought may be had in many ways, but rather in the fact that one and the same power of reason can produce thoughts having various formal objects – that is, various kinds of thought about one and the same subject matter – while it is up to us which kind of thought about birds we have. I can think about birds from various points of view, and the change of viewpoint is effected within the very power of my reason (whereas in the case of sight, it is not effected within the very power of sight).

The third difference concerns the role of freedom: for Oderberg, freedom of abstraction is the core of mental finality and intentionality. On my account, by contrast, it is the very change of formal object – the very change of the directedness of a power – that is the mark of mental intentionality. The change may be a matter of free choice, as it is in the case of thought, but it may also just be produced by some circumstances.

The fourth difference is that Oderberg's account of mental intentionality focuses on the realm of reason and will. On my account, by contrast, sensation and perception are mental phenomena having the marks of a mental sort of intentionality (displaying changes in their formal objects). So, in general, on my account the spectrum of mental sorts of intentionality is quite rich, and the rational intentionality based on freedom of abstraction occupies just a part of it.

Concluding Remarks

To sum up, I have claimed that the old notion of the formal object of something (and the distinction between formal and material objects of something), which is one of the main notions of the scholastic ontology of powers, is indispensable for a sound analysis of the contemporary issue of physical intentionality. The necessity of the distinction between formal and material objects is grounded in a certain obvious ambiguity pertaining to object and manifestation ascriptions. I have tried to show that the introduction of the distinction between a formal and a material object (i) paves the way for a sound analysis of the "marks of intentionality", (ii) sheds much light on some parallel errors in the ontology of intentionality, the ontology of action and the ontology of powers – namely, those which are committed when we *reify formal objects*, and, finally, (iii) allows one to demarcate the borderline between the physical and the mental in an interesting and, I would argue, correct way.

Acknowledgment: I would like to thank Christian Kanzian for his helpful remarks on a previous version of this paper.

Bibliography

Anscombe, G. E. M. (1981a). The intentionality of sensation: A grammatical feature. In *The collected philosophical papers of G.E.M. Anscombe: Vol. 2. Metaphysics and the philosophy of mind* (pp. 3–20). Oxford: Blackwell.
Anscombe, G. E. M. (1981b). Under a description. In *The collected philosophical papers of G.E.M. Anscombe: Vol. 2. Metaphysics and the philosophy of mind* (pp. 208-219). Oxford: Blackwell.
Armstrong, D. M. (1997). *A world of states of affairs*. Cambridge: Cambridge University Press.
Armstrong, D. M. (1999). The causal theory of properties: Properties according to Shoemaker, Ellis, and Others. *Philosophical Topics, 26*, 25–37.
Bird, A. (2007). *Nature's metaphysics. Laws and properties*. Oxford: Oxford University Press.

Cajetan, Thomas de Vio (1888). Commentaria in Summam Theologiae. In S. Thomas Aquinas, *Opera omnia: Vol. 4. Summa Theologiae, I, q. 1-49* (Editio Leonina). Rome.
Chrudzimski, A. (2013). Varieties of intentional objects. *Semiotica, 194*, 189–206.
Coope U. (2007). Aristotle on action. *Proceedings of the Aristotelian Society. Supplementary volumes, 81*, 109–138.
Geach, P. T. (1980). A medieval discussion of intentionality. In Geach, P. T., *Logic Matters* (pp. 129–138). Berkeley–Los Angeles, CA: University of California Press.
Głowala, M. (2015). Unity and plurality in joint manifestations of powers: A scholastic approach. *Revista Portuguesa de Filosofia, 71*, 873–894.
Głowala, M. (2018). Polygeny, pleiotropy, and two kinds of concurrentist ontology. In M. Szatkowski (Ed.), *Ontology of theistic beliefs*. Berlin: De Gruyter, 39–61. doi.org/9783110566512-004
Hauska, J. (2007). Dispositions and Meinongian objects. *Polish Journal of Philosophy, 1*, 45–63.
Heil, J. (2005). Dispositions. *Synthese, 144*, 343–356.
Heil, J. (2012). *The universe as we find it*. Oxford: Oxford University Press.
Lowe, E. J. (2004). Some formal ontological relations. *Dialectica, 58*, pp. 297–316.
Lowe, E. J. (2008). *Personal agency. The metaphysics of mind and action*. Oxford: Oxford University Press.
Lowe, E. J. (2011). How not to think of powers: A deconstruction of the "dispositions and conditionals" debate. *The Monist, 94*, 20–34.
Marmodoro, A. (2007). The union of cause and effect in Aristotle: Physics 3. 3. *Oxford Studies in Ancient Philosophy, 20*, 205–232.
Martin, C. B., (2008). *The mind in nature*. Oxford: Oxford University Press.
Martin, C. B, & Pfeifer, K. (1986). Intentionality and the non-psychological. *Philosophy and Phenomenological Research, 66*, 249–269.
McCann, H. (1975). Trying, paralysis, and volition. *The Review of Metaphysics, 28*, 423–442.
McKitrick, J. (2010). Manifestations as effects. In A. Marmodoro (Ed.), *The metaphysics of powers. Their grounding and their manifestations* (pp. 73–83). New York, NY: Routledge.
Molnar, G. (2003). *Powers: A study in metaphysics*. Oxford: Oxford University Press.
Mumford, S. (1999). Intentionality and the physical: A new theory of disposition ascription. *The Philosophical Quarterly, 49*, 215–225.
Nes, A. (2008). Are only Mental Phenomena Intentional? *Analysis, 68*, 205–215.
Nuchelmans, G. (1980). *Late-scholastic and humanist theories of the proposition*. Amsterdam–Oxford–New York: North Holland.
Oderberg, D. S. (2017). Finality revived: Powers and intentionality. Synthese. 194: 2387. doi: 10.1007/s11229-016-1057-5
Place, U. T. (1996). Intentionality as the mark of the dispositional. *Dialectica, 50*, 91–120.
Soncinas, P. (1579). *Quaestiones metaphysicales acutissimae*. Lugduni: Apud Carolum Pesnot.
Toledo, F. (1615). *Commentaria una cum quaestionibus in tres libros Aristotelis De anima*. Coloniae: In Officina Birckmannica, sumptibus Hermanni Mylii.
Tugby, M. (2013). Platonic dispositionalism. *Mind, 122*, 451–480.

Bartłomiej Skowron*, Tomasz Bigaj, Arkadiusz Chrudzimski, Michał Głowala, Zbigniew Król, Marek Kuś, Józef Lubacz, and Rafał Urbaniak

An Assessment of Contemporary Polish Ontology

Abstract: In this article we undertake a diagnosis of the state of ontology as it is currently practiced in Poland. We point to the strengths of Polish ontology and the aspects that should be improved in order for Polish ontology to flourish further. We cover different styles of thinking in Polish ontology, as well as different methodologies. We address threads influencing Polish ontology, both substantive, as well as those of an organizational, cultural and institutional nature. Apart from the ontologists, this paper also includes the opinions of scholars from the hard sciences, which allows us to shed light on Polish ontology from outside the philosophical community. The article presents concrete steps that could improve and strengthen Polish ontology in the future. We also present here thirty examples of important contributions to ontology by Polish ontologists.

Keywords: Polish ontology, organization of ontology, teaching of ontology, ontology and natural sciences

Introduction

In Warsaw, on 10 May 2016 during the conference entitled "Polish Contemporary Ontology" organized by the International Center for Formal Ontology (Faculty of Administration and Social Sciences, the Warsaw University of Technology), a

---

***Corresponding author: Bartłomiej Skowron,** International Center for Formal Ontology, Warsaw University of Technology, Poland.
Tomasz Bigaj, Institute of Philosophy, University of Warsaw, Poland.
Arkadiusz Chrudzimski, University of Szczecin, Poland.
Michał Głowala, Institute of Philosophy, Wrocław University, Poland.
Zbigniew Król, International Center for Formal Ontology, Warsaw University of Technology, Poland.
Marek Kuś, 1. International Center for Formal Ontology, Warsaw University of Technology, Poland; 2. Center for Theoretical Physics, Polish Academy of Sciences, Poland.
Józef Lubacz, International Center for Formal Ontology, Warsaw University of Technology, Poland.
Rafał Urbaniak, 1. Centre for Logic and Philosophy of Science, Ghent University, Belgium; 2. Institute of Philosophy, Sociology and Journalism, University of Gdańsk, Poland.

https://doi.org/10.1515/9783110669411-015

panel discussion entitled "The Future of Polish Ontology" took place. The panel was attended by Tomasz Bigaj, Arkadiusz Chrudzimski, Paweł Garbacz, Marek Piwowarczyk, Mirosław Szatkowski, Rafał Urbaniak, and was led by Bartłomiej Skowron. The discussion was attended by, among others, Michał Głowala, Józef Lubacz, and Marek Kuś.

The panellists were asked to answer the following four questions:

A. How do you see the state of ontology in Poland in terms of its research, educational, and organizational aspects?
B. How do you see the future of ontology in Poland over the next 5–10 years? What would the ideal situation in this area of study look like in 10 years or more?
C. Which topics within ontological research, possibly specific to Poland, could be suggested for the next 5–10, or perhaps more, years?
D. How do you see the relation between ontology and other academic disciplines (philosophy, logic, mathematics, computer science, physics, engineering, medicine, etc.)?

These questions were the starting point. However, the panellists were not limited to these questions. The panellists and those taking part in the discussion were then asked to prepare the comments that make up this joint paper. This is the origin of this work.

The International Center of Formal Ontology unites Polish and foreign ontologists who are not afraid of modern science and are able to use it. It is not a coincidence, however, that the place where the Center was established is a technical university, not an ordinary university. Of course, in our respect for science we are not without criticism. Philosophers are looking for science, philosophy without science is naive and science without philosophy is blind – that is why scientists turn to philosophy. It was precisely for this reason that the panel was attended not only by pure and untouched-by-science philosophers, but also by physicists, mathematicians, computer scientists and even biologists, who were present and who took an active part in the discussion.

Tomasz Bigaj

In my answer to the first question of this questionnaire I will limit myself to describing the current state of ontology at the Institute of Philosophy at the University of Warsaw, as I don't have sufficient knowledge about this topic in regard to other philosophical centers in Poland. Ontology has a strong presence in both teaching and research at the Institute of Philosophy. We have a number of faculty members whose research interests include topics that are traditionally categorized as belonging to ontology, or metaphysics. Among them perhaps Mariusz Grygianiec deserves to be singled out as a pure metaphysician in the analytical tradition of Jonathan Lowe and Peter van Inwagen. Grygianiec's extensive work touches upon such core metaphysical issues as the problem of persistence and identity in time, the problem of universals, and the problem of personhood. Other metaphysically oriented philosophers at the Institute combine their ontological investigations with different philosophical and extra-philosophical interests. Ontological reflections inspired by science and mathematics have always been strong at the Institute. In the past, Zdzisław Augustynek developed an original event-based ontology of the physical world derived from the postulates of special relativity, while Władysław Krajewski analysed the concept of causation in physics. Among the current followers of this tradition, Krzysztof Wójtowicz publishes extensively on the problem of realism in the philosophy of mathematics, and Tomasz Bigaj works on some metaphysical issues brought about by quantum theory. Many logicians working at the Institute also discuss problems that come close to traditional ontology. The most famous research program in logic and ontology carried out in Warsaw is unquestionably the development of the ontology of situations, inspired by Ludwig Wittgenstein's *Tractatus* and formalized with the help of non-Fregean logic. The main Warsaw contributors to this program were Roman Suszko, Bogusław Wolniewicz, Mieczysław Omyła and, most recently, Anna Wójtowicz. Finally, we should mention Joanna Odrowąż-Sypniewska, who is first and foremost a philosopher of language, but whose works include an analysis of the ontological notion of identity in the context of the vagueness problem.

Teaching ontology has always been an important part of the curriculum at the Institute of Philosophy. Currently a two-semester course in ontology, comprised of lectures and discussions, is obligatory for second-year students of the BA ("first cycle") program. This course taught variably by Mariusz Grygianiec, Tomasz Bigaj, and other faculty members, introduces students to the main themes of modern analytic metaphysics, such as the problem of universals and particulars, the notions of possibility and necessity, time and space-time, causation, free will, and determinism. On top of that several advanced seminars on various ontology-related topics are usually offered each year. Many students choose to write their Bachelor's or Master's theses on topics in ontology, and it is not uncommon to have PhD dissertations in this area as well (for instance, recently an excellent PhD thesis on the subject of the ontology of musical works was defended at the Institute).

Regarding the future of Polish ontology it is difficult, and in my opinion, impractical to try to predict or influence the direction of future ontological investigations in Poland. These things are better left for individual researchers to decide. However, there are certain systemic solutions that can be implemented in order to give ontology in Poland the best chance to flourish. The solutions which I would like to suggest are general enough to be applied to any subfield of philosophy, not limited to ontology or metaphysics. First and foremost, a necessary condition for any successful scientific enterprise is a constant and unimpeded flow of ideas between the involved parties. This means that the Polish philosophical community should be more open to international collaboration and exchange. As Poland is a medium-sized European country, its sheer demographics imply that the Polish community of metaphysicians cannot be particularly sizeable. In order not to stagnate within a close-knit circle of a few friends and colleagues, we need to go out and interact scientifically with the rest of the world. To a certain extent this interaction is already present; however, it is decidedly one-sided in that we are often influenced by international trends and debates in ontology, but rarely influence these trends in return. The main reason for this regrettable situation is that the overwhelming majority of philosophical works in Poland are published in Polish or – much less frequently – in English volumes which nevertheless are not in worldwide circulation. This is deplorable, as the research done by Polish philosophers is often of the highest quality and deserves to be noticed and discussed by the international community. And yet Polish authors are virtually absent from the main international journals in philosophy, such as *Journal of Philosophy*; *Mind*, *Philosophy and Phenomenological Research*; *Synthese* and many more. I believe that some administrative steps can and should be taken in order to change this sorry state of affairs.

Another important aspect in need of improvement is the connection between research and teaching. All the top-ranking departments of philosophy in the world put strong emphasis on their PhD programs, which are seen as a major force driving the research done by their faculties. This attitude is visible in that these programs are highly competitive and selective. Only a few carefully selected candidates are admitted every year; however, in return they are usually offered excellent financial conditions (fully paid tuition and living expenses), and high chances of employment in academia after successful completion of the program. To illustrate this with an example: every year Harvard University admits approximately ten candidates from around the world to their doctoral program in philosophy. In comparison, this number in the case of the University of Warsaw is above thirty, with virtually all the candidates coming from Poland. Consequently, the quality of the work done in Poland by PhD students in philosophy is on average lower than that of their counterparts abroad (with some notable exceptions, of course). And only a handful of graduates decide to pursue an academic career. I believe that we should aim to select only those PhD candidates who have clear potential to do high-quality research pushing the boundaries of their field. We should also avoid the situation in which one professor supervises more than three or four PhD dissertations, which makes it virtually impossible to give enough attention to individual students and the problems they encounter. And, in order to increase the number of well-qualified candidates, we should be more open to prospective students from abroad. This of course requires some changes in the way PhD programs are run (more courses offered in foreign languages and more financial help for international students coming to Poland).

Given my scientific interests, it should come as no surprise that I am a strong proponent of a close interaction between ontology and the special sciences. In my opinion no metaphysician can afford to ignore the radical advances in our knowledge of the world brought about by the development of particular disciplines such as physics, mathematics, biology, chemistry, psychology, and sociology. Of these disciplines I am most familiar with physics and its influence on modern metaphysics. This is of course an incredibly broad topic, so I can only mention that the number of new metaphysical conceptions of objects, properties, relations etc. directly inspired by modern physical theories is staggering. Also, some physicists admit that contemporary theoretical physics has recently run into so many roadblocks and conceptual quagmires that quite likely no further progress will be made without some help from philosophers and metaphysicians. I refer those who are interested in learning more about this topic to the volume *Metaphysics in Contemporary Physics* (Brill/Rodopi, 2015) that I edited together with Christian Wüthrich from the University of Geneva.

I am well aware of strong anti-scientific sentiments present in many approaches to ontology (including, perhaps surprisingly, analytic metaphysics done in the tradition of Peter Strawson). I don't want to arbitrarily reject these approaches out of hand, as I am reasonably sure that they have much to offer broader philosophical discussions. However, I would like to make an appeal to those who want to pursue metaphysics conceived as an autonomous field of study, independent and isolated from science. Please, try to explain why your arm-chair reflections on the nature of reality should be superior to the scientific analysis based on careful experimental studies. I am not prejudging this issue, and perhaps there are good arguments in support of the anti-scientific, a priori approach to metaphysics, but so far, I am not aware of any.

Rafał Urbaniak

On the State and Development of Ontology in Poland

Since my comments are somewhat orthogonal to the questions I was expected to address, I have decided not to answer them directly, but rather to elaborate on those issues which deserve emphasis. Because of my background and interests, I will be commenting on *formal ontology* and *mathematical philosophy* in general.

As Tomasz Bigaj correctly emphasizes, universities in Poland, when it comes to PhD programs, think in terms of numbers and not in terms of quality. What I would like to add here is that the problems begin even earlier in the students' career. Many courses in philosophy programs in Poland are not designed to, and do not improve, students' philosophical and analytic skills. A standard course in the philosophy program in Poland requires a student to quickly read a large number of texts of varying quality, not enough analytic attention is paid to any of them in class, and at the exam the student is mainly expected to remember dates, names, titles, and catch phrases supposedly summarizing those texts. Multiple courses are oriented toward teaching students to briefly summarize who said what, in what text and when, instead of teaching them to understand why they said it, in response to what problem, and to assess the quality of what has been said. Yes, learning some history of philosophy is important, but it should be taught with a philosophical education in mind, not as a mere *history of ideas*.

Another problem is that financial and organizational considerations led universities to adopt policies requiring rather high minimal numbers of students in programs and in particular courses. As a result, philosophy programs at universities, which aren't particularly popular anyway, are under threat of not having enough students. Accordingly, program administrators and instructors are pressed to keep the number of students high, usually at the cost of the quality of the program. God forbid we seriously require students to learn some difficult material (as material related to mathematical philosophy often is) and fail them at the exam if we honestly think they haven't mastered it!

As for the need to publish in leading journals, it's hard to expect someone to do that, if during their own education they haven't been exposed to multiple papers actually published in such journals and haven't been trained to develop their writing to a similar level. Yet, it's very unusual for a student in a Polish philosophy program to receive such support. Almost all the courses and almost all the texts read for classes are in Polish. Even if some writing on the part of the student is involved, the usual procedure is that a student writes a paper without much guidance in the process, submits it, and gets a grade, usually not even receiving feedback that they could learn from, and almost never being required to submit further versions of the paper developed in light of detailed remarks. The students that we have now will be the researchers that we will have soon, so there's no surprise in the faculty members publishing only in local Polish journals.

So, I submit some systematic solutions to the quality of formal ontology and mathematical philosophy in Poland in general. They would have to:
(i) eliminate incentives that departments currently have to keep their students at any price, and promote quality-based assessment of a department's teaching,
(ii) modify course content so that they focus more on the argumentative and analytic skills of the students rather than merely exercising their memory,
(iii) introduce more courses in English, where students are required to read, understand and discuss current research papers, and
(iv) force students to properly practice writing argumentative philosophical papers in English.

Józef Lubacz

As I do not consider myself a professional philosopher,[1] my answers to the questions pertain mainly to general circumstances which illustrate an impact on the current state and possible development of ontology in Poland. The achievements of the Lvov-Warsaw Philosophical School may be considered to be phenomenal, not only because of their impact on philosophy and mathematics, but also because they appeared shortly after Poland regained its independence and had to build its scientific potential practically from close to nothing. Poland is now in an incomparably better position from this point of view, so one may wonder why it is so difficult to succeed in following the Lvov-Warsaw tradition and achievements? Some might believe that the constellation of stars in some epochs favour great achievements in poetry, art and music, rather than philosophy. But somewhat more rational reasons seem more plausible.

Clearly, the general intellectual atmosphere has changed, not only in Poland, since the first half of the 20th century. On the one hand, we have witnessed the impressive success of science and technology, but on the other hand, belief in the omnipotence of science in explaining the world and in the human ability to

---

[1] Józef Lubacz is a professor of information and communication technology, increasingly engaging in philosophical issues. He currently chairs the Program Council of the International Center for Formal Ontology. In the past he chaired the General Council for Science and Higher Education in Poland.

control technology seems to be weakening. Philosophy should, in principle, be helpful in explaining this evolution of attitudes, if not in finding remedies. The real influence of philosophy on intellectual life, and on culture in general, seems however to have already been decreasing for several decades. This, hopefully, is a transient phenomenon, nevertheless presently it is reflected, in particular, in societal perception of the importance of philosophical education, not only at universities, but also in high-school education. Philosophy is considered to be just one more subject, on par with many other subjects taught, rather than a super-disciplinary (not to be confused with inter-disciplinary!) area of human reflection.

On top of general, worldwide tendencies, Poland has its specific problems. For many past decades Polish scientists, and philosophers in particular, were virtually separated from international academic societies. Over two decades ago the geopolitical situation changed radically, but this did not result in an equally rapid improvement in the internationalization of Polish science. The inertia towards desirable change was greater than expected, partially due to the fact that university professors became overloaded with teaching. This resulted from an almost 500% increase (!), in a very short time, in the number of university students. In consequence, professors were distracted from scientific work.

The above remarks pertain practically to all areas of current scientific activity in Poland, thus also to philosophy and ontology in particular. Therefore, the current general conditions for academic activity are not favourable. I believe at least two things have to be undertaken at the governmental level to change this state of affairs: first, education and science have to be considered a necessary investment rather than a cost for society. Second, the limited public resources available (Poland is still a relatively poor country) should be allocated according to diversified principles depending on the distinction between research universities and higher-education institutions which concentrate on teaching. Unfortunately, the scientific community has limited influence on such desirable government-level decisions. But there are of course things that our ontological community can undertake in its own capacity, without waiting for a positive change stemming from political decisions. An example of this is expressed in the mission statement of the International Center for Formal Ontology:[2]

> ICFO aspires to consider both natural beings and artefacts, i.e. human-made things. Of special interest are the relations of artefacts with natural things, relations which influence the evolution of different forms and aspects of human culture.

---

[2] To read the full text of the mission statement visit the ICFO home-page at http://www.icfo.ans.pw.edu.pl/en/.

ICFO considers how artefacts come into existence, i.e. the poetic aspects of human activity, in particular, the ethical aspects of the creation of artefacts, and the ontology of human and divine creators, and also the epistemic aspects of human activity and their influence on human behaviour.

The above citation is an answer to question D, and the preceding remarks, though not quite directly, to questions A–C.

Arkadiusz Chrudzimski

There are many good ontological papers produced in Poland but that is not to say that the situation, on a whole, cannot be improved. I agree with Tomasz Bigaj, that many papers which are written by our colleagues in Polish could easily be published in journals like *Husserl Studies* and *Grazer Philosophische Studien*. These are good things about Polish ontology, however there are cases of inferiority complexes and defensiveness. These are the reasons that make people publish only in Polish and that, in consequence, the papers are available only to a narrow group of recipients. Frankly speaking, they are usually only read by fellow lecturers from the same Institute.

I will agree again with the previous contributors, that opening up to the world is key for the development of Polish ontology and stems not only from the fact that – as has been mentioned before – we are a relatively small nation. Apart from that, in many ways we are behind. We've inherited many systematic carcinoses and I'm not only talking about the legacy of communism. A prodigious example is Polish Thomism which in its religious influence and its isolation from contemporary thought is a very Polish phenomenon. Thomism cured of those diseases could be an interesting branch of analytical ontology and in the world-at-large it is occasionally practiced that way. In Poland, however, practicing Thomism in this manner is almost inconceivable.

The consequence of the fact that Polish ontology is not open enough to what happens outside of texts available in Polish, is that apart from the good papers I mentioned before, many bizarre and archaic works are created. They would not stand a chance of being published elsewhere because, simply, they would not be understood.

There's no "particularly Polish" problem and thank God for that. Reason is universal and in ontology we deal with what the world is like and not what the Polish soul is like. It would be a true misfortune if we had such a "specifically Polish" philosophy like, for example, Russians who have a "specifically Russian" philosophy. Currently in Poland we have the phenomenological and analytical traditions, some interesting traits related to Marxism, we have Elzenberg. But they are all universal things. Well, we also have Thomism, which — if it drops the chains I mentioned before — has a chance of becoming a universally understood philosophy.

In this sense, there isn't a specifically Polish philosophy. Another issue is that you may take something that Poles have done in philosophy and research them partly historically, partly reconstructively, and partly by drawing some interesting consequences.

That is related to a postulate I would like to pose here which pertains to an adequate way of financing philosophical research. Currently the trend in Poland is to treat philosophy the way exact sciences are treated. Because of that approach, programs which finance purely historical research have almost completely vanished. In order to get a grant, you have to propose systematic research or convince the jury that our historical research will have some sort of value for systematic philosophy. That approach is wrong because philosophy is not, like chemistry or mathematics, a straightforwardly cumulative science. Philosophy always returns to the same problems, we still don't have answers to the questions asked by Plato or Aristotle. For these reasons an interesting historical reconstruction in many cases would be more valuable than systematic research which is done at a push.

Michał Głowala

The assessments of the present condition and prospects of Polish ontology may focus on institutional or on substantial aspects of research; I would like to focus on the latter (giving thus my answer to questions A and B). I certainly lack the prudence necessary to discuss the former; and what I offer focusing on the latter is a rather subjective sketch of what I take to be particularly helpful and promising for doing ontology.

Let me begin with two more general points on doing ontology which are relevant for my assessment of its present state. On the one hand, a basic requirement of a sound work here is participation in a vivid and detailed exchange of arguments (meeting possibly high standards of argumentation); one who withdraws from such an exchange is exposed to the danger of philosophical *torpor* (which is clearly described in Rafał Urbaniak's contribution to the present paper). I think it is just this danger of torpor that justifies the claim that scholars should publish papers in international journals. On the other hand, however, there is another danger of gradually losing one's grip on what one is talking about: one gets accustomed to some routine aspects of the tools one uses, focuses on some routine problems and puzzles; this produces a sort of philosophical *blindness* and one is in urgent need of fresh light to *see again* what one is talking about. What I mean here is by no means a generic scepticism about "conceptual schemes" or about using formal tools in general; what I mean is that there might be particular reasons against a method that proves workable and very popular in some communities. So, for example, I think Lowe is basically right in his objections to Kripkean modal metaphysics (Lowe, 2013, pp. 139–160) or against the "insouciance of the questions of serious ontology" produced by a particular use of the first-order predicate logic with identity (Lowe, 2013, pp. 50–55); I also think Anscombe is basically right in her criticism (in Anscombe, 1981) of the association (produced and petrified by various philosophers) of causation and necessitation. Finally, I think that Geach is right in his objections against a kind of formalism of deontic logic (Geach, 1991); and he says there that due to some faults of formalism "our mental vision, so to say, is prevented from coming to a proper focus" (Geach, 1991, p. 36). Certainly, all these examples are debatable; but they show the sort of objection (of a philosophical blindness) that I have in mind.

From this point of view, the Latin scholasticism of the 13th–17th centuries offers a unique example of a vast and long-lasting community of ontological/metaphysical discourse well equipped with antidotes both against philosophical *torpor* and against philosophical *blindness*. On the one hand, particular arguments were discussed in detail by a great number of authors, analysed through and through, again and again, refused, revisited, and refined in various ways; that is the best antidote against philosophical torpor. On the other hand, various domains of philosophy were interrelated strongly enough to ensure that some light from other domains was constantly being shed on what one wanted to discuss; that often is a good antidote against philosophical blindness. Moreover, philosophical research was closely connected to teaching (not only professional philosophers), and that also, I think, often cures philosophical blindness. As a matter of historical fact, I would claim, there is nothing else like scholasticism in these respects in the European intellectual history. Jan Salamucha claimed that Thomism offered a unique combination of maximalism both in scope and in method (positivism being marked by maximalism in method and minimalism in scope, and Bergson's philosophy – by minimalism in method and maximalism in scope); if that is true, the Thomist combination became possible not only thanks to what Thomists themselves did, but also to what used to be their natural milieu – scholasticism in general; that milieu both forced a detailed discussion of arguments and provided antidotes against philosophical blindness. The point is that the merits of a milieu that no longer exists are by their very nature difficult to reconstruct; what seems necessary is to make this past richness accessible again for contemporary debates in ontology (in particular, in the metaphysics of powers or metaphysics of mind and action); and what is crucial is first-hand access to various *details* of past debates, and not just to a general picture of it which is part of the cursory knowledge of the history of philosophy. It is these details that may shed light on contemporary issues.

From this perspective, there are two *big gaps* that seem to me crucial for the present condition and the prospects of ontology in Poland (and probably elsewhere).

The first is the gap between the Thomist tradition (as represented, for example, by the Lublin school) and analytical philosophy; the strength of *a* Thomist tradition in Poland makes that gap particularly striking. I am *not* saying that in Poland there are no interesting or fruitful attempts at combining the two perspectives; but the fact is that one who tries to combine them is doomed to a somewhat risky business that Klima (2004) aptly calls paradigm-straddling. It is very symptomatic here, I think, that Geach's work on Aquinas and scholasticism (which I think it is hard to overestimate, especially as far as the issues of actual existence, the concept of matter, the analogy between forms and Fregean *Begriffe*, or the concept of tendency are concerned) is not so popular as one might expect; the texts full of interesting and insightful hints or suggestions remain unnoticed. This big gap, I think, may be not just a matter of some idiosyncrasies; when, for example, Oderberg claims (2002, p. 125) that the defence of scholasticism is possible only "through the medium of analytical philosophy" and that analytical philosophy is "the only legitimate heir" of scholasticism, this must sound strange to the ears of existential Thomists in Poland who flatter themselves that they have never betrayed some fundamental merits of the Aristotelian legacy they had inherited.

The other big gap is the one between historical research and the systematic study of ontology. It seems sometimes to be taken as a special advantage of purely historical research in philosophy that it does not require any philosophical commitments or even full-fledged philosophical skills; what makes that research scientific is rather the use of the tools of history or philology. Purely exegetic work, however, that is not guided by philosophical interest in some particular problems or arguments, typically proves uninteresting for contemporary debates, because it is easy to overlook just these details that are crucial from a systematic point of view. In this way scholasticism becomes, to use Oderberg words, "the ossified material of an essentially tedious historical analysis" (2002, p. 125). Here again it is symptomatic that Geach's work on Aquinas is so underestimated (as supposedly not having serious merits for the study of past thought). From this point of view, it would be very promising, I think, to bridge these two gaps and to make the vast forgotten richness of scholasticism accessible again for contemporary debates in ontology, especially in the analytic tradition; it seems particularly promising for issues concerning the metaphysics of powers, causality, substantiality, mind and action, or in the ontology of intentionality. To repeat, it is crucial to make *details* of the past debates accessible. Here, in particular, even a cursory knowledge of the technical Latin of scholastic metaphysics/ontology (no doubt one of the most precise and sophisticated tools of philosophy in history, and an important source of the contemporary language of ontology) proves very helpful, as far as it enables an acquaintance with scholastic texts (which are now easily accessible in digital form).

That, of course, was the idea of analytical Thomism (and of analytical scholasticism put forward by the Czech school of Stanislav Sousedík). Szatkowski (2016, p. 25) and Pouivet (2009) are clearly right in their claim that analytical Thomism was born in Poland long before Haldane, in the works of Bocheński, Salamucha and other authors of the Cracow Circle. Some Polish existential Thomists may think it is a program for a doubtful synthesis of heterogeneous elements, a hybrid with no serious philosophical merits. Some Polish analytical philosophers may think that a time-consuming study of exotic debates enacted hundreds of years ago does not offer any substantial help for solving contemporary problems. And many historians of philosophy may think it means anachronistic speculation instead of a robust scientific study of the history of thought. The former, I think, would reflect just the way in which scholastic debates are often presented in contemporary research; the latter – just a clearly exaggerated scepticism about philosophy (as opposed to philology).

There are in Poland, however, some interesting and promising attempts to bridge these two gaps. To give just a few examples: Marek Piwowarczyk's work on the subject-properties structure is an excellent example of it; Jacek Wojtysiak discusses, in a number of works, various issues concerning existence both in the Thomistic and the analytical traditions; Arkadiusz Chrudzimski discusses (in a systematic context) a number of issues of the ontology of intentionality in the Aristotelian and scholastic traditions. As these examples also suggest, the phenomenological tradition inspired by Roman Ingarden that is quite strong in Poland (for example, in the work of Marek Rosiak) offers a unique chance to bridge the two abovementioned gaps.

I think that efforts to bridge the two gaps and to make late scholasticism accessible for contemporary debates do offer interesting prospects for ontology itself: a better grounding of contemporary debates in the classics of ontology, more challenges for the argumentation, more unexpected light on already known issues, a richer set of examples, more philosophical surprises and more chances to cure ontology of philosophical blindness.

Zbigniew Król

First of all, there are many great ontological problems and questions still unresolved. For instance, the problem of "epistemic access" to existence. What are the conscious tools in which the existence of something is given to the subject of cognition? How is the knowledge of being possible? What is the source of our knowledge of existence?

We have a few mainstream traditions or – rather isolated – ontological schools which are predominant nowadays. It means that ontological problems are approached in a highly specialized way with regards to the given school methods. Therefore, there are also three main ontological traditions, schools or philosophical styles in ontology, i.e. analytical, phenomenological, and Thomistic. As in world ontology, a similar situation also exists in Polish ontology.

In analytical philosophy, original ontological problems are often trivialized, flattened out, and oversimplified. Formal considerations, in many cases, use advanced logical and mathematical apparatuses but *ontological* results are mainly apparent. See, for instance, the quantificational criterion of being and Quine's prevalent theory of ontological commitments. (The last point needs a long explanation for analytical philosophers who are usually very happy and satisfied with their exact theories.) For example, for many philosophers, there is only one mode of being because only one type of existential quantifier is possible. However, there are many types of semantics possible; e.g. substitutional. "To be" means "to be" or "there is such x that ..." and nothing more. Therefore, there is no room for plenty of original ontological distinctions and questions. Nevertheless, there is no reasonable and fruitful treatment of ontological problems without the use of analytical and formal methods or without meaningful contact with science. The weak point is the lost contact with direct experience, conscious data, intellectual intuition, and being mindful that before "calculating" it is necessary to ascertain "for what?"

The phenomenological approach is more concerned with formal ontology than metaphysics (in Roman Ingarden's distinction), i.e. concerns more the ontology of essences than the ontology of being and existence. Thus, in many cases, collaboration with analytical philosophers is possible. Roman Ingarden tried to find the lost connection between ontology and metaphysics by considering modern physics. However, original phenomenology neglected the problem of being and the existence of something. In the original approach of Husserl, existence is bracketed from direct data. Moreover, one can prepare the same essence with the use of imaginative variation over the imagined and the given directly in sensual perception data. It means that existence can be, in a sterile way, separated (bracketed) from essences. Nevertheless, every direct given in transcendental cognition content of a conscious act is given *as existing*. Therefore, there is an open question concerning the transcendental constitution of existence. Heidegger-like hermeneutical ontology tries to find an answer to a great ontological problem, unsolved by original phenomenology: the problem of "epistemic access" to existence. What are the conscious tools in which the existence of something is given to the subject of cognition? Da-sein is a place where being is unveiled and manifests itself as its internal essential possibility of being. However, Heidegger's approach is isolated from science, especially mathematics. Many original ontological ideas, even from a purely historical point of view, are taken from mathematical investigations and the reflection upon it. Being in mathematics is given in some implicit, horizontal phenomena accessible with the use of hermeneutical methods such as the *reconstruction of the hermeneutical horizon for mathematics*. Such a reconstruction, in many cases forces the use of formal methods.

Thomistic ontology is in methodological isolation from science and epistemology. To the great advantage of Thomism, not used in other philosophical schools, the theory of the analogy of being. The formalization and inquiry of its formal properties as well as its incorporation within other types of ontology is a great challenge. There is also the possibility to analyse *intellection* with the use of phenomenological methods. On the other hand, Thomism offers contact with great forgotten historical traditions and original ontological problems.

Thus, the basic conclusion which follows from the above is to consider ontological problems which overlap between the schools, with different methods and not in isolation from the other approaches and sciences, especially in concert with physics, cosmology, logic, and mathematics. The same concerns the education of young philosophers.

Marek Kuś

Not being a professional philosopher[3] I would refrain from commenting on all issues raised in the questions. For me, as a physicist, a particularly acute problem is a lack of genuinely interdisciplinary or cross-disciplinary research encompassing ontology and the exact sciences (in particular physics, i.e. the area of my professional

---

[3] Marek Kuś is a profesor of theoretical physics but he is keenly interested in philosophical issues. He is a member of the the Program Council of the International Center for Formal Ontology and has served as the Center's Director since September 2016, as the successor of Mirosław Szatkowski.

expertise). We have many excellent philosophers of physics and mathematics at Polish institutions, doing excellent research, but their activities are hardly supported by physicists or mathematicians. At first glance it is easy to say who is to blame for this state of things. A deeper insight into philosophical problems of science is usually absent in the scientific activity of scientists. Obviously, a major part of, say physical research, does not require any philosophical input. On the other hand, as correctly stated by Tomasz Bigaj, contemporary theoretical physics desperately needs such input to surmount hindrances it has encountered. The fact that physical problems can be formulated and, to some extent, understood in purely ontological terms, usually escapes the notice of physicists. As an example, let me mention now-commonly-accepted but previously-not-imaginable explanations of some "paradoxes" of quantum mechanics[4] denying the existence of such things like "the value of physical quantity prior to a measurement". There are many ontological questions posed by even "elementary" quantum mechanics that are still unanswered (e.g. the ontological status of a state in quantum mechanics, ontological vs. epistemological status of randomness,[5] the nature of the indistinguishability of particles, and many others). The same can be said about many problems of quantum field theory, the theory of gravitation, and, especially topics on the border between the two, where we look for a unified description of all fundamental interactions.

So, why are ontological issues so rare in the reflections of physicists? My answer is simple. Philosophers gave up their role as leaders of all scientific activities, by not trying hard enough to make their discipline indispensable for higher education, and in particular for future scientists. Students of the best science departments in Poland are quite effectively "protected" from any deep encounter with philosophy. Even if curricula foresee some philosophical education, very rarely (if at all) are they adapted to a particular area of studies. As was stated by Rafał Urbaniak, students are taught (if at all) a kind of history of ideas, whereas what is needed, already at the undergraduate level, and indispensable for PhD students, are solid courses on the philosophy of mathematics, physics, technology, etc.

A consequence of this state of things is that common scientific programs at the border of science and philosophy are not common enough. One should admit that, apart from the abovementioned, there are some other obstacles along the way of those who would like to execute such projects, mostly of an organizational nature.

---

[4] Cf. Einstein et al. (1935). The example is now of a rather historical nature, but spectacular enough for having brought to ontological problems the foreground.

[5] Despite the common belief that quantum mechanics is genuinely, i.e. objectively probabilistic theory, the randomness of which is not caused by our ignorance, we do not have an ultimate proof of such a statement.

One of the problems is the attitude of e.g. funding agencies to interdisciplinary or cross-disciplinary projects. Although officially such projects are praised and encouraged, I cannot resist saying that often this is only lip service being paid to the idea. Thus, for example, the assessment of such projects is poor, while the number of specialists capable of making the evaluation fair and significant is low. Admittedly, these problems are not limited to our country, from my experience I can say that this situation is also similarly gloomy on a European level.

Now a few more concrete words concerning questions C and D. It should be clear from what I have written above that for me direct connections between ontological studies and science are of paramount importance and the main directions worth exploring. Although it is hard to state that such directions are nowadays specific for Poland, at least they are well founded in the Polish philosophical tradition. This is more important than pursuing some goals that are nation-specific. On my side I can only promise that my PhD students will not be deprived of good philosophical education (despite some common fears, young theoretical physicists are usually very excited by the philosophical problems of their discipline), and, what is obvious, and what I have been doing for some years, is to cherish my personal collaboration with philosophers.

Bartłomiej Skowron

A lot has already been said about Polish ontology by my colleagues. I agree with most of the criticisms and suggestions for improvement. Nevertheless, apart from a few critical remarks, I would like to focus on what is creative and what distinguishes Polish ontology. Firstly, in order to avoid unnecessary and meaningless discussions, I would like to point out that by Polish ontology I mean the ontology developed by Poles. End of definition.

I am not answering questions A–D, because after so many contributions it is difficult to say anything new and important. In my answer,[6] I would like to draw attention to the achievements of Polish ontology over the existing philosophical schools. I would like to mention those achievements which I myself consider important.

The main message of my intervention is to call for Polish ontologists to think for themselves. Independence of thought requires focusing on things themselves. An ontological thought, I think, should not be limited only to an exegesis of the writings of our predecessors, as a large part of Polish ontologists actually do. Since

---

[6] My contribution consists of fragments of a paper I published earlier in Polish (Skowron, 2017).

the umbilical cord of tradition obliges us in a strong sense, especially if it is as important as the legacy of the Lvov-Warsaw School, I believe that this umbilical cord should be shortened. Or perhaps it should be cut altogether. This is a way to continue, of course, creatively and independently, the legacy of our great Masters – not limiting ourselves to the history of philosophy. This was explicitly mentioned by Rafał Urbaniak above, but important points bear repeating.

The greatness of the achievements of Polish philosophers (the Lvov-Warsaw School plus Roman Ingarden and his successors, as well as Henryk Elzenberg and Ludwik Fleck) in the first half of the 20th century has paradoxical consequences – on the one hand, it encourages us to continue in the directions of the Masters' reflections, and on the other hand, the greatness of their achievements is paralyzing, as it makes it difficult to obtain independent and original results. In this sense, we are the victims of our great Masters, too often we choose safe historical work and do not think independently. We are afraid that our achievements are far from those of our predecessors. In addition, there are systemic factors connected with the criteria of academic career and the level and mechanisms of financing science, which support conservative academic attitudes and, in any case, do not encourage the undertaking of risky, primary philosophical undertakings.[7]

I think that in Poland we have a very good tradition and history of ontology. Below one can find a non-exhaustive list of Polish philosophers who have taken up ontological issues creatively[8]: Marcin Śmiglecki (1562–1618), Szymon Stanisław Makowski (1612–1683), Jan Morawski (1633–1700), Kazimierz Twardowski (1866–1938), Benedykt Bornstein (1880–1948), Tadeusz Kotarbiński (1886–1981), Roman Ingarden (1893–1970), Stanisław Leśniewski (1886–1939), Tadeusz Czeżowski (1889–1981), Józef Maria Bocheński (1902–1995), Jan Salamucha (1903–1944), Leopold Blaustein (1905–1944), Roman Suszko (1919–1979), Karol Wojtyła (1920–2005), Mieczysław Krąpiec (1921–2008), Jerzy Perzanowski (1943–2009), Leszek Nowak (1943–2009), Bogusław Wolniewicz (1927–2017).

I am also adding here a set of Polish philosophers and scientists active in 2019 (I mention scientists, because ontological issues in Poland are closely connected with science and not only with pure philosophy) who have creatively taken up ontological topics (in alphabetical order, the list is obviously not complete): Andrzej Biłat, Tomasz Bigaj, Piotr Błaszczyk, Izabela Bondecka-Krzykowska, Arkadiusz Chrudzimski, Janusz Czelakowski, Jan Czerniawski, Paweł Garbacz, Michał Głowala, Rafał Gruszczyński, Mariusz Grygianiec, Wojciech Grygiel, Piotr Gutowski, Michał

---

[7] Contemporary Polish philosophy is unfortunately practiced only at universities, hence it is influenced by the system of science in Poland, as already explained by Józef Lubacz.
[8] Cf. (Paź, 2011) and (Kaczmarek, 2012).

Heller, Stanisław Judycki, Janusz Kaczmarek, Tomasz Kąkol, Leszek Kopciuch, Zbigniew Król, Wojciech Krysztofiak, Piotr Kulicki, Marek Kuś, Damian Leszczyński, Józef Lubacz, Marek Łagosz, Dariusz Łukasiewicz, Marek Magdziak, Witold Marciszewski, Adam Olszewski, Jacek Paśniczek, Bogusław Paź, Maciej Piasecki, Andrzej Pietruszczak, Robert Piłat, Marek Piwowarczyk, Tomasz Placek, Witold Płotka, Paweł Polak, Andrzej Półtawski, Paweł Rojek, Marek Rosiak, Zbigniew Semadeni, Błażej Skrzypulec, Władysław Stróżewski, Mirosław Szatkowski, Kordula Świętorzecka, Robert Trypuz, Rafał Urbaniak, Jacek Wojtysiak, Jan Woleński, Wiesław Wójcik, Krzysztof Wójtowicz, Ireneusz Ziemiński, and Wojciech Żełaniec.

The legacy of Polish ontology, taking into account the size and complicated history of our country, can be considered very good, and in part even outstanding. There is no doubt that the ideas of Śmiglecki, Twardowski, Leśniewski, or Ingarden have influenced philosophy as such and will continue to influence it. The ontological ideas of Twardowski, the ideas Edmund Husserl engaged with in his *Logical Investigations*, and the works of his student Leśniewski, as well as the works of Ingarden, are recognizable beyond the borders of Poland.

Let me say it again. The indicated imperfections of Polish ontology, both by myself and by my colleagues, should not obscure many important contemporary achievements of Polish ontologists. The following are some of the themes, concepts, approaches, and analyses that I have learned and that I consider to be important contributions[9] – although this subjective criterion will not be accepted for many philosophers, it was the only one that was applied. In some cases, these are the latest ideas and analyses of a given author whose thoughts I have come to know, while in others, they are broader areas in which the author has worked creatively for many years. This is a list of contributions of Polish living ontologists:

1. Phenomenological ontology in the contemporary approach: the development of Roman Ingarden's ontology; formal analysis of moments of being by Jan Woleński; the ontological analyses in the spirit of Ingarden conducted by Marek Rosiak (ontology of the dispute: idealism-realism and the part-whole theory); the ontological studies of intentional objects by Arkadiusz Chrudzimski.
2. Analytical scholastic metaphysics: in particular the study of singleness by Michał Głowala.

---

[9] To give a full bibliography of the achievements of Polish ontological thought cited here would be ungrateful to both the reader and the author. The vast majority of the authors' works, which I refer to here, are easy to find – it does not take more than 2 seconds for the average Internet user. Some of the presented achievements of contemporary Polish ontology I know only from private conversations and seminars. If I feel that I understood them properly and I consider them important cognitively, they have added to the list.

3. Ontology of mathematics: in particular ontology of set theory and category theory. An ontology of mathematical objects proposed by Piotr Błaszczyk; the ontologies of mathematical development suggested by Jerzy Dadaczyński.
4. Ontology, ontologics, ontomethodics – an approach to ontology created by Jerzy Perzanowski and his followers.
5. Part-whole theory; mereology (Andrzej Pietruszczak); mereotopology (Rafał Gruszczyński); ontologies of space; an interpretation and new approach to the part-whole theory of Twardowski; new approaches to Leśniewski's systems (Rafał Urbaniak).
6. Ontology of the absolute: e.g. metaphysics of the absolute. Benedict Bornstein's architectonics of being in its contemporary formulation.
7. Engineering ontologies; ontologies of knowledge and beliefs; ontologies of virtual world; ontologies of large data systems; the Internet as a *mathesis universalis* (Paweł Garbacz, Piotr Kulicki, and Robert Trypuz).
8. Ontology of values: especially the ontology of values proposed by Władysław Stróżewski; an ontological study of values carried out by Leszek Kopciuch.
9. Research on ontological proofs (Mirosław Szatkowski, Edward Nieznański, Kordula Świętorzecka, and Stanisław Judycki).
10. Ontological analysis of the theory of ideas in the spirit of Roman Ingarden and Janusz Kaczmarek; ontology of concepts and ontology of individuals introduced by Janusz Kaczmarek; a study of platonism carried out by Zbigniew Król.
11. Ontologies of epistemic and poetic processes developed by Józef Lubacz.
12. Formalized ontologies of the world as a whole developed by Andrzej Biłat.
13. Ontologies in biology, ontology of life presented by Krzysztof Chodasewicz.
14. Phenomenological (Filip Kobiela) or analytical (Tomasz Placek, Władysław Krajewski) studies on causality.
15. Time-ontologies, recentivism of Józef Bańka. The event-theory created by Zdzisław Augustynek.
16. Formal ontology of truth as provided by Marek Magdziak.
17. Formal ontology of the rules of action and situational systems of action introduced by Janusz Czelakowski.
18. Ontology of the basic structure of the object studied by Marek Piwowarczyk.
19. Ontologies of the situation created by Bogusław Wolniewicz.
20. Ontology of science (and ontology in science) carried out by Michał Heller and his followers.
21. Formal ontology of objects, properties, and situation presented by Jacek Paśniczek. An ontology of dream-objects designed by Jacek Paśniczek.
22. Language-ontologies. Maciej Piasecki's *Word-Net* for the Polish language.

23. Jan Czerniawski's reism, as a legacy of the ontology proposed by Tadeusz Kotarbiński.
24. The application of the mathematical idea of the invariant (with respect to a transformation) in ontology and theology proposed by Wojciech Grygiel.
25. Metaphysics in contemporary physics, research on essentialism, symmetry, counterfactuals in physics undertaken by Tomasz Bigaj.
26. Ontological studies of objects of contemporary physics authored by Marek Kuś.
27. The method of the hermeneutical horizon in the studies of the ontological situation of mathematics and hermeneutical ontology of mathematics in Zbigniew Król's approach.
28. Ontologization of informal logic realized by Andrzej Kisielewicz, development of argument ontology and argumentation ontology (research on argumentation in the *Polish School of Argumentation*, see Budzynska et al. (2014)).
29. Theories of *continuum* philosophically explored by Wiesław Wójcik.
30. Topological ontology, i.e. analysis of spatiality using topological tools. Studies of spatiality conducted by Rafał Gruszczyński. Krzysztof Śleziński's analyses of the world's topology in the view of Benedict Bornstein. The full topologization of the trees of ideas and individuals proposed by Janusz Kaczmarek. The topo-ontology suggested by Bartłomiej Skowron. The topology of the branched space-time continuum in the view of Tomasz Placek.

I consider all these (as well as those omitted) cognitive projects to be important contributions by Polish ontologists to ontology in general. Some of them are completely unknown, because, as my colleagues have already mentioned, we often do not take enough care in Poland to ensure that our works are published in languages other than Polish. However, some of them have been published in prestigious international journals and are therefore available to an international audience.

In this context, it is worth mentioning three Polish contemporary initiatives of an organizational nature:
1. The Copernicus Center was initiated by Michał Heller and successfully managed by his followers. The Center deals creatively, actively, and on a large scale with issues at the intersection of science and philosophy. The philosophy of science, which is an important change, has been replaced by the philosophy *in* science. The Center's activities include, among others, broad research activities, valued publishing initiatives, and comprehensive educational and popularization activities.
2. A group of ontologists and philosophers at the John Paul II Catholic University of Lublin, such as Paweł Garbacz, Piotr Kulicki, and Robert Trypuz, is an example of a flourishing ontological initiative. This group deals both with

traditional ontological issues and with modern engineering ontologies and cooperation with natural sciences (e.g. *Laboratory for AgriFood Ontology*).
3. The International Center for Formal Ontology, which was established at the Faculty of Administration and Social Sciences of the Warsaw University of Technology in 2015, is another flourishing ontological initiative (over 150 ontological papers have been delivered at conferences organised by the Center, 5 ontological monographs on international circulation were published at the Center). The Center cooperates with renowned ontologists from all over the world. Research groups within the Center address both classical metaphysical issues, such as the existence of God, as well as modern ontologies of artefacts and creative processes, the ontology in physics, the ontology of mathematics, and mathematical ontologies.

To summarize, it seems to me that Polish ontology is doing quite well. I will present below a list of steps that could improve the condition of Polish ontology in the coming decades. Not all of the indications are direct from what I have said, but I join them together here, even though I have not given any reasons.
1. Wider use of natural, strict, and technical sciences in ontological considerations than has been the case so far and (!) integration of ontological considerations into these sciences. In particular, closer contact between ontologists, by organising joint seminars and carrying out interdisciplinary projects with cosmologists, theoretical physicists, mathematicians, computer scientists, engineers, molecular biologists, to mention a few examples.
2. Cultivating good practice, so that the diploma, doctoral, and postdoctoral theses are also oriented at attempting to solve original problems, and not only at reconstructive analysis of the concepts of other philosophers.
3. Spending, in the philosophical community, at least several months' worth of trips abroad (fellowships, study visits, etc.) at every stage of one's academic career.
4. Eliminating or ignoring these Polish philosophical journals which are unprofessional, profit-making, and redundant.
5. Launching a mentoring process in the field of ontology, selecting mentors (including foreign ones) due to the substantive fit, and not friendly dependencies.

Conclusion

Ontology cultivated in Poland is diverse and should remain so. Ontological research in Poland is conducted on an international level, although some Polish schools of thought remain isolated from the rest of the world. Many Polish ontologists, especially the ontologists whose work is present in this volume, work in contact with modern science, but still too many ontologists focus their forces on historical issues. Polish ontologists still have to be careful about the danger of philosophical topor and philosophical blindness. The works of Polish ontologists are more and more visible in international circles, which allows us to hope that Polish ontology will flourish in the future, just like Polish philosophy did in the first half of the 20th century.

Bibliography

Anscombe, G. E. M. (1981). Causality and determination. In *The collected philosophical papers of G. E. M. Anscombe, Vol. 2* (pp. 133–147). Oxford: Blackwell.

Budzynska, K., Araszkiewicz, M., Bogołębska, B. et al. (2014). The Polish School of Argumentation: A Manifesto. *Argumentation, 28*, 267. https://doi.org/10.1007/s10503-014-9320-8.

Einstein, A., Podolsky, B., & Rosen, N. (1935). Can quantum-mechanical description of physical reality be considered complete? *Phys. Rev., 47*, 777.

Geach, P. T. (1991). Whatever happened to deontic logic? In P. T. Geach (Ed.), *Logic and ethics* (pp. 33–48). Dordrecht: Kluwer Academic Publishers.

Kaczmarek, J. (2012). Ontologia formalna. Przykład: ontologia w Polsce. [Formal ontology in Poland] *Przegląd Filozoficzny – Nowa Seria, R. 21, 3*(83), 165–175.

Klima, G. (2004). On Kenny on Aquinas on being. *International Philosophical Quarterly, 44*, 567–580.

Lowe, E. J. (2013). *Forms of thought: A study in philosophical logic*. Cambridge: Cambridge University Press.

Bigaj, T., & Wüthrich, C. (Eds.) (2015). *Poznań Studies in the Philosophy of the Sciences and the Humanities: Vol. 104. Metaphysics in Contemporary Physics*. Leiden–Boston: Brill–Rodopi.

Paź, B. (2011). Ontologia versus metafizyka? Geneza, rozwój i różne postaci nowożytnej teorii bytu. [Ontology versus metaphysics? Genesis, development and various forms of modern theory of being.] *Filo-Sofija, 15*(2011/4), 817–847.

Oderberg, D. S. (2002). Hylomorphism and individuation. In J. Haldane (Ed.), *Mind, metaphysics and value in the thomistic and analytical traditions* (pp. 125-142). Notre Dame, Indiana: University of Notre Dame Press.

Pouivet, R. (2009). Jan Salamucha's analytical thomism. In S. Lapointe, J. Woleński, M. Marion, W. Miskiewicz (Eds.), *The golden age of Polish philosophy: Kazimierz Twardowski's philosophical legacy* (pp. 235–245). Dordrecht–Heidelberg–London–New York: Springer.

Skowron, B. (2017). Ojcobójcza i nieco stronnicza diagnoza stanu i uwarunkowań rozwoju ontologii w Polsce [Father-killing and a biased diagnosis of the state and determinants of the development of ontology in Poland]. *Studia Philosophica Wratislaviensia, 12*(2), 75–84.

Szatkowski, M. (2016). The recovery of St. Thomas Aquinas. Part II: What is analytically oriented Thomism. In M. Szatkowski (Ed.), *Analytically oriented Thomism* (pp. 1–53). Neunkirchen-Seelscheid: Editiones Scholasticae.

Author Index

Ajdukiewicz K. VIII, XI, XIII, 112, 113, 117, 118, 122, 130
Anscombe G. E. M. 248, 251, 259, 262, 264, 269, 281, 293
Antos C. 172, 178
Araszkiewicz M. 293
Aristotle VII, XI, 88, 91–93, 100, 102, 107, 149, 159
Armstrong D. M. 245, 250, 269
Arrigoni T. 172, 178
Arthan R. D. 187, 203
Augustine (St). X, 51, 53, 59, 61, 68
Augustynek Z. 130, 273, 290

Bailey A. M. 33, 36, 37, 46
Baker L. R. 31, 46
Balaguer M. 175, 178
Bańka J. 290
Bartels A. 20, 22
Batitsky V. 199, 200, 203
Beaney M. 97, 107
Belshaw C. 34, 35, 46
Benacerraf P. 190, 203
Bergson H. X
Besler G. 130
Bigaj T. VIII, IX, XIX, 1, 5, 17, 22, 271–273, 276, 279, 286, 288, 291, 293
Biłat A. X, XI, 87, 288, 290
Bird A. 247–250, 259, 260, 264, 269
Błaszczyk P. XVIII, 186–188, 191, 197, 202, 203, 288, 290
Blatti S. 33, 34, 46
Blaustein L. 288
Bocheński J. M. VIII, XIII, 117, 130, 155, 159, 283, 288
Bondecka-Krzykowska I. 288
Bonikowski Z. XIX, 130
Boolos G. 99, 107, 182, 203
Borkowski L. 129, 131
Bornstein B. XII, XIII, 133, 135, 137, 138, 140, 142, 143, 145–148, 288, 290, 291
Bottani A. XVIII, 68
Brading K. 16, 22

Brandom R. 206, 207, 209, 210, 218
Brighouse C. 20, 22
Brower J. E. 27, 46
Budzyńska K. 293
Bunge M. XIII, 155
Butterfield J. 20, 22

Cajetan T. XVII, 255, 269
Caplan B. XVIII
Carnap R. 114, 122, 130, 195, 203
Carrara M. 30, 46
Castellani E. 16, 22
Cayley A. 147
Chappell T. 42, 46
Chisholm R. M. 26, 31, 46
Chodasiewicz K. 290
Chrudzimski A. 52, 68, 69, 251, 256, 270, 271, 279, 283, 288, 289
Cieśliński C. 202
Citkin A. 130
Coope U. 264, 270
Copleston F. 154, 159
Costa D. 27, 46, 47
Crane T. 36, 47
Cresswell M. J. 206, 217
Crook S. 36, 47
Czelakowski J. 288, 290
Czerniawski J. 288, 290
Czeżowski T. 288

da Costa N. C. A. 174, 178
Dadaczyński J. 290
Daly Ch. XVIII
Doria F. A. 174, 178
Dummett M. 190, 203
Dunn J. M. 209, 218

Effingham N. 27, 47
Eilenberg S. XIII
Einstein A. 286, 293
Elzenberg H. 280, 288

Fara D. G. 5, 22

Feser E. 88, 108
Field H. 175, 178
Fine A. 167, 178
Fine K. 28, 47, 183, 203
Fleck L. 288
Fletcher S. XVIII, 21
Fox C. 205, 218
Frege G. 91, 108, 117, 122, 130, 181, 182, 190–192, 197, 203
Friedman H. 173, 178
Friedman M. 18, 22
Friedman Sy-D. 172, 178

Gaio S. 30, 46
Garbacz P. 272, 288, 290, 291
Geach P. T. 248, 270, 281, 282, 293
Gettier E. L. 73, 85
Gillett C. 36, 47
Giordani A. 27, 46, 47
Glick D. 2, 22
Głowala M. XVI, XVII, 245, 263, 264, 270, 271, 280, 288, 289
Godziszewski T.M. 202
Gołosz J. 2, 22
Gödel K. 161, 163, 166, 173, 175, 176, 178
Grädel E. 108
Greaves H. 2, 22
Griesel H. 193, 203
Gruszczyński R. 288, 290, 291
Grygianiec M. IX, 25, 31–33, 47, 273, 288
Grygiel W. 288, 291
Gutowski P. 288

Hale B. XIV, XVIII, 181–183, 197–199, 201–203
Halpern J. L. 73, 85
Hamkins J. D. 172, 178
Hardegree G. M. 218
Hartmann N. 92
Haslanger S. 27, 47
Hauska J. 249, 265, 270
Hawley K. 27, 47
Heck R. 195, 203
Heidegger M. 285
Heijenoort van J. 91, 108
Heil J. 252, 270
Heller M. 27, 47, 94, 108, 288, 290, 291

Hellman G. 161, 162, 178
Hempel C. G. 195, 203
Hershenov D. B. 37, 47
Hintikka J. 73, 85
Hodges W. 4, 22
Hoffman J. 38, 47
Honzik R. 172, 178
Horsten L. 28, 30, 47
Houser N. 131
Humphries C. VII, XIX
Huntington E. 138
Husserl E. VII, 52, 73, 75, 76, 79, 85, 92, 118, 130, 149, 159, 284, 289

Incurvati L. 184, 203
Ingarden R. VIII–X, XII, XIII, XV, 52, 54–59, 61, 62, 69, 75, 76, 85, 97, 108, 155, 283, 284, 288–290
Ingram D. 69

Jadacki J. J. 115–117, 125, 131
James W. 60, 69
Janowicz K. 73, 85
Jech T. 186, 204
Johansson J. 34, 43, 47
Jubien M. 31, 47
Judycki S. 289, 290

Kaczmarek J. XI, XIII, 149, 150, 159, 289–291, 293
Kahn Ch. H. 91, 93, 102, 108
Kanzian Ch. XVIII, XIX, 41, 47, 269
Kąkol T. 289
Kirsch A. 197, 204
Kisielewicz A. 291
Klima G. 282, 293
Kobiela F. X, 51, 69, 290
Kopciuch L. 289, 290
Koslicki K. 39, 47
Kotarbiński T. 288, 291
Krajewski W. 273, 290
Krąpiec M. 288
Kripke S. 206, 218
Król Z. X, XIX, 71, 80, 81, 85, 86, 107, 271, 284, 289–291
Krysztofiak W. XVIII, 289
Kulicki P. 289–291

Kuratowski K. 152, 159
Kurtz R. M. 27, 47
Kuś M. XIX, 107, 271, 285, 289, 291

Ladyman J. 1, 3, 22
Lakatos I. 83, 86
Lappin S. 205, 218
Leśniewski S. VIII, XII, XV
Leibniz G. W. F. VII, IX, XIII, 133, 142, 145, 153, 154, 157, 159
Leitgeb H. XIX
Lem S. X
Lem S. 53, 65–67, 69
Leszczyński D. 289
Leśniewski S. 117, 118, 131, 288–290
Levine J. 202
Lewis D. 5, 8, 9, 22
Licon J. A. 33, 47
Lowe E. J. 27–31, 40, 47, 48, 256, 263, 267, 270, 273, 281, 293
Lubacz J. X, XIX, 71, 107, 271, 277, 289, 290
Łagosz M. 289
Łukasiewicz D. 289

Mac Lane S. XIII
Mackie D. 34, 38, 48
Mackie P. 6, 22
Maddy P. 172, 178
Madell G. 31, 48
Magdziak M. XVI, 219, 289, 290
Makowski S. 288
Marciszewski M. 289
Marmodoro A. 264, 270
Martin C. B. 245, 247, 252, 270
Maudlin T. 20, 22
McCann H. 267, 270
McDaniel K. XVIII
McKitrick J. 263, 270
Meinong A. 91
Meixner U. XVIII
Melin R. 34, 37, 48
Mellor D. H. 36, 47
Melnyk A. 36, 48
Merricks T. 31, 39, 48
Mitscherling J. XIX, 52, 68, 69
Møller-Nielsen T. 3, 16, 18, 22
Molnar G. 245, 247, 253, 263, 270

Montero B. 36, 48
Morawski J. 288
Morgan C. 206, 218
Mormann T. XVIII
Mulligan K. XVIII
Mumford S. 249, 257, 270
Murawski R. XVIII
Müller T. XVIII, 21

Nakamori Y. 84, 86
Napolitano J. 16, 23
Nasieniewski M. XVIII, 243
Nes A. 245, 249, 270
Ney A. 36, 48
Nida-Rümelin M. 31, 48
Nieznański E. 290
Noonan H. W. 31, 34, 48
Norton J. D. 19, 22
Novak L. 88, 108
Novotny D. D. 88, 108
Nowak L. 288
Nuchelmans G. 248, 270

Oderberg D. S. 27, 37, 48, 245, 247, 249, 258, 267–270, 282, 293
Odrowąż-Sypniewska J. 273
Ogrodnik B. X, 53, 64, 65, 69
Olson E. T. 31–35, 37, 48
Olszewski A. 289
Omyła M. 273
Orilia F. XIX, 107
Owen G. E. L. 102, 108

Pabjan T. 94, 108
Paganini E. XIX
Papineau D. 36, 48
Paśniczek J. XV, 205, 208, 211, 212, 215, 217, 218, 289, 290
Pawlowski P. 202
Paź B. 289, 293
Peirce C. S. 114, 115, 131, 141
Pelc J. 131
Perzanowski J. VIII, XI, XII, XV, XIX, 110, 131, 219, 243, 288, 290
Pfeifer K. 245, 247, 270
Piasecki M. 289, 290
Pietruszczak A. 289, 290

Piłat R. 289
Piwowarczyk M. 27, 48, 272, 283, 289, 290
Place U. T. 245, 247, 270
Placek T. 289–291
Plato VII, XI, 86, 89, 91–93, 108
Płotka W. 289
Podolsky B. 286, 293
Polak P. 289
Poli R. 73, 86
Pooley O. 18, 20, 23
Popper K. R. X, 53, 63, 68, 69
Pouivet R. 283, 293
Półtawski A. 289
Presnell S. 22
Priest G. 206, 218
Pruss A. XIX, 107
Putnam H. 99

Quine W. V. 99, 102, 108, 175–179, 284

Rantala V. 206, 218
Reale G. 90, 108
Rescher N. 206, 207, 209, 210, 218
Rojek P. 289
Rosen N. 286, 293
Rosenkrantz G. S. 38, 47, 48
Rosiak M. 283, 289

Sakurai J. J. 16, 23
Salamucha J. 150, 159, 283, 288
Salmon N. U. 32, 48
Saunders S. 16, 23
Schnädelbach H. 88, 89, 97, 108
Semadeni Z. 289
Shapiro S. XIV, 98, 108, 170, 179, 181, 193, 204
Sharpe K. W. 34, 48
Sher G. 170, 179
Sider T. 27, 48
Sillari G. 206, 218
Simons P. XIV, XIX, 27, 49, 52, 69, 73, 86, 181, 190, 191, 204
Simpson S. G. 164, 167, 179
Skow B. 5, 23
Skowron B. VII, 107, 130, 138, 148, 150, 243, 271, 287, 291, 293
Skrzypulec B. 289

Słupecki J. 129, 131
Snowdon P. F. 33, 46, 49
Soncinas P. XVII, 254, 270
Sousedík S. 283
Spinoza B. 88
Stalnaker R. 220, 243
Stoljar D. 36, 49
Strawson P. 275
Stróżewski W. 289, 290
Suszko R. 117, 131, 273, 288
Swinburne R. 32, 49
Szatkowski M. XIX, 150, 272, 283, 285, 289, 290, 293
Śleziński K. XII, XIII, 133, 138, 148, 291
Śmiglecki M. 288, 289
Świętorzecka K. 289, 290

Tallant J. 69
Tarski A. XII, XV, 98, 108, 114, 131
Tatarkiewicz W. 89, 108
Ternullo 172, 178
Thomas W. 108
Thomasson A. 52, 69
Toledo F. 267, 270
Toner P. 35, 36, 49
Trypuz R. 289–291
Tugby M. 256, 263, 270
Twardowski K. VII, VIII, XII, XIX, 91, 135, 148, 288–290

Urbaniak R. XIV, 181, 184, 204, 271, 276, 280, 286, 289, 290

van Fraassen B. 211, 218
van Inwagen P. 38, 49, 273
van Rooij R. XIX
Vetter B. 43, 49
von Wachter D. 51, 69
von Wright G. H. XVI, 220, 243

Wallace D. 2, 22
Warren R. H. 151, 159
Wasserman R. 27, 49
Wierzbicka A. 93, 102, 108
Wierzbicki A. 84, 86
Wilke T. 108
Williamson T. 29, 30, 49

Wilson J. 36, 49
Wittgenstein L. 122, 131, 273
Wojtyła K. 288
Wojtysiak J. 283, 289
Woleński J. XIX, 90, 97, 108, 289
Wolff C. 88
Wolniewicz B. VIII, 273, 288, 290
Wood N. 202
Woodin H. 174, 179
Wójcik W. 289, 291

Wójtowicz A. XIX, 273
Wójtowicz K. XIV, 161, 273, 289
Wright C. 182, 192, 195, 196, 203, 204
Wüthrich C. 275, 293
Wybraniec-Skardowska U. XI, XII, 109, 116, 118, 120–122, 125, 131, 132

Zalta E. 182, 204
Zermelo E. 81, 86
Ziemiński I. 289
Żełaniec W. 289

www.ingramcontent.com/pod-product-compliance
Lightning Source LLC
Chambersburg PA
CBHW061933220426
43662CB00012B/1892